METALLOCENES

AN INTRODUCTION TO SANDWICH COMPLEXES

TO MUM AND DAD

METALLOCENES

AN INTRODUCTION TO SANDWICH COMPLEXES

NICHOLAS J. LONG DEPARTMENT OF CHEMISTRY

IMPERIAL COLLEGE OF SCIENCE, TECHNOLOGY & MEDICINE

LONDON

Blackwell
Science

© 1998 by
Blackwell Science Ltd
Editorial Offices:
Osney Mead, Oxford OX2 0EL
25 John Street, London WC1N 2BL
23 Ainslie Place, Edinburgh EH3 6AJ
350 Main Street, Malden
  MA 02148 5018, USA
54 University Street, Carlton
  Victoria 3053, Australia

Other Editorial Offices:
Blackwell Wissenschafts-Verlag GmbH
Kurfürstendamm 57
10707 Berlin, Germany

Blackwell Science KK
MG Kodenmacho Building
7–10 Kodenmacho Nihombashi
Chuo-ku, Tokyo 104, Japan

First published 1998

Set by Semantic Graphics, Singapore
Printed and bound in Great Britain
by MPG Books Ltd, Bodmin, Cornwall

The Blackwell Science logo is a
trade mark of Blackwell Science Ltd,
registered at the United Kingdom
Trade Marks Registry

DISTRIBUTORS

Marston Book Services Ltd
PO Box 269
Abingdon, Oxon OX14 4YN
(*Orders*: Tel: 01235 465500
         Fax: 01235 465555)

USA
Blackwell Science, Inc.
Commerce Place
350 Main Street
Malden, MA 02148 5018
(*Orders*: Tel: 800 759 6102
           781 388 8250
         Fax: 781 388 8255)

Canada
Copp Clark Professional
200 Adelaide St West, 3rd Floor
Toronto, Ontario M5H 1W7
(*Orders*: Tel: 416 597-1616
           800 815-9417
         Fax: 416 597-1617)

Australia
Blackwell Science Pty Ltd
54 University Street
Carlton, Victoria 3053
(*Orders*: Tel: 3 9347 0300
         Fax: 3 9347 5001)

A catalogue record for this title
is available from the British Library

ISBN 0-632-04162-5

Library of Congress
Cataloging-in-publication Data

Long, Nicholas J., 1965–
    Metallocenes : an introduction to
    sandwich complexes / Nicholas J. Long.
          p.   cm.
    Includes bibliographical references
    (p.   –) and index.
    ISBN 0-632-04162-5
    1. Metallocenes.   I. Title.
QD411.L84   1997
547′.05–dc21
                                       97–23272
                                          CIP

Contents

Contents

Preface

Organometallic complexes, in which low oxidation state metals are co-ordinated to a range of organic-based ligands, have been known for over two hundred years. However, since the 1950s, there has been an exponential increase in research into this area of chemistry and the field continues to be one of great excitement and activity. The explosion of interest can be traced to the discovery of ferrocene and the recognition of its *sandwich* structure (featuring two parallel η^5-cyclopentadienyl rings). Nowadays, primary chemical journals publish extensive investigations into metallocenes and metallocene derivative research. Metallocene chemistry has become an integral part of the majority of undergraduate courses at university and sections of this area of chemistry are found in virtually every general inorganic, organometallic and even organic chemistry textbook. Since Rosenblum's text on *The Chemistry of the Iron Group Metallocenes* in 1965 (the year of my birth!), the field has expanded enormously with a wide variety of sandwich complexes now being known. It is my hope that chemists from all branches of the subject will be interested in this timely and up-to-date monograph on the subject.

The main focus of this book is to present a broad overview of this important area. It is surprising for a field of this size, relevance and interest, that a monograph has not been produced to link the aspects of all the various metallocenes (such as parallel, bent, main group, multi-decker and multi-cyclopentadienyl species) together – I hope that I have done so here. The book aims to be a valuable source of information for undergraduates, postgraduates and senior researchers but is not meant to be a complete literature review. Therefore, to guide the interested reader, key primary papers plus relevant review articles, monographs and textbooks have been referenced and listed at the end of each chapter and in the Bibliography. The book could not have been written without drawing on information contained in earlier articles and treatises and these, along with copyrighted material, are acknowledged throughout the book.

The contents of this book encompass all the relevant aspects of metallocene compounds. Chapter 1 provides a general and historic introduction to organometallic chemistry and, in particular, metallocene chemistry. This is followed by a brief discussion of the various types of metallocene complexes that are featured in the book and finalised by a description of 'the 18-electron rule' and its application to metallocenes. Chapter 2 describes the general synthetic methods applied in the synthesis of metallocenes and then details the cyclopentadienyl (and the most important 6- and 8-membered ring) complexes of many of the elements in the Periodic Table, enabling the reader to attain an appreciation of the extent of the field. A rigorous examination of the electronic structure and bonding of metallocenes is given in Chapter 3 via application of molecular orbital theory. This is finalised by a discussion on the ionic and covalent character of the bonding in metallocenes. Chapter 4 initially illustrates the wide range of chemistry of metallocenes and then details the spectroscopic properties of the complexes. Some important and topical derivatives of metallocenes are selected for discussion in Chapter 5 and Chapter 6 puts the field into context by discussing the uses and importance of metallocenes. Present-day and applied subjects are mentioned, focusing on the important catalytic and material applications of metallocenes and their derivatives.

Clearly, in a book of this size, it is impossible to include and do justice to everything in a field with an almost infinite amount of material. The scope of the book also does not allow for a detailed examination of other cyclic conjugated ring systems or 'half-sandwich' species. There may be those experts who disagree with my choice of salient features, examples and references but care has been taken in the allocation of emphasis and the examples discussed, with topicality and appeal to a wide readership being the important factors. I have tried to keep the book as error free as possible but there are, no doubt, some omissions and inaccuracies. For these I am solely responsible and please forgive any oversights that I might have made.

It is impossible to give credit to everyone who has been of help but I would like to single out certain people who have been especially instrumental, directly or indirectly, in the writing of this book. Firstly, it has been a pleasure to work with Anne Stanford, Anna Woodford, Karen Moore, Anna Rivers and Simon Rallison at Blackwell Science who have kept the project moving forward through their friendly and helpful comments and advice. I would like to thank Melissa Levitt (formerly of Blackwell Science) for encouraging me to write this book and getting the project started, before 'defecting'! What confidence I have in offering this book to the world stems largely from the helpful and much appreciated criticism, comments and suggestions of Scott Ingham, John Gallagher and Paul Beer, who each read the whole text, and Phil Dyer who read Chapter 6. Their remarks have, I hope, ensured clarity and interest. Thanks are also due for the many encouraging words and wishes from my family, friends, students (particularly those dedicated proof-readers) and colleagues over the last couple of years, who have all helped to keep me going through 'thick and thin'. As an undergraduate and postgraduate, I was lucky enough to have exceptional teachers such as Ken Wade, Vernon Gibson, Eddie Abel and Tony Osborne who stimulated an interest in organometallic chemistry and, in particular, sandwich compounds. Finally, it is perhaps fitting that I have written this book in the Department of Chemistry, Imperial College of Science, Technology and Medicine, University of London where Geoffrey Wilkinson spent most of his career. Unfortunately he passed away before I could show him the manuscript but, without him, all the other pioneers in this field and the many current researchers in this topical, enlightening and often remarkable area, this book would not have been written. Thanks to you all!

NICK LONG
E-mail: n.long@ic.ac.uk
Web Site: http://www.ch.ic.ac.uk/long/n/intro1.htm

1 An Introduction to Metallocenes

1.1 The birth of organometallic chemistry

Organometallic chemistry, a broad interdisciplinary field between the classical subdivisions of organic and inorganic chemistry, has become one of the most interesting and important areas of modern-day chemistry. It has developed into a thriving discipline with a wide variety of compounds, a host of strange and beautiful structures and a gamut of reactions and applications in organic synthesis and industrial catalysis. Research study continues to expand rapidly in this area through the synergetic convergence and the beneficial interaction of inorganic, organic and physical chemists and biochemists. Bonding theories to account for and predict novel compounds and amazing structures have been important in the expansion of the area, as has the increased use of organometallic compounds in industrial processes. Few areas of chemistry have created as many challenges and surprises as has occurred in organometallic chemistry in the past few decades and the field continues to be one of great excitement and activity.

Convention has it that an *organometallic compound* contains at least one direct metal–carbon (M—C) bond, though this definition is open to interpretation. For instance, elements such as boron, silicon, phosphorus and arsenic are not true metals yet their organic derivatives constitute a major part of organometallic chemistry. The term organoelement is, therefore, sometimes employed to include compounds formed with non- or semi-metals, and whose characteristics mimic those analogues formed with metallic counterparts.

The field of organometallic chemistry is relatively new, with comprehensive systematic study not beginning until the 1950s. The development of the area, with particular emphasis on the organometallic chemistry of the transition elements (or *d*- and *f*- block elements), dates from midway through this century although the initial organometallic compounds were synthesised and characterised to some degree in the nineteenth century. The first of these, an ethene complex of platinum(II) (Form **I.1**) was prepared by heating $PtCl_2$–$PtCl_4$ in ethanol, evaporating and treating the residue with aqueous KCl (Equation 1.1) (Zeise, 1827).

$$C_2H_4 + K_2PtCl_{4\,(aq)} \longrightarrow K^+[(C_2H_4)PtCl_3]^- + KCl \tag{1.1}$$

I.1

In 1849, during attempts to prepare an '*ethyl radical*', zinc alkyls were formed and consequently investigated (Frankland, 1849).

$$3\ C_2H_5I + 3\ Zn \quad\begin{cases} \xrightarrow{\times} ZnI_2 + 2\ C_2H_5 \\ \\ \longrightarrow [(C_2H_5)_2Zn] \quad \text{(a pyrophoric liquid)} \\ \qquad + C_2H_5ZnI \quad \text{(a solid)} \\ \qquad + ZnI_2 \end{cases} \tag{1.2}$$

Frankland swiftly followed this by preparing some important alkylmercury compounds and, via *alkyl transfer reactions*, used R_2Hg and R_2Zn in the synthesis of many main-group organometallics.

$$2\ CH_3X + 2\ Na/Hg \longrightarrow [(CH_3)_2Hg] + 2NaX \tag{1.3}$$

These studies led him to introduce the first theory of valency where he suggested that each element has a definite limiting combining capacity (Frankland, 1852). The next major discovery was the first binary metal carbonyl, nickel tetracarbonyl (**I.2**) (Mond, 1890). There followed a substantial

$$(I.2)$$

amount of research undertaken on metal carbonyls during the first half of the twentieth century. From a more practical standpoint, Grignard's Nobel Prize-winning synthesis and exploitation of organomagnesium halides (Grignard, 1900) provided easily handled and versatile intermediates for a variety of organic and organometallic syntheses.

$$Mg + C_2H_5Br \xrightarrow{\text{ether}} C_2H_5MgBr \tag{1.4}$$

However, of greatest importance to the more recent developments in organometallic chemistry was the synthesis and discovery of *ferrocene*.

1.2 The history of ferrocene

1.2.1 The initial formation of a new cyclopentadienyl–iron compound

In 1951, the compound dicyclopentadienyliron was discovered and the field of organometallic chemistry was thereby transformed. Two independent groups of chemists almost simultaneously arrived at the same conclusions, albeit accidentally. A reaction was carried out on cyclopentadienyl magne-

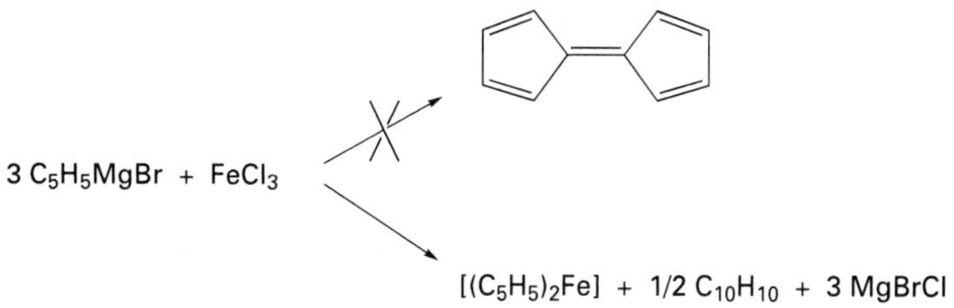

$3 C_5H_5MgBr + FeCl_3$

$[(C_5H_5)_2Fe] + 1/2 C_{10}H_{10} + 3 MgBrCl$

Scheme 1.1 An attempted formation of fulvalene.

sium bromide with anhydrous iron(III) chloride in ether in an attempt to synthesise fulvalene via the oxidation of the cyclopentadienyl Grignard reagent (Kealy and Pauson, 1951) (Scheme 1.1). However, via reduction of the iron(III) to (II) by the Grignard species, they instead obtained a crop of orange crystals that analysed for $C_{10}H_{10}Fe$. At the same time, Miller *et al.* (1952), who were investigating the preparation of amines, reported forming the orange compound $C_{10}H_{10}Fe$ by direct reaction of cyclopentadiene with iron in the presence of aluminium, potassium or molybdenum oxides at 300°C.

$$2 C_5H_6 + Fe \longrightarrow [(C_5H_5)_2Fe] + H_2 \qquad (1.5)$$

Both groups noted that the new compound was an air-stable, sublimable, orange solid that melted at 173°C with excellent solubility in organic solvents but was insoluble in water. The structures originally proposed for dicyclopentadienyliron (Fig. 1.1) featured firstly two, flat, planar cyclopentadienide rings, where one of the five carbon atoms of each ring was linked by a single σ-bond to a central iron atom. Then, by resonance, another structural representation indicated a divalent iron moiety (Fe^{2+}) coordinated to two molecules of cyclopentadienide anion (Cp^-). The structure, held together by σ-bonds, showed a high degree of ionicity and was stabilised by ionic–covalent resonance.

1.2.2 The 'breakthrough' by Wilkinson and Fischer

It was not long, however, before the correct structural formulation was identified. On reading the initial papers, it was clear to Geoffrey Wilkinson and Ernst Fischer that the unusual thermal and chemical behaviour observed for $C_{10}H_{10}Fe$ could not be explained by σ-bonding. A simple σ-bond be-

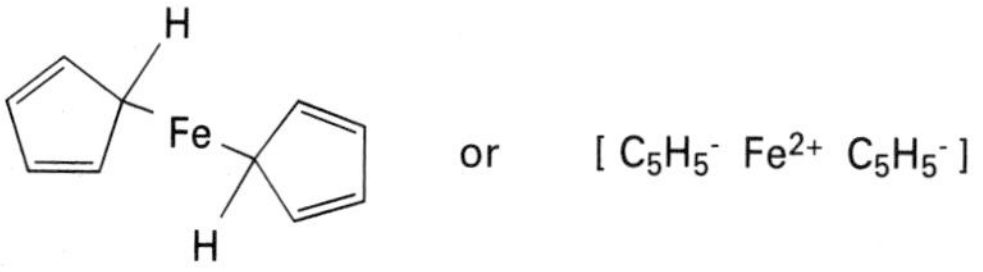

or $[C_5H_5^-\ Fe^{2+}\ C_5H_5^-]$

Fig. 1.1 Early representations of dicyclopentadienyl iron.

tween the organic and metallic fragments would be relatively unstable and the observed volatility could not be a result of an ionic interaction. There was obviously some unusually strong bonding between the cyclopentadienyl ligands and the iron centre. At Harvard University, Wilkinson used chemical, physical and spectroscopic methods to elucidate the correct structure of dicyclopentadienyl iron, whilst independently (and unaware of each others work), Fischer used X-ray crystallography to structurally characterise the compound. Wilkinson realised that with five electronically equivalent carbons (five in each cyclopentadienyl ring), they must all contribute, in an equal way, to the bonding to iron. He sketched a structure where the iron atom was placed between the two cyclopentadienyl (Cp) ligands. From this, the bonding appeared very strong due to excellent overlap of the metal *d* orbitals and the π-electrons in the *p* orbitals of the Cp ligands. From an interesting recollection (Wilkinson, 1975), he was particularly excited by the thought that if iron behaved in this way then surely other transition metals should form '*sandwich*' compounds.

A Harvard organic chemist, R.B. Woodward, was also interested in the topic and the two colleagues agreed to collaborate in order to prove the postulated structure. Some of the iron compound was prepared and Wilkinson, Woodward and co-workers began to characterise the material by infrared and ultraviolet spectroscopy, Gouy magnetic susceptibility and by measuring the dipole moment. There was just one C—H stretch in the infrared spectrum, indicating one type of C—H bond and magnetic susceptibility experiments showed the compound to be diamagnetic with no unpaired electrons. Together with an effectively zero dipole moment, this information indicated that the π-complexed sandwich structure had to be correct (**I.3**) (Wilkinson *et al.*, 1952). Wilkinson discovered that the iron

(**I.3**)

centre of the compound could be readily oxidised from +2 to +3 and formed a number of $[(C_5H_5)_2Fe]^+X^-$ derivatives from the blue cation. Wilkinson and Woodward first submitted their results in the middle of March 1952 followed by a series of papers concerning dicyclopentadienyl iron and analogues during June of that year. Meanwhile, Fischer and Pfab were working on their first publication until the end of June and unaware of Wilkinson's initial paper. Fischer's X-ray diffraction studies (Fischer and Pfab, 1952) gave unequivocal evidence of the 'sandwich' structure and postulated a 'double-cone' shape (**I.4**). The iron atom was located between the two planar,

(**I.4**)

parallel cyclopentadienyl rings but these rings were not mutually coincident and, in fact, were staggered in conformation. The structure gave a pentagonal antiprism, which Wilkinson had considered distinctly possible in his paper. It became obvious that the sandwich structure was due to good orbital overlap between the π-electrons in the *p* orbitals of the Cp rings and the *d* orbitals of the iron with the compound's high stability resulting from the π-complexation. Whilst this type of bonding had been suspected before, there had been no proof of it before this time. The discovery and recognition of this new type of bonding between metals and organic unsaturated molecules gave organometallic chemistry a whole new lease of life.

Fischer's group then went on to prepare and carry out crystal structure analysis on other analogues, i.e. sandwich structures involving cobalt and nickel, and they demonstrated that this type of structure was not unique to iron. Back in Harvard, Woodward discovered that the cyclopentadienyl rings were of a sufficiently aromatic nature (i.e. similar to benzene) to allow for such classic electrophilic aromatic substitution reactions as the Friedel–Crafts reaction to be carried out. This, and a number of other aromatic similarities of the Cp rings to the benzene moiety, led one of Woodward's postdoctoral fellows, Mark Whiting, to coin the name '**ferrocene**' (Woodward *et al.*, 1952). The entire class of transition metal dicyclopentadienyl compounds became quickly known as the '**metallocenes**', and this has since been expanded for compounds $[(\eta^5\text{-}C_5H_5)_2M]$ in general. For this and subsequent pioneering work in organometallic chemistry, Wilkinson and Fischer shared the Nobel Prize for Chemistry in 1973. (NB: Woodward was awarded the Prize in 1965 for his 'contributions to the art of organic synthesis'.)

1.3 Classification of metallocenes

In metallocene chemistry, ferrocene is still the main focus of research primarily due to its remarkable stability and ease of preparation. For example, the formation of the analogous ruthenocene (**I.5**) (Wilkinson, 1952) (formed

(**I.5**)

from ruthenium(III) acetylacetonate and cyclopentadienyl magnesium bromide) and osmocene (**I.6**) (Fischer and Grubert, 1959) (via osmium tetra-

$$(\text{I.6})$$

chloride and excess sodium cyclopentadienide) quickly followed on from the ferrocene studies. However, the chemistry of both of these heavier metallocenes has still not been thoroughly studied mainly because of the high cost and the lack of convenient synthetic routes which produce both high yields and substantial quantities. The enormous role that the recognition of the 'sandwich' concept of bonding has played in the advancement of organometallic chemistry cannot be disputed. The momentum generated by the discovery and analysis of ferrocene continues as a potent driving force, both via synthetic studies and in a theoretical sense, as the understanding of chemical bonding was significantly enhanced by its formulation.

To date, virtually every element and its complexes have been reacted with various cyclopentadienyl reagents (some with more success than others) and this has culminated in an enormous number of η^5-C_5H_5–metal complexes now being known. They can be classified in a number of ways:

[(η^5-C_5H_5)$_2$M]—symmetrical complexes with a classical 'sandwich' structure;

[(η^5-C_5H_5)$_2$ML$_x$]—metallocene species where the cyclopentadienyl rings are 'bent' or 'tilted', and where L represents a ligand such as H$^-$, R$^-$, halide, olefin, NO or other cyclopentadienyl ligands with variable hapticity (i.e. η^1, η^3,η^5);

[(η^5-C_5H_5)ML$_x$]—these complexes contain only one cyclopentadienyl ligand in addition to other ligands, L.

Within these classifications, there are five different structural types of compounds that can occur.

1 Parallel sandwich complexes.
2 Multi-decker sandwich complexes.
3 Half-sandwich complexes.
4 Bent or tilted sandwich complexes.
5 Complexes with more than two cyclopentadienyl ligands.

There now follows a brief introductory discussion on each of these five structural types. Most of the points and compounds mentioned now are featured and discussed more expansively later in the book.

1.4 Parallel sandwich complexes

1.4.1 Dicyclopentadienyl compounds of 3*d* elements

These are the most common but not always the most stable of the metallocene series of compounds. Following on from ferrocene the di-π-cyclopentadienyl compounds of the 3*d* elements have been obtained as neutral molecules and, excepting the manganese and titanium species, they are isomorphous with remarkably consistent melting points (see Table 1.1). They possess the same structure and essentially the same bonding as in ferrocene but with valence electron counts ranging from 15 to 20. (NB: A discussion of the '18-electron rule' and its relevance to metallocenes follows at the end of this chapter.)

With all the bonding and non-bonding orbitals filled for the group 8 metallocenes (Fe, Ru, Os) it is not surprising that these are the most stable members of the series. The metallocenes featuring cobalt and nickel as the central atoms have one and two electrons, respectively, in degenerate anti-bonding e^*_{1g} (d_{xz}, d_{yz}) orbitals and are thus paramagnetic. The occupancy of the anti-bonding orbitals weakens the metal–ring carbon bond and thereby increases the metal–carbon distance. Chromocene and vanadocene are also paramagnetic (possessing fewer than 18 electrons) and are liable to welcome additional ligands that will contribute extra electrons. This makes them sensitive to decomposition or oxidation by air in the order $Ni > Co > V \gg Cr$. With d^5 ions having no crystal-field stabilisation in their high-spin form Cp_2Mn is very reactive and considered strongly ionic in character. The other species in the series (i.e. Cr, Fe, Co) are considered to possess strong covalent bonding between the metal and the rings, illustrated by the dissociation energies of the metallocene cations (Table 1.2) (Cais and Lupin, 1970) (the data are for +1 cations as appearance potentials in the mass spectra of these compounds were used). There is some polarity in these metal–ring bonds but the compounds do not react like polar organometallic species

Table 1.1 Some di-η^5-cyclopentadienyl complexes of the first row transition metals.

Compound	Colour	Melting point (°C)
'Cp$_2$Ti'	Dark green	> 200 (decomposition)
Cp$_2$V	Purple	168
Cp$_2$Cr	Scarlet	173
Cp$_2$Mn	Amber	173
Cp$_2$Fe	Orange	173
Cp$_2$Co	Purple	174
Cp$_2$Ni	Green	173

Table 1.2 Dissociation energies of metallocene cations.

Reaction			ΔH (kJ mol^{-1})
Cp_2Mg^+ $\longrightarrow$	$CpMg^+$	$+ Cp$	310
Cp_2V^+ $\longrightarrow$	CpV^+	$+ Cp$	515
Cp_2Cr^+ $\longrightarrow$	$CpCr^+$	$+ Cp$	633
Cp_2Mn^+ $\longrightarrow$	$CpMn^+$	$+ Cp$	364
Cp_2Fe^+ $\longrightarrow$	$CpFe^+$	$+ Cp$	641
Cp_2Co^+ $\longrightarrow$	$CpCo^+$	$+ Cp$	754
Cp_2Ni^+ $\longrightarrow$	$CpNi^+$	$+ Cp$	524

(cf. the Grignard reagent).

$$RMgX + H_2O \longrightarrow RH + MgXOH \tag{1.6}$$

$$Cp_2Fe + H_2O \longrightarrow \text{No reaction} \tag{1.7}$$

1.4.2 Dicyclopentadienyl compounds of 4d and 5d elements

Few heavy transition metals (4d and 5d) form stable sandwich metallocene structures and the bulky Cp* (η-C$_5$Me$_5$) ligand is often used to obtain stable systems. (NB: The chemistry of bulky or substituted cyclopentadienyl groups has been reviewed elsewhere (Janiak and Schumann, 1991).) 'Cp$_2$Nd', 'Cp$_2$Mo' and 'Cp$_2$W' have similar structures to that of 'Cp$_2$Ti', whilst Cp$_2^*$Re is isolatable and Cp$_2$Re is unstable but can be formed at low temperatures. Of the lower congeners, only ruthenium and osmium form stable ferrocene analogues. 'Cp$_2$Rh' and 'Cp$_2$Ir' are unstable radicals and, to date, there are no Pd or Pt metallocenes.

1.4.3 Dicyclopentadienyl compounds of main group elements

Of course, the polyhapto bonding of π-ligands is not restricted to transition elements and there are many examples of main group elements acting as the central atoms. Initial examples date back to the late nineteenth century when arene complexes of gallium and antimony were reported by Lecoq de Boisbaudran and Smith, respectively. Over the years there has been a wealth of main group–cyclopentadienyl chemistry and it is now possible to classify the complexes via structural and bonding features, these features being dependent on the position of the element in the Periodic Table. Many of the structures for the transition metal complexes are mimicked by the range of main group element species.

Bonding in most main group element–cyclopentadienyl ligand species can be thought of as primarily covalent but with some additional degree of ionic character. Intermediate bonding situations are usually found in Cp$_n$E species, where $n = 1$, E = In, Tl; $n = 2$, E = Be, Mg, Pb; $n = 3$, E = Bi, whilst compounds of the alkali metals and heavier alkaline earth metals (e.g. Ca, Sr, Ba) feature ionic bonding and their structures are not defined by electron counting rules. With regard to parallel sandwich structures, magnesocene and its derivatives are the lightest compounds of this type and possess no d orbitals available for bonding. Although magnesocene is structurally very similar to ferrocene it is thought of as essentially ionic though this is still unresolved. In the crystalline state, a regular sandwich structure is observed with a staggered conformation (cf. the gas phase, where the eclipsed conformation seems to be apparent) and the metal–ring bonding distance is much longer in magnesocene than ferrocene. The former's lower force constant of the metal–ring stretch indicates a weaker metal–ring bond in magnesocene.

1.4.4. Dibenzenechromium (a benzene sandwich compound)

Whilst the cyclopentadienyl group is the best known aromatic ligand in organometallic chemistry there are several other cyclic conjugated ligands of considerable importance. Compounds featuring ligands $C_nH_n^{+,0,-}$ (where $n = 4, 6, 7$ or 8) tend not to be as stable as cyclopentadienyl species and their chemistry is more limited but in this 'parallel, sandwich structure' section a couple of important examples should be mentioned. Dibenzenechromium was first synthesised rationally in 1955 (Fischer and Hafner, 1955) and was the initial compound recognised as having η^6 bonding. It is an isoelectronic analogue of ferrocene and was formed by the reduction of $CrCl_3$ with Al metal in the presence of C_6H_6 using $AlCl_3$ as a catalyst and isolated as the air-stable cation.

$$3CrCl_3 + 2Al + AlCl_3 + 6C_6H_6 \longrightarrow 3[(\eta^6\text{-}C_6H_6)_2Cr]^+[AlCl_4]^- \qquad (1.8)$$

For some years after the synthesis, the structure and the symmetry of the complex (D_{3d} or D_{6h}) was a controversial subject. Finally, electron diffraction and low temperature X-ray diffraction studies gave a molecular structure comprising planar, parallel rings in eclipsed conformation above and below the chromium atom (D_{6h}) (**I.7**).

(**I.7**)

1.4.5 Uranocene (a cyclooctatetraene sandwich compound)

Arguably the best known metallocene not featuring cyclopentadienyl rings is 'uranocene', which possesses a parallel, sandwich structure (D_{8h}) involving

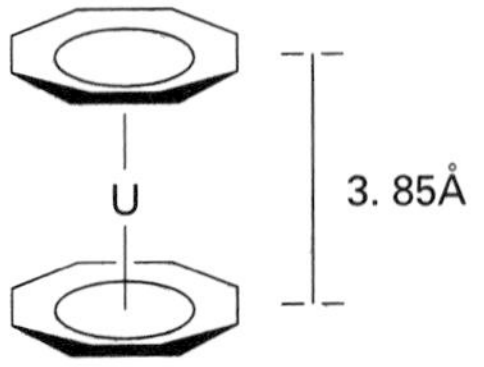

(I.8)

two η^8-cyclooctatetraene ligands about a central uranium atom (**I.8**). The rings are eclipsed and in many respects the complex is remarkably similar to ferrocene and other d-block cyclopentadienyls, hence the name uranocene. There are differences, however, the main one being that $[\eta^8\text{-}C_8H_8)_2U]$ is a 22-electron species and for sandwich complexes of the d-block this would lead to occupation of anti-bonding molecular orbitals (MOs). In this case, due to the available $5f$ orbitals, more bonding MOs can be constructed from interaction of symmetries e_2 and e_3 and the additional electrons accepted into a bonding situation (for full discussion, see Chapter 3).

1.5 Multi-decker sandwich complexes

The first triple-decker sandwich compound featuring cyclopentadienyl ligands was isolated and characterised in 1972. On treating nickelocene (paramagnetic with 20 valence electrons) with HBF_4 in propionic anhydride the 34 valence electron cation $[\eta^5\text{-}C_5H_5)_3Ni_2]^+$ (**I.9**) was obtained. Since this

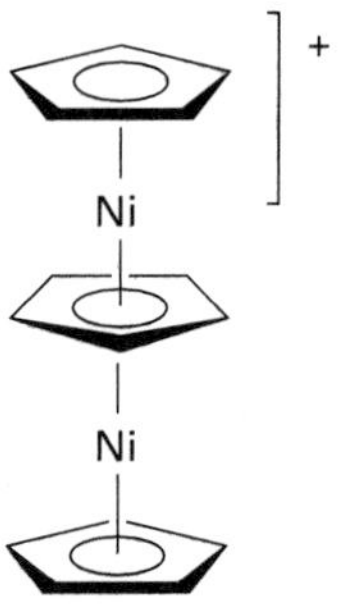

(I.9)

initial discovery, a range of multi-decker sandwich compounds containing heavier metals, larger rings or heterocycles (such as carborane, azacarborane or thiacarborane ligands) have been synthesised. Most triple-decker systems have 30 valence electrons but as different numbers of electrons can be accommodated systems with 26 ranging to 34 valence electrons are known.

1.6 Half-sandwich complexes

Half-sandwich complexes of the general type $[(\eta^5\text{-}C_5H_5)ML_n]$ ($n = 2$, 3 or 4) represent a major class of organometallic derivatives. As the cyclopentadie-

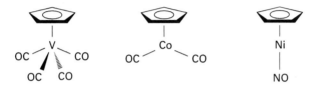

Fig. 1.2 Some examples of half-sandwich cyclopentadienyl compounds.

nyl group is very firmly bound and generally inert to both nucleophilic and electrophilic reagents it is often used as a stabilising ligand. There are a whole series of half-sandwich complexes sometimes referred to as two-, three- or four-legged 'piano stools'. When L is an excellent π-acid ligand (e.g. CO or NO) the 18-electron rule can be applied and stoichiometries predicted (e.g. Fig. 1.2). However, with ligands such as NH_3, PR_3 and SMe_2 (poor π-acceptor species) the 18-electron rule has less predictive value and complexes with 16 and 17 valence electrons are prevalent. These complexes frequently possess geometries with various degrees of distortion, e.g. in solution $[(C_5H_5)Cu(PR_3)]$ distorts to a 14-electron monohapto (η^1) geometry **(I.10)** rather than an 18-electron pentahapto (η^5) geometry **(I.11)**. The

(I.10)

(I.11)

chemistry of these 'piano stool' species is far more plentiful than that of the metallocenes mainly because fewer ligands can bind to the metallocenes without overstepping the 18-electron rule. The scope of this book deals primarily with those compounds featuring *at least* two metal-bound cyclopentadienyl ligands; therefore, half-sandwich compounds will not be discussed in any detail.

1.7 Bent or tilted sandwich complexes

Compounds featuring two η^5-C_5H_5 rings that are not parallel are also numerous and those featuring *d*-block species generally involve group 4 and heavier groups 5–7 elements. The structures arise out of the electron-deficient nature of the complexes, and in order to achieve the desired stable

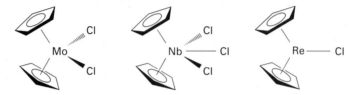

Fig. 1.3 Some examples of bent metallocenes.

18-electron configuration additional ligands which can contribute extra electrons are included. Up to three ligands are capable of being bound and when these ligands co-ordinate to the metal, a geometry is enforced whereby the cyclopentadienyl ligands are not planar (Fig. 1.3).

Group 5 metals can bind three additional ligands, e.g. Cp_2MX_3 where the metal is d^0, as all three valence orbitals of the fragment are used to bind the three σ-ligands, e.g. Cp_2NbCl_3 and Cp_2TaH_3. Due to the 18-electron rule, most metallocene complexes are restricted to metals with a low number of d electrons. In contrast, group 4 metallocenes prefer to form stable 16-electron species and bind only two singly donating ligands. This leaves an unpaired central orbital which can act as a Lewis acid, where the Ti interacts with a lone pair of a π-basic ligand such as alkoxide or aroxide (OR) **(I.12)**.

$$\text{(I.12)}$$

This is in contrast to molybdocene dihalides where this orbital is filled and acts as a Lewis base binding π-acceptor ligands such as ethene **(I.13)**. The

$$\text{(I.13)}$$

bent 'pseudo-tetrahedral' $[Cp_2MX_2]$ (M = Ti, Zr, Hf) complexes have been very well studied in recent years because of their utilisation in the Ziegler–Natta polymerisation of olefins. In 1957 it was discovered that $[Cp_2TiCl_2]/AlCl_3$ can form a heterogeneous catalysis system for the polymerisation of ethylene and this has recently been re-examined. One such observation was that homogeneous isotactic polymerisation of α-olefins could be effected by a catalyst featuring chiral zirconocene dihalide and methylaluminoxane (full details in Chapter 6).

Of the compounds featuring predominantly main group element–cyclopentadienyl ligand covalent bonding, bent sandwich structures tend to be the most common. This is exemplified by the metallocenes of silicon, germanium, tin and lead (Jutzi, 1986; Jutzi *et al.*, 1989) and follows from the lone pair requirements of the elements' + II state. The distorted sandwich structure allows for the presence of a stereochemically active lone pair on the metal (**I.14**).

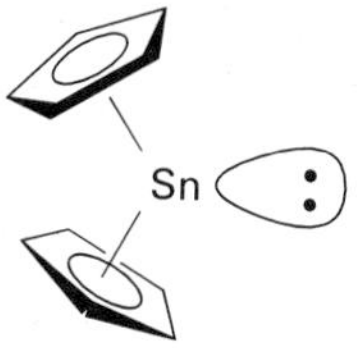

(**I.14**)

1.8 Complexes with more than two cyclopentadienyl ligands

1.8.1 Transition metal species

It is important to realise that C_5H_5 can also act as an η^1-ligand and forms a σ M—C bond giving rise to mixed compounds where two, three or four cyclopentadienide or cyclopentadienyl ligands are present. The 18-electron rule can help to induce one electron donor, σ-bonded arrangements where, for instance, one cyclopentadienyl ring is σ-bonded and another is π-bonded (**I.15**).

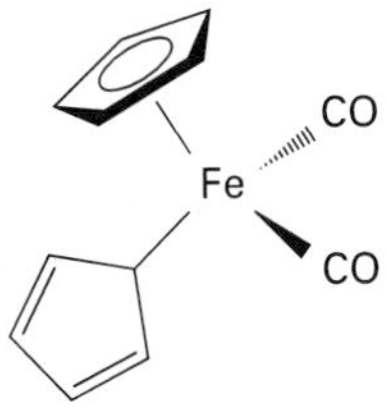

(**I.15**)

However, this can also be the case in situations that do not conform to the 18-electron rule, e.g.

$$MoCl_5 \xrightarrow{NaC_5H_5} [(\eta^5\text{-}C_5H_5)Mo(\eta^1\text{-}C_5H_5)_3] \qquad (1.9)$$

$$NbCl_5 \xrightarrow{NaC_5H_5} [(\eta^5\text{-}C_5H_5)_2Nb(\eta^1\text{-}C_5H_5)_2] \qquad (1.10)$$

Other examples with more than one Cp ligand are the tri- and tetra-cyclopentadienyl titanium and zirconium species. The number and type of bonding of these Cp ligands is dependent on the size of the metal rather than the 18-electron rule.

1.8.2 Lanthanide and actinide species

Another class of compounds featuring a number of Cp ligands are the lanthanide and actinide complexes (Baker *et al.*, 1976). In general, tricyclopentadienides are formed and for syntheses of all the lanthanides and many of the actinides ligand exchange reactions have been effective, e.g.

$$2PuCl_3 + 3[(C_5H_5)_2Be] \longrightarrow 2[(C_5H_5)_3Pu] + 3BeCl_2 \qquad (1.11)$$

The compounds tend to be air and moisture sensitive, thermally stable and display properties characteristic of ionic materials. This is especially true of the lanthanides as the covalency of the metal–carbon bond increases in the order lanthanide—C < actinide—C < transition metal—C. In the actinide complexes, the $5f$ orbitals are subject to less shielding than $4f$ orbitals, thereby overlapping more extensively and contributing more covalent interaction. The cyclopentadienyls form three main types: $[(\eta^5\text{-}C_5H_5)_3An^{III}]$, $[(\eta^5\text{-}C_5H_5)_4An^{IV}]$ and $[(\eta^5\text{-}C_5H_5)_3An^{IV}X]$ (where X = halogen, alkyl and alkoxy group, An = actinide element). The tricyclopentadienyls (An = U → Cf) are seemingly isostructural with $[(C_5H_5)_3Sm]$ which features both η^5 and bridging ring systems. Each ring is pentahapto towards one metal atom but some additionally can bridge by presenting a ring vertex (η^1) or edge (η^2) towards an adjacent metal atom thereby building up a chain structure. On the other hand, the $[(C_5H_5)_4An^{IV}]$ species (An = Th → Np) contain four identical η^5-rings arranged tetrahedrally around the metal atom. Four η^5-cyclopentadienyl ligands can only be accommodated by the actinide metals due to the size considerations, the availability of $5f$ orbitals and the ability of the actinides to engage in covalent bonding.

1.9 Simple, predictive bonding theories: 'the octet rule' and 'the 18-electron rule'

In organometallic chemistry there are predictive functions that can act as simple 'rule-of-thumb' starting points for discussion of structure and bonding. There are exceptions but main group elements following the '*octet rule*' and, more importantly, transition metals fulfilling the '*18-electron rule*' result in compounds that normally represent relative energy minima.

1.9.1 The octet rule

A main group element, e.g. silicon, can use its four valence electrons in the formation of four σ-bonds ('*2-centre, 2-electron*' bonds) thereby giving rise to a stable system such as $SiMe_4$. The stability of $SiMe_4$ (and the instability of other species) can be predicted by the '*octet rule*' which states that 'a sufficient number of bonds are formed to surround the central element by an octet of electrons'. Therefore, the element reaches the electron configuration of the noble gas with the next highest atomic number.

1.9.2 The 18-electron rule

As stated earlier, there are exceptions to this rule but, in general, transition metal species that do follow it tend to be particularly stable. It states 'a stable complex, with an electron configuration of the next highest noble gas, is obtained when the sum of metal d electrons, electrons donated from the ligands and the overall charge of the complex equals 18'. It is also known as the '*inert gas rule*' or '*effective atomic number rule*' and the application of the rule to metallocenes will be discussed. With regard to electron counting conventions, it is useful to treat anionic (or cationic) ligands as neutral radicals and the metal centre as zerovalent (thereby avoiding discussions about formal oxidation states). The other option is to distribute charges in an 'ionic' fashion and consider metals as cations and ligands as anions (Scheme 1.2). Considering formal oxidation states, since $Na^+C_5H_5^-$ exists as an ionic substance, it is reasonable to regard C_5H_5 in ferrocene as anionic as well and therefore assign a formal oxidation state of +2 to iron. It should be remembered though that bonding in complexes such as ferrocene is predominantly covalent.

The rule has considerable predictive value in that the composition of many transition metal complexes may be anticipated from the combination of sets of ligands with transition metals of appropriate d electron count (Scheme 1.3). The rule is satisfied by many of the stable, diamagnetic d-block complexes, but exceptions to the rule are reasonably common since antibonding orbitals are often readily accessible and vacancies in the nonbonding orbitals do little to stabilise the complex. It generally works well for

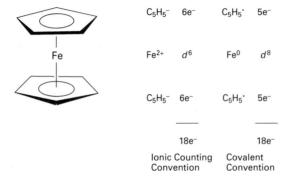

	$C_5H_5^-$	$6e^-$	$C_5H_5 \cdot$	$5e^-$
Fe	Fe^{2+}	d^6	Fe^0	d^8
	$C_5H_5^-$	$6e^-$	$C_5H_5 \cdot$	$5e^-$
		$18e^-$		$18e^-$
	Ionic Counting Convention		Covalent Convention	

Scheme 1.2 Electron counting conventions using ferrocene as the example.

15

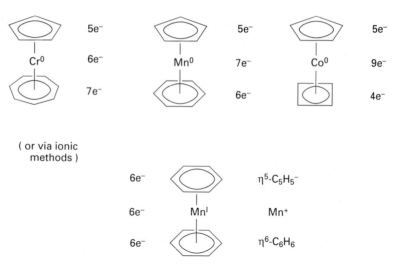

Scheme 1.3 Examples of electron counting for various metal sandwich complexes.

organometallic compounds but not for co-ordination compounds. A perspective view of co-ordination chemistry suggests a division into three classes (Table 1.3).

Class I

The splitting Δ_o is relatively small for $3d$ metals as well as for ligands at the lower end of the spectrochemical series. NB: The **spectrochemical series** orders ligands according to their ability with respect to crystal field splitting, e.g. halides are *weak field* ligands and phosphines and cyanides are *strong field* ligands:

$$I^- < Br^- < [NCS]^- < Cl^- < F^- < [OH]^- < [ox]^{2-} \sim H_2O < [NCS]^- < NH_3 < en < [CN]^- < PR_3$$

t_{2g} is non-bonding and can be occupied by 0–6 electrons.
e_g^* is weakly anti-bonding and can be occupied by 0–4 electrons.

Table 1.3 The classes of co-ordination compounds.

Class	No. of valence electrons	18-valence electron rule
I	←16, 17, 18, 19→	Not obeyed
II	←16, 17, 18	Not obeyed
III	18	Obeyed

Therefore, 12–22 valence electrons (VE) can be accommodated, i.e. the 18 VE rule is not obeyed. Tetrahedral complexes, with small Δ_t values, also belong to this class. Examples are: $[TiF_6]^{2-}$ (12 VE), $[Cr(NCS_6)]^{3-}$ (15 VE), $[Fe(C_2O_4)_3]^{3-}$ (17 VE), $[Cu(NH_3)_6]^{2+}$ (21 VE).

Class II

Δ_o is larger for $4d$ and $5d$ metals (especially in high oxidation states) and for σ-ligands in the intermediate and upper ranges of the spectrochemical series.
t_{2g} is essentially non-bonding and can be occupied by 0–6 electrons.
e_g^* is more strongly anti-bonding and thus no longer available for occupancy.
Consequently, the valence shell contains 18 VE or less. A similar splitting Δ_o is also observed for complexes of the $3d$ metals with ligands possessing extremely high ligand field strength, such as CN^-. Examples are: WCl_6 (12 VE), $[OsCl_6]^{2-}$ (16 VE), $[PtF_6]^-$ (17 VE), $[W(CN)_8]^{4-}$ (18 VE).

Class III

Δ_o is largest for ligands at the high end of the spectrochemical series (good π-acceptors like CO, PF_3, olefins, arenes).
t_{2g} becomes bonding due to the interaction with π-orbitals of the ligands and should be occupied by six electrons.
e_g^* is strongly anti-bonding and remains empty.
Therefore, the 18 VE rule is obeyed unless steric reasons prevent the attainment of an 18 VE shell. Examples are: $[CH_3Co(CO)_4]$, $[Fe(CN)_6]^{4-}$, $[Ni(CNR]_4]$, $\{[(C_5H_5)Cr(CO)_3]_2\}$ (all 18 VE). Organometallic compounds of the transition metals almost exclusively belong to Class III.

It should be noted that organometallic compounds with 16 VE are common on the right of the d-block because the ligand field stabilisation energy of d^8 complexes favours a low-spin, square-planar configuration, with Δ being large and the $d_{x^2-y^2}$ orbital empty and the d_{z^2} stabilised. Exceptions to the 'Rule' are also sometimes seen to the left of the d-block. Here, steric and electronic factors are in competition and it is not always possible to crowd enough ligands around the metal either to satisfy the rule or to permit dimerisation. In general, the 18-electron rule is not obeyed as consistently by metallocenes as by carbonyls and nitrosyls and their derivatives. When dealing with cyclopentadienyl sandwich complexes, ferrocene is the ultimate example of a compound following the 18-electron rule. Most transition metal sandwich complexes do follow the rule but it is also the case that the favourable metallocene structure can frequently override the requirements of the 18-electron rule. For instance, first-row metallocenes feature electron counts ranging from 15 to 20, e.g. (see p. 18)

$[(\eta\text{-}C_5H_5)_2V] = 15e^-,\ [(\eta\text{-}C_5H_5)_2Cr] = 16e^-,\ ([\eta\text{-}C_5H_5)_2Mn] = 17e^-,$
$[(\eta\text{-}C_5H_5)_2Fe] = 18e^-,\ [(\eta\text{-}C_5H_5)_2Co] = 19e^-,\ [(\eta\text{-}C_5H_5)_2Ni] = 20e^-$

and 'uranocene' $[(\eta^8\text{-}C_8H_8)_2U]$ (**I.8**) is formally a 22-electron compound.

Although the 18-electron structure is not a prerequisite for formation of a sandwich structure, only ferrocene shows exceptional thermal stability (up to 500°C) and is not air sensitive but oxidation of cobaltocene to $[(C_5H_5)_2Co]^+$ then gives a cation which possesses much of the thermal stability of fer-rocene. Cobaltocene and nickelocene are 19- and 20-valence electron com-plexes respectively and their chemistry is influenced by their tendency to attain an 18-valence electron configuration (also detailed in Chapter 4) (Scheme 1.4). The ease of oxidation of cobaltocene makes it an attractive one-electron reductant and the rate constants for one-electron oxidation to the 18-electron cationic species are about an order of magnitude faster than the corresponding rate constants for ferrocene. Nickelocene can undergo up to a one-electron reduction and a two-electron oxidation. Additionally, the 18 VE ions $[Cp_2Rh]^+$ and $[Cp_2Ir]^+$ are also very stable, as opposed to the neutral monomers Cp_2Rh and Cp_2Ir which as yet have only been observed by *matrix isolation*. (NB: Matrix isolation is a technique whereby unstable compounds, radicals and intermediates are trapped in an inert or reactive solid matrix by co-condensing the matrix, e.g. argon, and the species to be studied at low temperatures of *c.* 4 to 20 K.)

In bent sandwich complexes that normally feature additional ligands other than the cyclopentadienyls, the 18-electron rule tends to restrict the complexes to metals possessing a low number of *d* electrons (Scheme 1.5).

These examples show that the compounds do not rigidly conform to the 18-electron rule and there are a number of structurally related complexes with 16, 17 and 18 electrons. Coupled with this, the previously mentioned 22-electron compound $[(\eta^8\text{-}C_8H_8)_2U]$ is also a stable entity. The number of electrons in these complexes and consequently their stability and behaviour

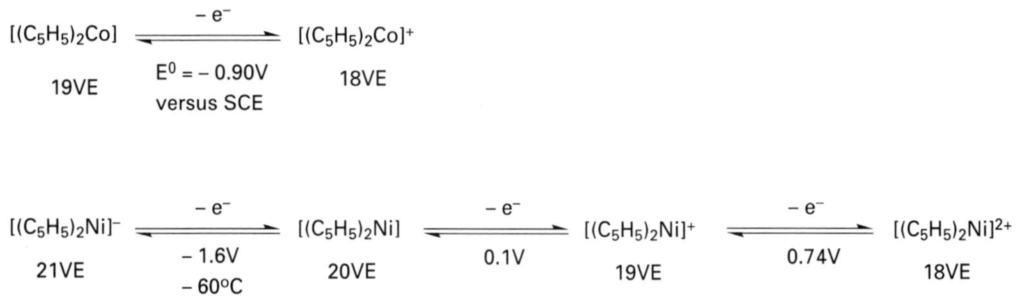

$[(C_5H_5)_2Co]$ $\xrightleftharpoons{\ -e^-\ }$ $[(C_5H_5)_2Co]^+$

19VE $E^0 = -0.90V$ 18VE
        versus SCE

$[(C_5H_5)_2Ni]^-$ $\xrightleftharpoons{\ -e^-\ }$ $[(C_5H_5)_2Ni]$ $\xrightleftharpoons{\ -e^-\ }$ $[(C_5H_5)_2Ni]^+$ $\xrightleftharpoons{\ -e^-\ }$ $[(C_5H_5)_2Ni]^{2+}$

21VE $-1.6V$ 20VE   0.1V   19VE   0.74V   18VE
      $-60°C$

Scheme 1.4 Oxidation and reduction of cobaltocene and nickelocene derivatives.

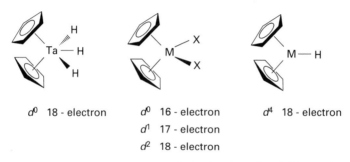

d^0 18 - electron d^0 16 - electron d^4 18 - electron

d^1 17 - electron

d^2 18 - electron

Scheme 1.5 The 18-electron rule applied to 'bent' metal sandwich complexes.

can be explained by considering the simple molecular orbital picture for each type of metallocene structure and this is fully discussed in Chapter 3.

References

Baker, E.C., Halstead, G.W. & Raymond, K.N. (1976) *Struct Bond*, **25**, 23.

Cais, M. & Lupin, M.S. (1970) *Adv Organomet Chem*, **8**, 211.

Fischer, E.O. & Pfab, W. (1952) *Z Naturforsch*, **B7**, 378.

Fischer, E.O. & Hafner, W. (1955) *Z Naturforsch*, **B10**, 665.

Fischer, E.O. & Grubert, H. (1959) *Chem Ber*, **92**, 2302.

Frankland, E. (1849) *J Chem Soc*, **2**, 263.

Frankland, E. (1852) *Phil Trans*, **142**, 417.

Grignard, V. (1900) *C R Acad Sci*, **130**, 1322.

Janiak, C. & Schumann, H. (1991) *Adv Organomet Chem*, **33**, 291.

Jutzi, P. (1986) *Adv Organomet Chem*, **26**, 217.

Jutzi, P., Hoffmann, H., Kanne, D. *et al.* (1989) *Chem Ber*, **122**, 1629.

Kealy, T.J. & Pauson, P.L. (1951) *Nature*, **168**, 1039.

Miller, S.A., Tebboth, J.A. & Tremaine, J.F. (1952) *J Chem Soc*, 632.

Mond, L. (1890) *J Chem Soc*, **57**, 749.

Wilkinson, G. (1952) *J Am Chem Soc*, **74**, 6146.

Wilkinson, G. (1975) *J Organomet Chem*, **100**, 273.

Wilkinson, G., Rosenblum, M., Whiting, M.C. & Woodward, R.B. (1952) *J Am Chem Soc*, **74**, 2125

Woodward, R.B., Rosenblum, M. & Whiting, M.C. (1952) *J Am Chem Soc*, **74**, 3458.

Zeise, W.C. (1827) *Ann Phys*, **40**, 234.

2 Synthesis and Physical Properties of Metallocenes

2.1 The methods of synthesis of binary cyclopentadienyl transition metal complexes

There are three main routes that are normally employed in the formation of these types of compounds.

2.1.1 Using a metal salt and cyclopentadienyl reagents

The normal starting point is the 'cracking' of dicyclopentadiene. This involves a retro Diels–Alder reaction to produce the monomeric and fairly unstable C_5H_6 (Scheme 2.1). Being a weak acid ($pK_a = 15$), cyclopentadiene can be deprotonated by strong bases or alkali metals. Sodium cyclopentadienide (NaCp) is the preferred reagent for this type of reaction, the most versatile synthetic method of forming simple metallocenes.

$$MCl_2 + 2NaC_5H_5 \xrightarrow[-\text{NaCl}]{} [(C_5H_5)_2M] \tag{2.1}$$

(M = V, Cr, Mn, Fe, Co; Solvent = THF, DME, $NH_{3(l)}$)

NaCp acts as a reducing agent (e.g. in Equation 2.2(i)) but depending on the reaction conditions sometimes forms cyclopentadienyl–metal hydrides or complexes containing σ-bonded cyclopentadienyl rings when added to metal (4d, 5d) salts (e.g. in Equation 2.2(ii)):

$$CrCl_3 + 3NaC_5H_5 \longrightarrow [(C_5H_5)_2Cr] + 1/2C_{10}H_{10} + 3NaCl \tag{2.2(i)}$$

$$TaCl_5 \xrightarrow{\text{NaCp}} [(\eta^5\text{-}C_5H_5)_2TaCl_3] \xrightarrow{\text{NaCp}} [(\eta^5\text{-}C_5H_5)_2Ta(\eta^1\text{-}C_5H_5)_2] \tag{2.2(ii)}$$

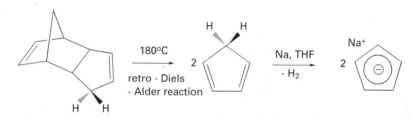

Scheme 2.1 The 'cracking' of cyclopentadiene and formation of sodium cyclopentadienide.

2.1.2 Using a metal and cyclopentadiene

$$M + C_5H_6 \longrightarrow MC_5H_5 + 1/2H_2 \ (M = Li, Na, K)$$

$$M + 2C_5H_6 \xrightarrow{500\,°C} [(C_5H_5)_2M] + H_2 \ (M = Mg, Fe) \tag{2.3}$$

In this context, the '*metal vapour synthesis*' or '*co-condensation method*' is employed extensively with the less electropositive metals. It is possible to use vapours such as atoms of transition metals, carbon atoms and molecules that may only exist in the gaseous phase (e.g. SiF_4, TiO) as routine reagents in synthesis and catalysis. Pimentel, Linevsky and Skell conducted pioneering work in the technique of vapour synthesis and this was logically extended to include transition metals by Timms and Turney (1977). The highly reactive atoms or molecules are generated at high temperatures in a vacuum and then brought together with the chosen co-reactants on a cold surface. Complexes are then formed on warming the system to room temperature (although metal aggregation is sometimes possible). The technique provides synthetic advantages of using metal atoms in the vapour phase rather than the solid metal, i.e. Cr or Ti atoms brought together with benzene form di(η^6-benzene)chromium in 60% yield or di(η^6-benzene)titanium in 40% yield, when they cannot be made directly from the solid metal. Although metal vapour synthesis is simple, installation and maintenance of the apparatus is not trivial. Therefore, it must have definite, recognisable advantages over other methods and this can be achieved in: (i) the synthesis of new and interesting compounds; (ii) improved synthesis of known compounds; and (iii) providing a quick, convenient research tool (Blackborow and Young, 1974).

2.1.3 Using a metal salt and cyclopentadiene

If the salt anion has poor basicity and cannot deprotonate cyclopentadiene, an auxiliary base can be utilised to generate the cyclopentadienyl anions *in situ* which can sometimes be more convenient.

$$Tl_2SO_4 + 2C_5H_6 + 2OH^- \xrightarrow{H_2O} 2TlC_5H_5 + 2H_2O + SO_4^{2-} \tag{2.4}$$

$$FeCl_2 + 2C_5H_6 + 2C_2H_5NH \longrightarrow [(C_5H_5)_2Fe] + 2[(C_2H_5)_2NH_2]Cl \tag{2.5}$$

Alternatively, a reducing agent is required:

$$RuCl_3(H_2O)_x + 3C_5H_6 + 3/2Zn \xrightarrow{C_2H_5OH} [(C_5H_5)_2Ru] + C_5H_8 + 3/2Zn^{2+} \tag{2.6}$$

The following sections will discuss the formation and chemical properties of the common cyclopentadienyl–element compounds, focusing on the dicyclopentadienyl species but also including the important tri- or tetra-cyclopenta-dienyl compounds and the well-known larger polyalkenyl ring sandwich compounds. Detailed information and further references can be obtained from consulting relevant sections in *Comprehensive Organometallic Chemistry I* (1982), *Comprehensive Organometallic Chemistry II* (1995) and the *Dictionary of Organometallic Compounds* (1995) (see Bibliography).

2.2 Cyclopentadienyl compounds of group 2 elements

2.2.1 Dicyclopentadienylberyllium (beryllocene)

Dicyclopentadienylberyllium can be prepared by the reaction of NaCp and $BeCl_2$ and, following sublimation, is obtained as colourless crystals (melting point 59–60°C) that decompose above 70°C. It is unstable to air and moisture (reacting violently with water), highly toxic and possesses a sizeable dipole moment (2.46(6) D in C_6H_6 and 2.24(9) D in C_6H_{12}). The compound features unusual modes of co-ordination of the Cp group to the metal atom. Several possible structures have been proposed and these have been the subject of controversy. It soon became clear that there was no possible symmetrical sandwich structure (like magnesocene or ferrocene) as the compound is polar in solution. Early electron diffraction data of the gaseous species indicated a molecule possessing C_{5v} symmetry featuring the beryllium atom situated unsymmetrically between the two parallel, staggered rings on a fivefold rotation axis (the perpendicular Be–ring distances being 0.1472(6) and 0.1903(8) nm). There followed other postulations supported by spectroscopic studies before a low temperature X-ray diffraction experiment concluded that the structure is disordered and the two rings are related by a centre forming a '*slip-sandwich*' or '*peripherally-bonded*' structure (Beattie and Nugent, 1992) (Fig. 2.1). The striking feature is that although the

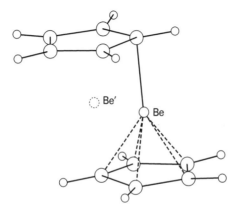

Fig. 2.1 The molecular structure of beryllocene determined by low temperature X-ray crystallography. (Wild *et al.* (1982), with permission.)

rings are parallel, one has slipped sideways by about 0.12 nm. The beryllium atom is 0.153(3) nm from the plane of one ring on the fivefold symmetric axis with all the Be–C distances 0.194(4) nm and is therefore thought of as π-bonded. The second ring which has 'slipped' features the shortest Be–C distance perpendicular to the plane of the ring and is therefore considered as σ-bonded to the beryllium centre. It is also interesting to note that in variable temperature ^{1}H nuclear magnetic resonance (NMR) spectra between − 100 and 50°C, there is only one single sharp peak indicating that all the hydrogen atoms are equivalent as the rings rotate relative to each other with the beryllium alternating rapidly between the two possible positions.

An alternative theory to the slipped situation is that there is pentahapto- and trihapto-bonding present (**II.1**). Raman and He(I) photoelectron spectra

$$(\text{II.1})$$

support this view along with electron diffraction data obtained on the vapour above 100°C. The unusual structure of Cp_2Be compared to symmetrically π-bonded metallocenes is a result of the ionic radius of Be^{2+} (0.03–0.04 nm). This is the smallest of any known metal cation and the metal forces the two rings into an unsymmetrical arrangement to minimise interplanar repulsion. In general, the 'peripheral' bonding in metallocenes is often thought of as monohapto (η^1) but it can also encompass trihapto (η^3) types of structure too. In cases where only one metal orbital is available for bonding with the ring, there are no essential bonding features that indicate a difference between mono- and tri-hapto bonding. The rationalisation for these and other types of metal–cyclopentadienyl bonding is the polarity of the M—C bond and the number of co-ordination positions on the metal atom available for bonding with the cyclopentadienyl ring.

2.2.2 Dicyclopentadienylmagnesium (magnesocene)

The direct reaction of halogeno-organic compounds with magnesium in polar solvents gives Grignard reagents, but in hydrocarbons the products are halogen deficient and chlorocarbons give species that are essentially MgR_2. A direct reaction between cyclopentadiene or alkyl-substituted cyclopenta-dienes and magnesium metal occurs at 450–600°C. The cyclopentadiene vapour is passed over heated magnesium turnings in an inert atmosphere to form Cp_2Mg (see Equation 2.7).

$$2C_5H_6 + Mg \longrightarrow [(C_5H_5)_2Mg] + H_2 \qquad (2.7)$$

Magnesocene is also formed following pyrolysis of [Mg(Br)Cp] and sublimation *in vacuo*. It is an air- and moisture-sensitive colourless solid (m.p. 176°C) and forms complexes with co-ordinated solvent or ligands such as ethers and amines. The thermally stable species possesses a typical monomeric sandwich structure in the solid state with staggered parallel C_5H_5 rings (Bünder and Weiss, 1975). The Mg–C distance (0.2304(8) nm) is relatively long compared to iron and cobalt dicyclopentadienyl analogues (although shorter than in calcocene) and this suggests significant ionic bonding character, i.e. $Cp^-Mg^{2+}Cp^-$ (Fig. 2.2). Infrared (IR) and Raman studies on magnesocene in the solid or the melt support the ionic situation; however, a covalent contribution to the bonding has not been discounted. Dicyclopentadienylmagnesium in ethers appears to exist essentially as a contact triple ion; however, on addition of hexamethylphosphorus triamide (HMPT), the formation of solvent-separated ions occurs in a two-step process, the fully-

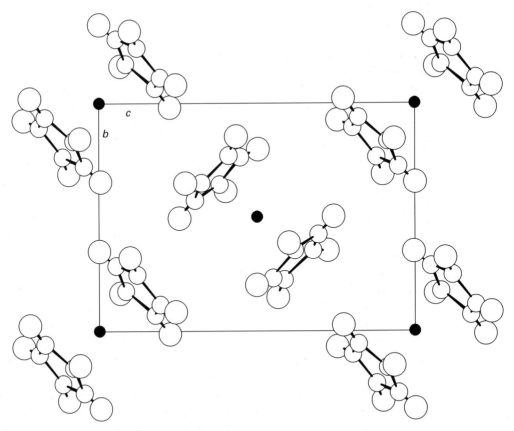

Fig. 2.2 The structure and unit cell packing of dicyclopentadienylmagnesium. (Bünder and Weiss (1975), with permission.)

separated species arising when the ratio [HMPT] : [Cp$_2$Mg] is >3. Electron diffraction studies in the gas phase support the metallocene sandwich structure with an Mg—C bond distance of 0.234 nm. However, the cyclopentadienyl rings appear to adopt an eclipsed geometry as opposed to the staggered form in the crystal. The compound has found use in the formation of a range of η^5-cyclopentadienyl metal compounds and it can be reacted with a transition or main group metal precursor in a suitable solvent.

2.2.3 Dicyclopentadienylcalcium (calciocene or calcocene)

The first well-characterised organo-alkaline earth metal compounds formed were dicyclopentadienylcalcium and analogous ring-substituted derivatives. The methods used were either reaction of CaCl$_2$ with cyclopentadiene in liquid NH$_3$ or direct reaction of the metal with cyclopentadiene in tetrahydrofuran (THF), liquid NH$_3$ or dimethylformamide (DMF).

$$2C_5H_6 + CaCl_2 \longrightarrow [(C_5H_5)_2Ca] + 2HCl$$
$$2C_5H_6 + M \longrightarrow [(C_5H_5)_2M] + H_2$$
$$(M = Ca, Sr)$$

(2.8)

The compound is obtained as a faint yellow powder or colourless crystals after sublimation at 200°C *in vacuo*. It is sensitive to oxygen and water and not particularly soluble and easily forms adducts with N- and O-donor ligands, e.g. *N,N,N',N'*-tetramethylethylenediamine (TMEDA), 2,2'-bipyridine or 1,2-dimethoxyethane. IR studies indicate an essentially ionic (C$_5$H$_5^-$ and M^{2+}) species and the usual singlets are observed in ^{1}H NMR spectra even at −100°C. Another indication of the ionicity are the high yields of Cp$_2$Hg, Cp$_2$Fe and Cp$_2$TiCl$_2$ formed from Cp$_2$Ca and HgCl$_2$, FeCl$_2$ and TiCl$_4$, respectively. An even better cyclopentadienylating agent is the bis-THF adduct [(C$_5$H$_5$)$_2$Ca·2THF] and this reacts readily in polar organic solvents with metal halides. The two solvent molecules can be removed by heating under high vacuum.

The unsolvated [(C$_5$H$_5$)$_2$Ca] species displays a significant increase in the interaction between the cyclopentadienyl rings and the calcium atom, as opposed to the situation in solution. X-ray analysis of crystalline Cp$_2$Ca shows a polymeric structure with bridging cyclopentadienyl ligands (Fig. 2.3). The coordination sphere around the calcium atom consists of two η^5-groups (A and B) (Ca—C c. 0.28 nm), one η^3 (C) (Ca—C 0.295 nm) and one η^1 (D) with a Ca—C bond of 0.31 nm. Three rings (A, B, C) are arranged in a trigonal orientation whilst the fourth lies perpendicular to the plane of the centroids of the other rings (Fig. 2.4). The long mean Ca—C distances of η^5-C$_5$H$_5$ rings (0.28 nm) indicate mainly ionic bonding but some covalent contribution from the Ca 4s, 4p and 3d orbitals is likely. The geometry of the

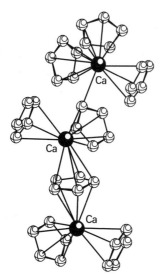

Fig. 2.3 Part of the infinite structure of $(Cp_2Ca)_\infty$.

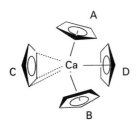

Fig. 2.4 A representation of the molecular structure of crystalline Cp_2Ca.

Cp rings around the metal atom is due to (i) steric interactions which are directly dependent on the size of the metal atom radius and (ii) maximum interaction between the metal atom and the Cp groups either by metal orbital–ligand orbital overlap, electrostatic interaction or both.

Di(pentamethylcyclopentadienyl)calcium can be synthesised by the reaction of pentamethylcyclopentadiene with calcium metal (m.p. 207–210°C) and many adducts of the species can be formed, including bis-THF, bipy, PEt_3 and pyradazine complexes. Base-free di(pentamethylcyclopentadienyl)calcium crystallises with two crystallographically independent but identical molecules in the unit cell. As in the gas phase structure, but unlike the polymeric Cp_2Ca metallocene structure described above (Fig. 2.3), Cp_2^*Ca contains two η^5-Cp* rings flanking a Ca core in a monomeric bent metallocene geometry. The ring centroid–Ca–ring centroid angles are 147.7° with average Ca—C distances of 0.264(1) nm. The structure backs up the strongly linear relationship between the bending angle in the complexes and the ionic radius of the metal centre. It also indicates that addition of substituents to the Cp ring (and/or Lewis base to the metal) increases the bulk around the metal centre, encouraging the formation of monomeric complexes.

2.2.4 Dicyclopentadienylstrontium (strontiocene or strontocene)

Strontiocene $[(C_5H_5)_2Sr]$ has been known for over 30 years and is thermally stable but has proved difficult to obtain. Initial preparations featured SrH_2 and C_5H_6 reacted at high temperatures to give the colourless strontiocene after sublimation. More recently, metal–vapour synthesis techniques have been employed with success along with the solution synthesis from lithium or potassium cyclopentadienes and strontium dihalide in THF. The colourless solid is soluble in THF and DMF, forming a THF complex $[(C_5H_5)_2Sr(THF)]$ which is a white solid and melts at > 400°C. Due to the high sublimation temperatures and the similarity of the IR spectrum with that of the Ca analogue, it is expected that strontiocene is polymeric in the crystalline phase.

2.2.5 Dicyclopentadienylbarium (bariocene or barocene)

This compound is not easily formed by normal solution techniques. It has been synthesised in low yields from BaH_2 and cyclopentadiene at 350–400°C. However, co-condensation of Ba vapour with cyclopentadiene at −196°C and then warming does produce Cp_2Ba in high yields. It is a colourless, crystalline solid, sparingly soluble in THF and soluble in DMF. The THF-adduct $[(C_5H_5)_2Ba(THF)_2]$ can be prepared in high yields from the reaction of lithium or potassium cyclopentadienes and a barium dihalide in THF. No structural data are available as yet but the IR spectra have been interpreted to illustrate essentially ionic $C_5H_5^-$ and Ba^{2+} bonding to be present. ^1H NMR spectra show the expected singlet resonance at temperatures down to −100°C.

2.3 Cyclopentadienyl and octatetraenyl compounds of group 3, lanthanide and actinide elements

2.3.1 Tricyclopentadienylscandium

The tricyclopentadienylscandium species was amongst the first series of organometallic lanthanide compounds to be reported. It can be synthesised from $ScCl_3$ and NaCp in THF and, following sublimation, isolated as a straw-coloured crystalline solid, melting at 240°C.

$$ScCl_3 + 3NaC_5H_5 \xrightarrow{\text{THF}} [(C_5H_5)_3Sc] + 3NaCl \qquad (2.9)$$

It is soluble in pyridine, THF and dioxane but insoluble in toluene. It is sensitive to air and decomposes rapidly in the presence of water to yield cyclopentadiene and the metal hydroxide. It was one of the first tricyclopen-

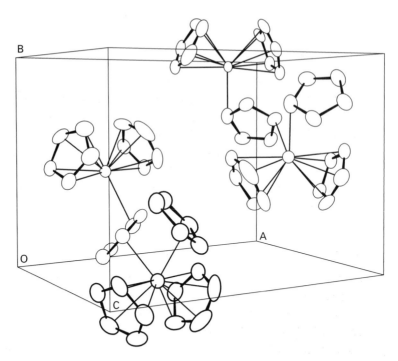

Fig. 2.5 The solid state structure and unit cell packing of [(C$_5$H$_5$)$_3$Sc]. (Atwood and Smith (1973), with permission.)

tadienyl species to be analysed accurately by X-ray diffraction (Atwood and Smith, 1973) (Fig. 2.5). The polymeric structure consists of [(C$_5$H$_5$)$_2$Sc] units bridged together by the remaining cyclopentadienyl groups. The environment of the scandium atom is such that two rings are coordinated in a pentahapto fashion (Sc—C average distance is 0.249 nm) and two rings are associated through essentially only one carbon atom (Sc—C bond distance average is 0.257 nm), i.e. [Sc(η^5-Cp)$_2$(μ-η^1:η^1-Cp)]$_x$ thus forming an infinite zig-zag chain arrangement. It is thought that the small size of the scandium ion (in relation to other lanthanide ions) produces the unusual monohapto linkages, as a formal co-ordination number of 9 would be required for scandium if all three Cp ligands were η^5-bound. There seems an indication of a considerable amount of covalency in this organoscandium compound.

2.3.2 Tricyclopentadienylyttrium

This compound can be synthesised from NaCp and YCl$_3$ and, after sublimation, isolated as a pale yellow crystalline solid. It has a melting point of 295°C and is soluble in THF but insoluble in most other common organic solvents. It is unstable to air and moisture and from magnetic susceptibility and chemical reactivity data can be thought of as a primarily ionic system.

$$YCl_3 + 3NaC_5H_5 \xrightarrow{\text{THF}} [(C_5H_5)_3Y] + 3NaCl \tag{2.10}$$

With NaH, it acts as a good catalyst for hydrogenation and/or bromination of alkenes and can effect the hydrometallation of 1-hexene. Molecular weight measurements in THF solution indicate that Cp_3Y is monomeric in solution and the THF adduct $[(\eta^5\text{-}C_5H_5)_3Y.OC_4H_8]$ has been investigated by X-ray diffraction. The $Y-C(\eta^5)$ bond distances average 0.271 nm which are longer than expected and appear to result from a severe steric problem due to the crowding of three η^5-cyclopentadienyl groups and one THF ligand around the yttrium atom.

2.3.3 Dicyclopentadienyllanthanides

A summary of the properties of **dicyclopentadienyleuropium, dicyclopentadie-nylytterbium** and **dicyclopentadienylsamarium** is listed in Table 2.1.

A molecular structure determination has been carried out on the analo-gous $[(MeC_5H_4)_2Yb(THF)_2]$ species (Zinnen *et al.*, 1980). In the solid state, each ytterbium(II) ion is bonded to one terminal MeC_5H_4 ligand, a single THF ligand and two shared, bridging MeC_5H_4 ligands resulting in a chain polymer structure (Fig. 2.6). The $Yb-C$ distances are longer than in related $Yb(III)$ species and the angles between the ring-centroids and the ytterbium ion range from 114.6 to 118.0°.

2.3.4 Tricyclopentadienyllanthanides

The first tricyclopentadienyl complexes of the lanthanides were reported in 1956 and they are generally formed from the reaction of the lanthanide

Table 2.1 Properties of some dicyclopentadienyllanthanides.

Complex	$[(C_5H_5)_2Eu]$	$[(C_5H_5)_2Yb]$	$[(C_5H_5)_2Sm]$
Synthesis	Eu + C_5H_6 in liquid NH_3	Yb + C_5H_6 in liquid NH_3	$[(C_5H_5)_3Sm]$ + $KC_{10}H_8$ in THF
Colour	Yellow	Green	Purple
Stability	Air and moisture sensitive	Air and moisture sensitive	Air and moisture sensitive
Solubility	Soluble in THF	Soluble in THF	Insoluble
Sublimation	400°C	400–420°C	Decomposes
Miscellaneous	High ionic character		Reducing agent in organic chemistry

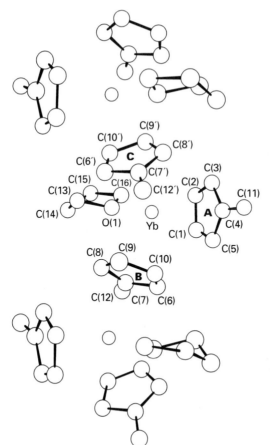

Fig. 2.6 The crystal structure of [(MeC$_5$H$_4$)$_2$Yb(THF)$_2$] showing a polymeric chain structure. (Zinnen *et al.* (1980), with permission.)

trichloride with sodium cyclopentadienide in THF or from the trichloride with potassium cyclopentadienide in benzene or toluene. Solvent adducts are normally first formed which can be converted to the solvent-free Cp$_3$Ln species by vacuum sublimation at *c.* 200–250°C.

$$LnCl_3 + 3NaC_5H_5 \xrightarrow{\text{THF}} [(C_5H_5)_3Ln] + 3NaCl$$

$$LnCl_3 + 3KC_5H_5 \xrightarrow{\substack{C_6H_6 \text{ or} \\ C_6H_5CH_3}} [(C_5H_5)_3Ln] + 3KCl \qquad (2.11)$$

Virtually all the lanthanides form tricyclopentadienyl compounds, and information on these species is summarised in Table 2.2.

A gadolinium–cyclopentadienyl compound structural analysis has been made on the THF adduct [(η^5-C$_5$H$_5$)$_3$Gd.OC$_4$H$_8$] (Rogers *et al.*, 1980) (Fig. 2.7). The molecule has a co-ordination sphere of three η^5-cyclopentadienyl ligands and one σ-bonded tetrahydrofuran ligand. The THF molecule is firmly bound with a Gd—O distance of 0.2494(7) nm whilst the three η^5-

Table 2.2 Properties of the tricyclopentadienyllanthanides.

Complex	Cp$_3$Ce	Cp$_3$Dy	Cp$_3$Er	Cp$_3$Eu	Cp$_3$Gd	Cp$_3$Ho	Cp$_3$La	Cp$_3$Lu	Cp$_3$Nd	Cp$_3$Sm	Cp$_3$Yb
Synthesis	CeF$_3$ + Cp$_2$Mg	DyCl$_3$ + NaCp	ErCl$_3$ + NaCp	Cp$_2$EuCl + NaCp	GdCl$_3$ + NaCp	HoCl$_3$ + KCp	LaCl$_3$ + KCp or NaCp	LuCl$_3$ + NaCp	NdCl$_3$ + NaCp	SmCl$_3$ + NaCp	YbCl$_3$ + NaCp or KCp
Colour	Orange	Yellow	Pink	Golden-brown	Yellow	Yellow	Colourless	Colourless	Red–blue	Orange	Green
Melting point	435°C	202°C	285°C	Decomposition > 70°C	Decomposition 350°C	295°C	395°C	264°C	380°C	360°C	273°C
Solubility*	Soluble in THF	Soluble in THF, pyridine	Soluble in THF, pyridine	Soluble in THF, benzene, CHCl$_3$	Moderately soluble in THF	Soluble in THF	Moderately soluble in THF, pyridine	Sparingly soluble in THF, CH$_2$Cl$_2$, toluene	Soluble in THF, MeCN, pyridine	Soluble in THF	Soluble in THF, MeCN, pyridine
Stability	Air and moisture sensitive	Air and moisture sensitive	Air and moisture sensitive	Air and moisture sensitive	Air and moisture sensitive	Air and moisture sensitive	Air and moisture sensitive	Air and moisture sensitive	Air and moisture sensitive	Air and moisture sensitive	Air and moisture sensitive
Miscellaneous	Structure is oligomeric (cf. Cp$_3$Sc)	Highly ionic character	Ionic, used to dope Er into InP and GaS	μ_{eff} = 3.74 BM†			Available commercially, used in synthesis of prostaglandins	Catalytic uses in alkene isomerisations and hydrogenations			Highly ionic character

* Formation of adducts normally affects solubility.

† BM = Bohr magnetons.

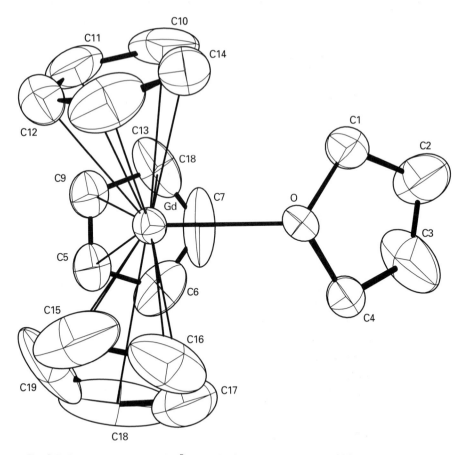

Fig. 2.7 The crystal structure of [(η⁵-C₅H₅)₃Gd.OC₄H₈]. (Rogers *et al.* (1980), with permission.)

bound cyclopentadienyl ligands are co-ordinated at an average Gd—C length of 0.274(3) nm and indicates that the overall geometry is very similar to [(η⁵-C₅H₅)₃Zr(η¹-C₅H₅)].

In the solid state **tricyclopentadienyllanthanum** possesses a non-linear polymeric chain structure. The zig-zag chains of distinct [(C₅H₅)₂La(μ-η⁵:η²-C₅H₅)] units involve two non-equivalent terminal Cp ligands (Eggers *et al.*, 1986a) (Fig. 2.8). The bridging Cp ring lies more remote from its η⁵-bonded La atom than the two terminal η⁵-Cp ligands and the individual La—C distances are predominantly about 0.26, 0.28, 0.29 and 0.3 nm and within the unexpectedly wide range of 0.256 to 0.303 nm. The co-ordination is irregular probably due to optimal intra- and inter-chain packing. **Tricyclopentadienyllutetium** also exists as a polymeric chain structure in the solid state (Eggers *et al.*, 1986b) (Fig. 2.9). The zig-zag chains are of the type [(η⁵-Cp)₂Lu(μ-η¹:η¹-Cp)]ₙ and are apparently isomorphous with the corresponding tricyclopentadienylscandium(III) species (see Fig. 2.5). The Lu—C

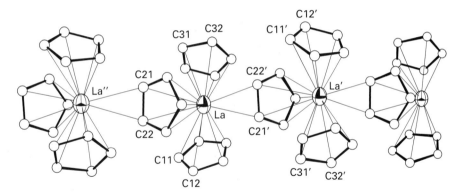

Fig. 2.8 The crystal structure of [(η⁵-C₅H₅)₃La]. (Eggers *et al.* (1986), with permission.)

distances for the η^5-co-ordinated Cp rings are 0.2602 nm and the centroid1–Lu–centroid2 angle is 126.9(8)°.

Studies on **tricyclopentadienylytterbium** indicate that like all the other Cp_3Ln species the bonding between the ytterbium and the $C_5H_5^-$ rings is essentially ionic in character. Due to the lack of low-lying vacant orbitals in the Yb^{3+} $(4f^{10})$ ion, the formation of covalent-type bonding would require the promotion of normally unoccupied high-energy $5d$, $6s$ and $6p$ orbitals, which is unlikely. This is reinforced by the high volatility, low melting points and solubilities of this and the other tricyclopentadienides of the lanthanides.

2.3.5 Tricyclopentadienylactinides

The properties of some tricyclopentadienylactinides are summarised in Table 2.3. **Tricyclopentadienylberkelium**, the first organometallic compound of berkelium, shows the expected properties common to the lighter tricyclopentadienyls of actinides and lanthanides. The visible spectrum of

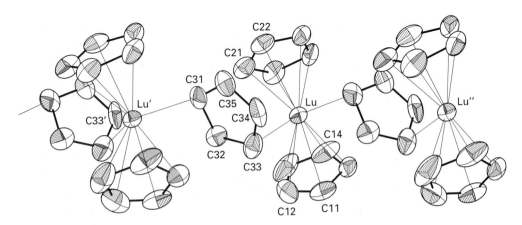

Fig. 2.9 The solid state structure of [(η⁵-C₅H₅)₃Lu].

Table 2.3 Properties of some tricyclopentadienylactinides.

Complex	$[(C_5H_5)_3Bk]$	$[(C_5H_5)_3Cf]$	$[(C_5H_5)_3Cm]$	$[(C_5H_5)_3Th]$	$[(C_5H_5)_3Pu]$	$[(C_5H_5)_3U]$
Synthesis	$BkCl_3$ + molten $[(C_5H_5)_2Be]$ in microgram quantities	$CfCl_3$ + molten $[(C_5H_5)_2Be]$ in microgram quantities	$CmCl_3$ + molten $[(C_5H_5)_2Be]$ in microgram quantities	(i) Reduction of $[(C_5H_5)_3ThCl]$ in THF; (ii) from Th(IV) alkyl derivatives and use of photochemistry	$Pu + Cl_2 + CCl_4$ + molten $[(C_5H_5)_2Be]$	(i) UCl_3 + KCp in benzene; (ii) reduction of Cp_4U with K metal; (iii) reduction of Cp_3UCl with NaH
Colour	Amber	Ruby-red	Colourless	(i) Purple (ii) Dark green	Green	Bronze
Melting point	Decomposition > 250°C	Decomposition > 250°C	Decomposition > 300°C	Decomposition > 170°C	Decomposition > 195°C	Decomposition > 200°C
Solubility			Soluble in THF	Soluble in THF		Soluble in THF, diethylether

$[(C_5H_5)_3Bk]$ was found to be in agreement with calculations and other spectra featuring the Bk^{3+} species, and involves f–f transitions. The structural data are also in close agreement with those of the other known tricyclopentadienyls of the lighter rare earths and actinides and therefore the character of the metal–ligand bonds can also be assumed to be ionic. This bonding is slightly more covalent than in the late lanthanide analogues but is considerably less covalent than in typical transition metal cyclopentadienyls.

Tricyclopentadienyluranium can undergo protolysis of the metal–ring bonds using HCN, leading to a dicyclopentadienylcyano complex and is thought to be a result of significant ionic character in the metal–ligand bonding. The THF adduct $[(\eta^5\text{-}C_5H_5)_3U.OC_4H_8]$ has been structurally analysed (Fig. 2.10), and in the solid state exists as a monomer with a distorted tetrahedral arrangement of THF and $(\eta^5\text{-}C_5H_5)$ ligands with Cp—U—Cp angles in the range 110–122° and Cp—U—O angles between 90 and 106° (Wasserman *et al.*, 1983).

2.3.6 Tetracyclopentadienylactinides

The first tetracyclopentadienyl actinides were prepared using the metal tetrachloride and potassium cyclopentadienide, e.g.

$$MCl_4 + 4KC_5H_5 \longrightarrow [(C_5H_5)_4M] + 4KCl \tag{2.12}$$

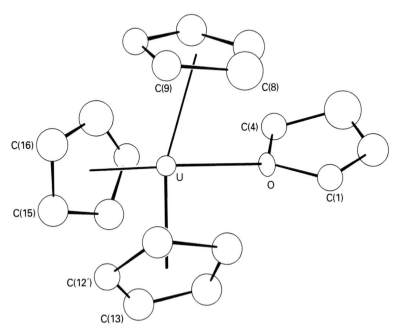

Fig. 2.10 The molecular structure of $[(\eta^5\text{-}C_5H_5)_3U.OC_4H_8]$. (Wasserman *et al.* (1983), with permission.)

Tetracyclopentadienylthorium is prepared in this fashion and is only sparingly soluble in common organic solvents but could be purified by Soxhlet extraction. Following dipole moment and IR studies it was proposed that the compound was a pseudo-tetrahedral tetra(η^5-cyclopentadienyl) structure.

Tetracyclopentadienylprotactinium was the first complex of this element to contain purely aromatic ligands, it being surprisingly tetravalent. It was synthesised by chlorinating Pa_2O_5 in a $Cl_2/CCl_4/Ar$ mixture at 600°C and reacting the product at 65°C with beryllocene.

$$Pa_2O_5 \xrightarrow[CCl_4]{Cl_2} PaCl_4 \xrightarrow{Cp_2Be} [(C_5H_5)_4Pa] \qquad (2.13)$$

Orange–yellow crystals were formed, decomposing between 210 and 230°C; however, sublimation was not possible. It is moderately soluble in benzene and apparently possesses a molecular structure analogous to that of the other tetracyclopentadienyl complexes of the 5f elements.

Tetracyclopentadienyluranium can be isolated as an air-sensitive, red crystalline solid that decomposes at 250°C. It is only sparingly soluble, but experimental measurements indicate a pseudo-tetrahedral tetra(η^5-cyclopentadienyl) structure which was confirmed by an X-ray diffraction study (Burns, 1974) (Fig. 2.11) and shows the compound to possess S_4 symmetry. About the central uranium atom, the co-ordination is nearly tetrahedral with the planar pentahapto-bonded rings and the ring centroid–U–ring centroid angles all *c.* 109°. The average U—C distance is 0.281 nm, which is slightly longer than usually found in U^{4+} cyclopentadienyl complexes and this is attributed to the pronounced crowding of the ligands about the metal ion and the concomitant ligand–ligand repulsion. The polyhapto bonding is indica-

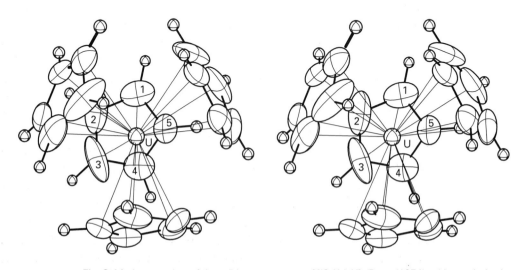

Fig. 2.11 A stereoview of the solid state structure of $[(C_5H_5)_4U]$. (Burns (1974), with permission.)

tive of the covalent character of this tetravalent uranium compound as opposed to the greater ionicity found in trivalent lanthanides (and actinides) where a mixture of η^5 and η^1 bonding is observed.

2.3.7 Dicyclooctatetraenyllanthanides

Dicyclooctatetraenyllanthanide can be formed as a very air- and moisture-sensitive green solid from $K_2(cot)$ and Cp_3La in THF and is only soluble in THF and dioxane.

$$LaCl_3 + 2 K_2(C_8H_8) \longrightarrow K^+[(C_8H_8)_2La^-] + 3KCl \qquad (2.14)$$

The compound is isostructurally related to the previously discovered cerium analogue $K[(C_8H_8)_2Ce]$ (see below) and chemical and spectral data indicate that there is significant covalent interaction between the rings and the metal. The ease of hydrolysis and formation of characteristic hydrolysis products indicate ionic salts of the C_8H_8 dianion and that this is a far more ionic situation relative to the analogous actinide complexes. This is undoubtedly due to the inability of the lanthanide $4f$ orbitals to contribute to covalent bonding relative to the $5f$ orbitals. Nephelauxetic ratio data suggest that the covalent character in Cp_3Ln compounds does not exceed 2.5% and, accordingly, the instantaneous and qualitative reaction with water and iron(II) chloride yields C_5H_6 and Cp_2Fe, respectively. Magnetic moments also indicate a similarity to free ion values, again consistent with a high degree of ionic character.

Dicyclooctatetraenylcerium is an unusual Ce^{IV} compound being isomorphous with and having an electronic structure similar to Th and U analogues. There are several methods of synthesis: (i) reduction of $Ce(OPr^i)_4$ by Et_3Al in the presence of excess cyclooctatetraene; (ii) by electrochemical oxidation of $[(C_8H_8)_2Ce]^-$; or (iii) oxidation of $[(C_8H_8)_2Ce]^-$ by excess AgI. It exists as black crystals or a red–violet powder, being pyrophoric and sparingly soluble in aromatic hydrocarbons, chlorinated aromatic hydrocarbons and ethers. Weak absorption bands in the infrared indicate aromatic character (i.e. π-bonded C_8H_8 rings) and the ^1H NMR spectrum once again shows the fluxionality of the cyclooctatetraene rings. Single crystals of $K(diglyme)^+$ $[(C_8H_8)_2Ce]^-$ have been analysed by X-ray diffraction (Hodgson and Raymond, 1972) (Fig. 2.12). The rings are approximately staggered with respect to each other with Ce—C bond distances of 0.274 nm. Some linear combination of atomic orbitals (LCAO) calculations have calculated the degrees of f-orbital participation, in comparison to the uranium analogue. The f-orbital mixing is greatest for uranium (22%) but for cerium the analogous $4f$ orbitals are much contracted and there is only a 3% mixing with the ligand molecular orbitals.

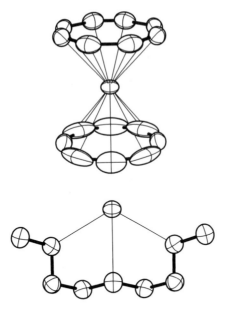

Fig. 2.12 The crystal structure of K(diglyme)$^+$[(C$_8$H$_8$)$_2$Ce]$^-$. (Hodgson and Raymond (1972), with permission.)

2.3.8 Dicyclooctatetraenylactinides

The synthesis of the uranium sandwich complex **dicyclooctatetraenyluranium** [(C$_8$H$_8$)$_2$U] 'uranocene' was an important point in organoactinide chemistry and it was first prepared in 1968 by the reaction of UCl$_4$ and K$_2$C$_8$H$_8$ (2.15). It can also be prepared by reacting cyclooctatetraene with finely divided uranium (generated electrolytically or by reducing UCl$_4$ with Na/K or using Mg(C$_8$H$_8$) and uranium tetrafluoride (2.16)).

$$UCl_4 + K_2(C_8H_8) \longrightarrow [(C_8H_8)_2U] + 4KCl \qquad (2.15)$$

$$UF_4 + 2Mg(C_8H_8) \longrightarrow [(C_8H_8)_2U] + 2MgF_2 \qquad (2.16)$$

After sublimation at 180°C *in vacuo* it can be isolated as green plates and it is flammable in air but stable towards H$_2$O. The molecular structure of uranocene has been determined by X-ray diffraction (Avdeef *et al.*, 1972). The molecule has D_{8h} symmetry with the eight-membered rings in an eclipsed configuration. The mean U—C bond length is 0.265 nm and the cyclooctatetraene dianion rings are planar with average C—C bond lengths of 0.139 nm (Fig. 2.13).

The bonding description of uranocene and other (cot)$_2$M organoactinides has been an area of intense study and in terms of orbital symmetry uranocene can be thought of as an *f*-orbital analogue of ferrocene. This will be discussed in Chapter 3 but it is clear that there is considerable metal *f*-ligand orbital

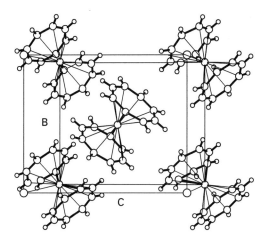

Fig. 2.13 The molecular structure and unit cell packing of [(C₈H₈)₂U]. (Avdeef *et al.* (1972), with permission.)

mixing and covalency in the bonding, which is a marked departure from the principally ionic organolanthanide species.

A summary of the properties of some other dicyclooctatetraenylactinides is listed in Table 2.4. Although the chemistry of **dicyclooctatetraenylthorium** has not been as well studied as that of uranocene, it seems to be more ionic; however, in the solid state the molecular structure of thorocene is isomorphous with that of uranocene (Avdeef *et al.*, 1972) (Fig. 2.14). The mean Th—C bond distance is 0.2701(4) nm and the mean C—C bond distance (0.1386(4) nm) and the differences in these values from those in uranocene are due to the difference in ionic radii of the metals.

Table 2.4 Properties of some dicyclooctatetraenylactinides.

Complex	[(C₈H₈)₂Np]	[(C₈H₈)₂Pa]	[(C₈H₈)₂Pu]	[(C₈H₈)₂Th]
Synthesis	K₂C₈H₈ + NpCl₄ in THF	K₂C₈H₈ + PaCl₄ in THF	K₂C₈H₈ + [(Et₄N)₂–PuCl₆] in THF	K₂C₈H₈ + ThCl₄ in THF
Colour	Orange	Golden-yellow	Cherry-red	Yellow
Melting point		Decomposition > 155°C		Decomposition > 190°C
Solubility	Soluble in benzene, toluene, CCl₄, CHCl₃	Soluble in benzene, THF	Soluble in benzene, toluene, CCl₄, THF	Soluble in DMSO
Stability	Stable in H₂O; reacts in air	Very air sensitive	Air sensitive	Air and water sensitive

DMSO = dimethylsulphoxide.

39

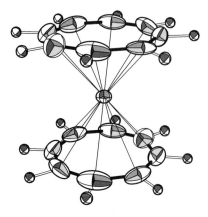

Fig. 2.14 The molecular structure of [(C₈H₈)₂Th].
(Avdeef *et al.* (1972), with permission.)

2.4 Cyclopentadienyl compounds of group 4 elements

2.4.1 Dicyclopentadienyltitanium 'titanocene'

The subject of 'titanocene' is still controversial and yet to be resolved, although it does seem clear that the monomeric metallocene structure is probably only observed as a reactive intermediate. Various methods of preparation resulting in complexes with different colours and spectroscopic properties have been published. The problems arise due to the extremely

(i) $TiCl_2$ + NaC_5H_5 $\xrightarrow{\text{THF}}$ (petrol extract) green, diamagnetic

$\xrightarrow{}$ (sublime) THF adduct, green paramagnetic then brown, diamagnetic

(ii) $[(C_5H_5)_2TiCl_2]$ + sodium naphthalenide $\xrightarrow{}$ green, dimer

(iii) $[(C_5H_5)_2TiCl_2]$ + sodium sand $\xrightarrow{}$ green dimer with a complex ^{1}H NMR spectrum

(iv) $[(C_5H_5)_2TiCl_2]$ + H_2 $\xrightarrow{\text{hexane, 20°C}}$ green complex via violet intermediate

(v) $[(C_5H_5)_2TiCl_2]$ $\xrightarrow{\text{pyrolysis}}$ hexane, green product

$\xrightarrow{\text{90°C}}$ THF, brown product

(vi) $[(C_5H_5)_2Ti(CH_3)_2]$ $\xrightarrow{h\nu}$ black product

(vii) $[(C_5H_5)_2TiPh_2]$ $\xrightarrow{h\nu,\ THF}$ green product

Scheme 2.2 Some preparations of 'titanocene'.

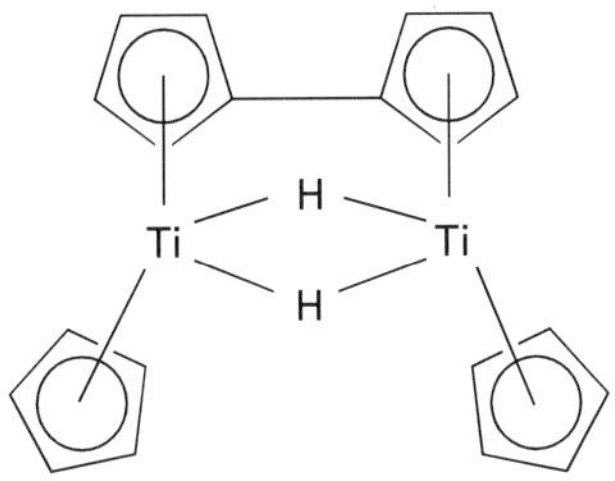

Fig. 2.15 The structure of μ-(η^5:η^5-fulvalene)-di-μ-hydridodi(cyclopentadienyl)titanium.

reactive nature of titanocene and its ability to easily abstract hydrogens from the coordinated cyclopentadienyl ligands. Some examples of 'titanocene' preparations are detailed in Scheme 2.2. Unfortunately, isolation of all these products proved so difficult that no crystals could be formed that were suitable for X-ray diffraction studies though the green dimeric species (reactions ii, iii, iv) has generally been accepted as being a bridging hydride structure. This was due to the IR data being too complex for titanocene [(η-C$_5$H$_5$)$_2$Ti] or its simple dimer. It is thought that titanocene initially forms but quickly decomposes through abstraction of a hydrogen atom from a cyclopentadienyl ligand by titanium, followed by dimerisation leaving a fulvalene ligand bonded to both titanium centres (Fig. 2.15).

   [(η^5-C$_5$H$_5$)$_2$Ti] is predicted to be very unstable with a carbene-like reactivity, leading to abstraction of one of the ring hydrogens and its shift to the titanium centre in [(C$_5$H$_5$)(C$_5$H$_4$)TiH]. The co-ordinated fulvalene and two hydride bridges were shown via ^{13}C NMR studies and confirmation was obtained via X-ray diffraction studies on analogous hydroxyl and chloro derivatives and recently on the species itself. Various reactivity studies to elucidate the reactive intermediates have been carried out and one interesting analogous preparation was that of decamethyltitanocene. This was thought to be more stable than titanocene due to the impossibility of a σ-hydrogen shift from the rings to the metal. Pure decamethyltitanocene was formed by reacting the dihydride complex with dinitrogen to form a purple dinuclear dinitrogen complex, readily isolatable at $-80°$C. Then on warming to room temperature, nitrogen was evolved to leave decamethyltitanocene which was then isolated as an orange crystalline solid. It was found to be monomeric in solution and very soluble in diethylether and hydrocarbons. The two unpaired electrons expected in a Ti(II) complex resulted in a magnetic moment of 2.6 BM (Bohr magnetons). However, at room temperature it was shown to exist as an equilibrium mixture of two isomers, the orange, 14-electron parallel sandwich species and a green, diamagnetic 16-electron 'bent' compound formed by insertion of titanium into the C—H bond of a methyl group of the pentamethylcyclopentadienyl ligand.

2.4.2 Dicyclopentadienylzirconium(II) and dicyclopentadienylhafnium(II)

Some properties of these two metallocenes are listed in Table 2.5 but no true metallocene (i.e. a monomeric dicyclopentadienyl metal species) has been isolated to date.

2.4.3 Group 4 di(η^5-cyclopentadienyl)metal(IV) dihalides (metallocene dihalides)

These compounds are best prepared by the reaction of either the cyclopentadienyl–Grignard or the sodium or lithium cyclopentadienide with the metal tetrahalide and are normally isolated in excellent yields. They belong to the general class of bent sandwich complexes of the type $[(\eta^5\text{-L})_2MX_2]$ with the geometry about the metal centre that of a distorted tetrahedron and where each η^5-L group is considered as a bulky unidentate ligand. A large number of Ti(IV) compounds of this general type have been reported. The area is of great current interest (as will be described later) due to the ready availability of convenient starting materials, i.e. $[Cp_2TiCl_2]$, ease of preparation, the excellent thermal, air and moisture stabilities of many of the derivatives and the use of them as catalysts in alkene polymerisation systems and as anti-tumour agents. The most important compound is bright-red, air-stable **di(cyclopentadienyl)titanium dichloride (titanocene dichloride)** (m.p. 289–291°C). IR spectroscopy has proved informative, with four major bands invariant for a wide range of η-cyclopentadienyltitanium(IV) compounds, i.e. ~ 3100 cm^{-1} (CH stretch), ~ 1435 cm^{-1} (CC stretch), ~ 1020 cm^{-1} (CH deformation in plane) and ~ 820 cm^{-1} (CH deformation out of plane). The ^{1}H NMR of $[Cp_2TiCl_2]$ in solution shows a sharp singlet, as expected. An X-ray crystal structure determination reveals a distorted tetrahedral coordination about the titanium atom by the two chlorine atoms and the centroids of the cyclopentadienyl rings. The average Ti—Cl bond distance is 0.2364(3) nm whilst that of the Ti—Cp (centroid) is 0.2058 nm.

Table 2.5 Some properties of dicyclopentadienylzirconium(II) and dicyclopentadienylhafnium(II).

Complex	$[(C_5H_5)_2Zr]$	$[(C_5H_5)_2Hf]$
Synthesis	Reduction of $[(C_5H_5)_2ZrCl_2]$ by $NaC_{10}H_8$ in THF	(i) Reduction of $[(C_5H_5)_2HfCl_2]$ by $NaC_{10}H_8$ in THF; (ii) photolysis of hafnocene dialkyls or diaryls gives grey '$[(C_5H_5)_2Hf]$'
Colour	Dark purple, diamagnetic	Very dark purple, diamagnetic
Sensitivity	Pyrophoric, water sensitive	Pyrophoric, water sensitive
Structure	Isomorphous with the dinuclear Ti species $[(C_5H_5)_2Ti]_2$	Isomorphous with the dinuclear Ti species $[(C_5H_5)_2Ti]_2$

[Cp$_2$TiCl$_2$] is a convenient starting material for a wide range of other organometallic titanium derivatives (some reactions are detailed in Chapter 4). It has recently become important with regard to catalysis in hydrometallation and carbometallation reactions and in DNA–metal binding processes exhibiting anti-tumour activity (see Chapter 6).

Dicyclopentadienylzirconium(IV) dihalides (zirconocene dihalides) are all colourless → yellow crystalline solids, with melting points between 240 and 300°C and ^{1}H NMR indicating freely rotating π-bound cyclopentadienyl groups. An important trend of these [(η^5-C$_5$H$_5$)$_2$ZrX$_2$] species is the variation in the X–M–X angle with change in the metal d-electron configuration, whilst the centroid–M–centroid angle remains invariant at c. 130°C. The trend is in accordance with molecular orbital calculations and generally values are 94–97° for d^0, 85–88° for d^1 and 76–82° for d^2. The only anomalies are for bulky ligands such as CH(SiMe$_3$)$_2$ or CH$_2$PPh$_2$. The stereochemistries have also been explained by inter-ligand repulsions where the largest X—Zr—X angles are associated with the bulkiest ligands, i.e. CH(SiMe$_3$)$_2$ and C$_5$H$_4$MeCH(SiMe$_3$). The mutual repulsion of the hydrogen atoms of the cyclopentadienyl ligands seems to result in a constant centroid–M–centroid angle. In these dihalide species there is an eclipsed conformation of the Cp rings in [(η-C$_5$H$_5$)$_2$ZrX$_2$] (X = F, I) (**II.2**), whilst in [(η-C$_5$H$_5$)$_2$ZrCl$_2$] these rings are staggered (**II.3**). In [(η-C$_5$H$_5$)$_2$ZrCl$_2$], the Zr—η—carbon distances

(**II.2**)

(**II.3**)

are arranged at 0.249 nm whilst the Cl—Zr—Cl angle is 97.8° and the centroid–Zr–centroid angle is 126°. The structure of [(η-C$_5$H$_4$CH$_2$Ph)$_2$ZrCl$_2$] is slightly different with a discrepancy between the Zr–ring normal distance (0.216 nm) and the Zr–centroid distance (0.235 nm). This seems to be because of steric interaction between the eclipsed methylene benzyl substituents with the observed unsymmetric bonding to the rings.

The **dicyclopentadienylhafnium(IV) dihalides** tend to be off-white or pale yellow solids melting between 200 and 300°C and are the starting points for a whole range of cyclopentadienyl–hafnium complexes. Two of the most interesting reactions have been (i) the displacement of a halide by a salt elimination route (2.17) and (ii) the formation of oxo-bridged dimers in the presence of water and/or base (2.18).

$$[(C_5H_5)_2HfCl_2] + Li[M'Ph_3] \longrightarrow [(hC_5H_5)_2HfCl][M'Ph_3] + LiCl \qquad (2.17)$$
(M = Si, Ge or Sn)

$$2[(C_5H_5)_2HfCl_2] + 2PhNH_2 + H_2O \longrightarrow \{[(C_5H_5)_2HfCl]_2O\}$$
$$+ 2PhNH_2 + HCl \qquad (2.18)$$

$[(\eta\text{-}C_5H_5)_2HfCl_2]$ has been shown to be isostructural with $[(\eta^5\text{-}C_5H_5)_2ZrCl_2]$ where the cyclopentadienyl rings are staggered (both in the vapour phase and solid state).

2.4.4 Group 4 tricyclopentadienyls

Tricyclopentadienyltitanium(III) is formed from the reaction of $[(Cp_2TiCl)_2]$ with sodium cyclopentadienide, or by the vacuum pyrolysis of Cp_4Ti. It is isolated after sublimation as a green crystalline solid with a melting point of 138–140°C. A monomeric structure is indicated by a magnetic moment close to the spin-only value and this was established by an X-ray structure determination. This indicated that two of the cyclopentadienyl groups were bound in a η^5-fashion whilst the third was attached to the titanium atom by only two carbon atoms, suggesting a three centre–four electron bond.

Few tricyclopentadienyls of zirconium have been studied. Starting from the dichlorides, the monochloride species **tricyclopentadienylzirconium(IV) chloride** can be formed using stoichiometric quantities of sodium cyclopentadienide (to prevent formation of the tetracyclopentadienyl species).

$$[(\eta\text{-}C_5H_4R)_2ZrCl_2] + NaC_5H_4R \longrightarrow [(\eta\text{-}C_5H_4R)_3ZrCl] + NaCl \qquad (2.19)$$
(R = H or Me)

For the methylcyclopentadienyl derivative, NMR data indicated a formulation of $[(\eta^1\text{-}C_5H_4Me)(\eta^5\text{-}C_5H_4Me)_2ZrCl]$. Only one tricyclopentadienyl species has been characterised by X-ray diffraction to date, namely $[(\eta\text{-}C_5H_5)_3Zr(HAlEt_3)]$ (formed from $[(\eta\text{-}C_5H_5)_3ZrH]$ and $AlEt_3$). The co-ordination about zirconium is pseudo-tetrahedral with elongated $Zr\text{—}\eta\text{—}carbon$ distances compared to mono- and di-cyclopentadienyl complexes. In solution, the complex is in equilibrium with the starting material and adduct formation by $AlEt_3$ is exhibited in Lewis base solvents such as THF.

2.4.5 Group 4 tetracyclopentadienyls

These compounds are usually prepared from the metal tetrahalide or metallocene dihalide and sodium cyclopentadienide in an inert solvent. **Tetracyclopentadienyltitanium(IV)** or **([di(η^1-cyclopentadienyl)di(η^5-cyclopentadienyl)titanium])** is a very air- and moisture-sensitive green–black crystalline solid (m.p. 128°C) (although it appears violet in transmitted light). From various reactions and spectroscopic data, it was postulated that the formula was $[(\eta^5\text{-}C_5H_5)_2Ti(\eta^1\text{-}C_5H_5)_2]$ (Calderon *et al.*, 1971) and the X-ray diffraction data agreed with a structure containing two monohapto and two pentahaptocyclopentadienyl rings (Fig. 2.16). The compound is fluxional in solution, undergoing both ring rotation and $\eta^1 : \eta^5$ interchange. At temperatures below ambient spectral changes indicate sigmatropic ring rearrangements, whilst at temperatures above ambient spectral changes due to interchange of the monohapto- and pentahapto-cyclopentadienyl rings were observed. The separate $\eta^1\text{—}C_5H_5$ and $\eta^5\text{—}C_5H_5$ lines in the ^{1}H NMR broaden and coalesce to a single line which narrows on raising the temperature. The ring interchange is attributed to the 16-valence electron configuration of the Ti atom and the low-lying empty valence shell orbital, characteristic of this configuration, can stabilise the transition state in which one ring is intermediate in character between η^1 and η^5. Sigmatropic rearrangement of the σ-cyclopentadienyl rings can also occur for the molecule in the crystalline state.

The NMR spectrum of **tetracyclopentadienylzirconium(IV)** at $-150°C$ shows only a single proton resonance. It is clear though that from dipole moment and IR studies not all the cyclopentadienyl ligands are bound in the same way, indicating that there are both σ- and π-bound ligands in these tetracyclopentadienyls. A structural determination shows the bonding as one η^1 and three η^5 rings (Rogers *et al.*, 1978) (Fig. 2.17). Compared to $[(\eta\text{-}C_5H_5)_2ZrX_2]$ (X = F, Cl, I), the $Zr\text{—}\eta^5\text{—}C$ distances are approximately 0.01 nm longer and their spread within each ring is longer but the symmetry is closer to tetrahedral geometry. The long $Zr\text{—}\eta^5\text{—}C$ distances are thought to arise from each ring contributing less than five electrons to give an 18- rather than 20-electron species.

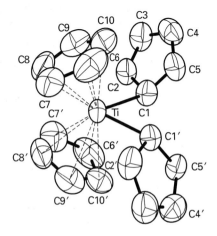

Fig. 2.16 The molecular structure of tetracyclopentadienyltitanium(IV). (Calderon *et al.* (1971), with permission.)

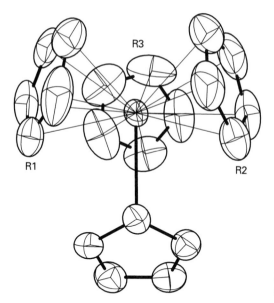

Fig. 2.17 The molecular structure of $[(\eta\text{-}C_5H_5)_3Zr(\eta^1\text{-}C_5H_5)]$. (Rogers *et al.* (1978), with permission.)

Physical measurements on **tetracyclopentadienylhafnium** also indicate that the species contains both σ- and π-bound ligands. From X-ray crystallography, this hafnium species is isostructural with $[(\eta^5\text{-}C_5H_5)_2Ti(\eta^1\text{-}C_5H_5)_2]$ possessing two η^1 and two η^5 rings as opposed to the zirconium species $[(\eta^5\text{-}C_5H_5)_3Zr(\eta^1\text{-}C_5H_5)]$. This is explained by the slightly smaller radius of Hf compared with Zr and the Zr structure features very long $Zr\text{—}\eta^1\text{—}C$ distances. The structure of the complex has also been determined at 160°C but shows little change with two η^5 rings at 0.25 nm and two rings at 0.234 nm.

2.5 Cyclopentadienyl compounds of group 5 elements

2.5.1 Dicyclopentadienylvanadium (vanadocene)

Vanadocene, an electron-deficient species (15 VE), was the first group 5 di(η^5-cyclopentadienyl) metal complex isolated and is often thought of as possessing carbene-like character. Vanadocene is usually prepared from the reaction of VCl_3 and NaCp in THF, followed by sublimation at 0.1 mmHg and at temperatures of 200°C. Other successful routes have included the reaction of $[VCl_2(THF)_2]$ with NaCp in THF where the important intermediate is $[VCl_3(THF)_3]$ which is generated *in situ* by $LiAlH_4$ or zinc reduction. It is a violet crystalline solid, paramagnetic, melting at 167–168°C and is very air sensitive in solution and in the solid state. It is soluble in organic solvents such as benzene and THF and it has found industrial use as a catalyst for acetylene polymerisation, resulting in linear, high molecular weight species with extended conjugation and good semiconducting behaviour. Vanadocene undergoes a reversible one-electron reduction at a very negative potential to give the stable anion $[Cp_2V]^-$. On oxidation, the parent metallocene is subjected to two one-electron transfer reactions; the first is reversible but the second irreversible, thus indicating a possible severe structural change.

The molecular structure of vanadocene almost certainly features a staggered arrangement of the cyclopentadienyl rings but X-ray and electron diffraction experiments are different and it appears that the measurements are distinctly temperature dependent. The X-ray study at room temperature shows a staggered structure with approximate D_{5d} symmetry with expected distances of 0.225 nm and 0.192 nm for the average V—C and V—Cp distances, respectively. The gas phase electron diffraction results at 120°C point to an eclipsed D_{5h} structure with the average V—C bond length of 0.228 nm (Fig. 2.18). Disorder and temperature/phase are major factors when trying to compare these results but the gas phase study, although favouring an eclipsed structure, does also indicate that the staggered conformation (D_{5d}) cannot be ruled out.

The species shows a strong parent ion of $[Cp_2V]^{2+}$ in the mass spectrum and the mean dissociation energy of $[Cp_2V]$ to vanadium atoms and C_5H_5 radicals is higher than for ferrocene. In fact, the order found for bond dissociation energies of $[Cp_2M]$ species is $M = V > Sc > Fe > TiCl_2 > Cr > Co > Ni > Mn$. Therefore, it is not surprising that there is no ring exchange reaction with $[(\eta-C_5D_5)_2Ni]$ in heptane at 51.3°C and little formation of ferrocene from reaction of vanadocene with $FeCl_2$.

2.5.2 Dicyclopentadienylniobium (niobocene)

Typically for second- and third-row transition metals, the reaction of niobium pentachloride with excess sodium cyclopentadienide does not give the metallocene Cp_2Nb and higher oxidation state complexes form instead. For instance, addition of excess NaCp to $NbCl_5$ gives $[Cp_2Nb(\eta^1-Cp)_2]$ whilst if $NaBH_4$ is added to the reaction mixture $[Cp_2NbH_3]$, an 18-electron hydride,

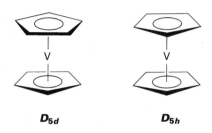

D_{5d}   **D_{5h}**

Fig. 2.18 Schematic drawing of vanadocene in the solid state (D_{5d} symmetry) and in the gas phase (D_{5h} symmetry).

47

is formed. Evidence for the formation of niobocene comes from the thermolysis of Cp_2NbH_3 which is accompanied by evolution of H_2. However the NMR shows that the product cannot be a simple monomeric form. 'Niobocene' is paramagnetic and a low frequency signal of metal-bonded hydrogen is observed (δ, –2.07) with several signals displayed in the Cp region. It was postulated that this indicates the presence of either a fulvenide-containing structure (**II.4**) (similar to that found for 'titanocene') or a (η^5,η^1-C_5H_4)-bridged structure (**II.5**). Crystallography has indicated that **II.5** is the ob-

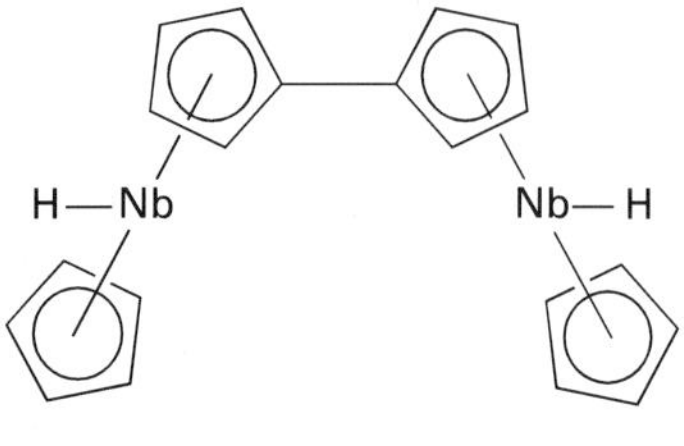

(**II.4**)

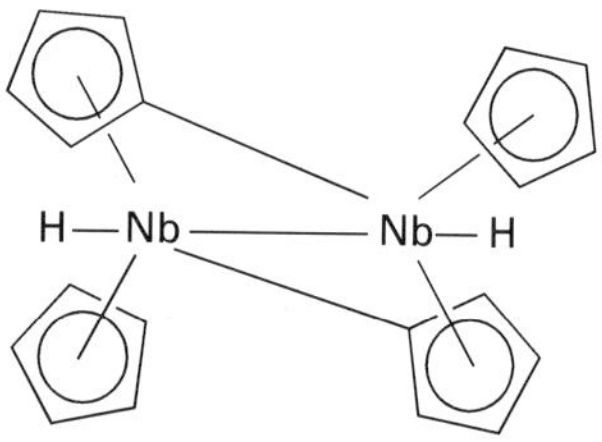

(**II.5**)

served species (Guggenberger, 1973). Each Nb centre can be thought of as a (Cp_2NbHR) group, a 17-electron species, and the two are joined by a metal–metal single bond to form an 18-electron dimer. In **II.4**, there is a more unfavourable situation with each centre being 16-electron Nb(III). Fulvalenide structures are known for niobium, e.g. the one unpaired electron per dimer species (**II.6**) obtained from either controlled reduction of

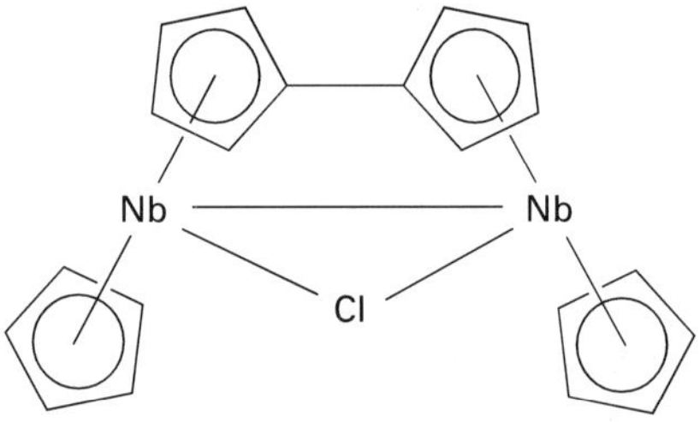

(**II.6**)

[Cp_2NbCl_2] or oxidation of **II.5**. Monomeric niobocene seems to be observed in electron paramagnetic resonance (EPR) studies. A signal observed for [Cp_2NbH_2] was replaced by a new signal with hyperfine coupling only to ^{93}Nb. The same signal was again observed following the treatment of [Cp_2NbCl_2] with a range of reducing agents. Unfortunately, the product was not isolatable but reactions performed on it suggest the presence of Cp_2Nb.

2.5.3 Dicyclopentadienylniobium dihalides

There are a number of stable dicyclopentadienyl systems involving niobium, namely: (i) [(Cp$_2$NbX$_3$], pentavalent complexes with three one-electron ligands; (ii) [Cp$_2$NbXL], trivalent with one one-electron and one two-electron ligands; and most commonly (iii) [Cp$_2$NbX$_2$]. The propensity of these species is a subtle balance between steric and electronic effects. Heavy early transition metals exhibit a preference for higher oxidation states but the coordination sphere is crowded with three ligands in addition to the two Cp groups. There are numerous examples of dicyclopentadienyl complexes containing other organic ligands (Labinger, 1982), but [Cp$_2$NbCl$_2$] is the compound that has featured in most synthetic studies. It is a black, crystalline paramagnetic solid being relatively air stable, mostly insoluble, easily isolated and handled and is the starting point for virtually all [Cp$_2$NbX$_n$] species. It was first reported from the direct reaction of NbCl$_5$ and NaCp in benzene-glyme but other methods have included the addition of HCl to an *in situ* preparation of Cp$_4$Nb followed by sublimation at 270°C (0.25 mmHg), by the reduction of NbCl$_5$ with SnCl$_2$ or by using CpSnR$_3$ reagents.

$$\text{NbCl}_5 \xrightarrow[\text{MeCN}]{\text{Al}} \xrightarrow{\text{THF}} [\text{NbCl}_4(\text{THF})_2] \xrightarrow{2\text{LiCp}'} [\text{Cp}_2'\text{NbCl}_2] \qquad (2.20)$$

All these compounds are bent metallocenes and have been the subject of many theoretical studies (see Chapter 3). NMR (for the penta- and trivalent compounds) and EPR studies (for the paramagnetic tetravalent species) have proved to be important. In the NMR, the Cp rings are usually equivalent and a single peak is observed with the normal range of shifts being dependent on the other ligands present and a decrease in electron density results in a more downfield shift. EPR studies show that there is a trend towards low hyperfine coupling constants to the ^{93}Nb nucleus as the ligands become less electronegative.

2.5.4 Tantalum metallocenes and related complexes

The chemistry of cyclopentadienyl complexes of tantalum (though less studied) follows very closely that of niobium. Reaction of tantalum pentachloride with excess sodium cyclopentadienide does not result in the formation of Cp$_2$Ta but gives instead [Cp$_2$Ta(η^1-Cp)$_2$] or, with addition of NaBH$_4$, the 18-electron hydride [Cp$_2$TaH$_3$]. Compared to niobium, little is known about tantalocene and its intermediate or bridged species; however the (η^5,η^1-C$_5$H$_4$)-bridged structure {[CpTaH(μ-C$_5$H$_4$)]$_2$} has been studied. From similar EPR and ^{1}H NMR studies, there seems no obvious reason why tantalocene should not be similarly formed (as niobocene, see earlier) and isolated.

As with niobium, there are several classes of compounds possible with tantalum: (i) [Cp$_2$TaX$_3$], pentavalent complexes with three one-electron ligands; (ii) [Cp$_2$TaXL], trivalent with one one-electron and one two-electron ligands; and most commonly (iii) the paramagnetic tetravalent [Cp$_2$TaX$_2$]. Once again, the compound [Cp$_2$TaCl$_2$] is the most important and well-studied species in this area.

Dicyclopentadienyltantalum dichloride is best prepared from TaCl$_5$ and CpSnBu$_3$ in CH$_2$Cl$_2$ and is the starting point for nearly all [Cp$_2$TaX$_n$] species. It is a brown, crystalline solid, moderately stable in air, sublimable at 222°C (0.001 mmHg) and soluble in toluene, CH$_2$Cl$_2$ and Me$_2$CO. It undergoes a wide variety of reactions, one of which gives **di(η^5-cyclopentadienyl)tantalum trihydride**. This species is formed from TaCl$_5$, NaCp and NaBH$_4$ in THF at reflux under N$_2$ or by prior preparation of [Cp$_2$TaCl$_2$]. It is a white crystalline solid (m.p. 187–189°C), sublimed at 110°C (0.1 mmHg) and soluble in benzene.

$$MX_5 + NaCp + NaBH_4 \xrightarrow{\text{THF}} [Cp_2MH_3] \qquad (2.21)$$
$$(M = Nb, Ta)$$

The most readily observed reactions of the trihydride are (i) the loss of H$_2$ to give the dimeric products (**II.7**) after heating to 130°C without any added

$$\text{(II.7)}$$

ligand and (ii) formation of trivalent derivatives [Cp$_2$TaHL] with a good two-electron donor ligand present. These reactions proceed via the co-ordinatively unsaturated [Cp$_2$TaH]. The hydride species exhibit acidic, basic and radical reactions. For instance, [Cp$_2$TaH$_3$] can be deprotonated by butyl-lithium and then reacted with benzyl chloride to form [Cp$_2$TaCl(CH$_2$Ph)]. Hydridic or nucleophilic character is shown by formation of adducts with Lewis acids such as BF$_3$ which binds to a hydride ligand.

Once again, these [Cp$_2$TaX$_n$] species are bent metallocenes. Most data have been directed toward establishing where the non-bonding d-electron density is localised in the compounds. It has been shown that the non-bonding orbital lies primarily outside of the region between the two ligands with the angle X—Ta—X decreasing on going from $d^0 \rightarrow d^1 \rightarrow d^2$, e.g. [Cp$_2TaH_3$] (based on a neutron diffraction study) has $\theta = 139.9$ and $\phi = 63.0$. The central TaH$_3$ fragment is planar with the three hydridic atoms equi-

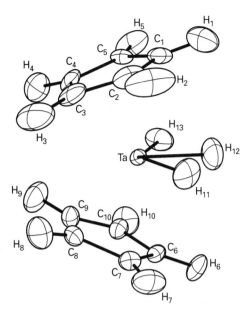

Fig. 2.19 The molecular structure of $[(C_5H_5)_2TaH_3]$ from neutron data. (Wilson et al. (1977), with permission.)

distant from the Ta and the central Ta—H bond in a bisecting position (Wilson *et al.*, 1977) (Fig. 2.19).

2.6 Cyclopentadienyl compounds of group 6 elements

2.6.1 Dicyclopentadienylchromium (chromocene)

This compound is normally prepared from chromium(II) or chromium(III) halides with NaCp in dimethoxyethane or tetrahydrofuran. Other cyclopentadienylating reagents such as TlCp or CpMgBr can also be used. It can be isolated as very air-sensitive, red needles that melt at 173°C and are soluble in THF. The material is paramagnetic with μ_{eff} = 3.27 BM which is considerably above the spin-only value (2.83 BM) for two unpaired electrons and suggests significant orbital contribution. This situation leads to a configuration arising from the removal of one electron from each of the completely filled e_{2g} and a_{1g} levels in the closed shell arrangement of ferrocene.

X-ray crystallographic studies on chromocene show a Cr—C bond length of 0.213 nm, whilst electron diffraction results indicate an M—C bond length which is 1.1×10^{-3} nm longer than in ferrocene. This is thought to be due to the increase in atomic size of chromium, due to its lower nuclear charge which results in a size difference between Cr(II) $(3d^4)$ and Fe(II) $(3d^6)$ of 0.005 nm. Secondly, the elongation on going from Cp_2Fe to Cp_2Cr can be attributed to the loss of bonding electrons. The electron diffraction studies indicate an eclipsed (D_{5h} symmetry) configuration of the rings (Fig. 2.20) but,

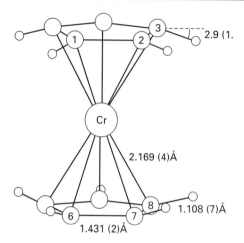

2.9 (1.

2.169 (4)Å

1.108 (7)Å

1.431 (2)Å

Fig. 2.20 The molecular model of $[(C_5H_5)_2Cr]$ with eclipsed rings. (Gard *et al.* (1975), with permission.)

as with Cp_2Co, a model with staggered rings (D_{5d} symmetry) has also to be considered. The M—C bond and C—C bond distances are 0.217 and 0.143 nm, respectively, and the internal barrier to rotation is small at 3.4 kJ mol^{-1} (Gard *et al.*, 1975).

Chromocene undergoes a wide range of reactions and the chromocenium ion $[Cp_2Cr]^+$ can be generated from chromocene by chemical and electro-chemical methods but having three unpaired electrons makes it unstable and difficult to handle.

2.6.2 Dicyclopentadienylmolybdenum (molybdocene)

The simple mononuclear complex (similar to the chromium analogue) is unknown mainly due to its highly electron-deficient nature. Attempted syntheses have resulted in the formation of di- or poly-nuclear products. For example, reaction of $[Cp_2MoCl_2]$ with sodium amalgam in THF yielded $[CpMo]_x$ ($x > 2$) whilst heating $[Cp_2MoMe_2]$ at 150°C for 24 hours also failed to produce $[Cp_2Mo]$. Photolysis of $[Cp_2MoH_2]$ at −78°C or elevated temperatures results in $[Cp_2Mo]_x$ but when the same pyrolysis is performed in an argon matrix at 10 K the transient species $[Cp_2Mo]$ can be observed. The process seems to involve a concerted elimination of H_2 to form the metallocene, and an IR spectrum indicates a structure containing parallel rings as it is very similar to that of $[Cp_2Cr]$, although it is also thought to exist in a similar form to that found for 'Cp$_2$Ti' (a fulvalenediyl- and dihydrido-bridged system). Dimeric isomers of composition $[C_{10}H_{10}M]_2$ are known and possess η^5–Cp bridges or η^5:η^5–fulvalenediyl bridges and terminal hydrido ligands (Fig. 2.21). The transient presence of $[Cp_2Mo]$ can be intercepted by trapping agents and in many reactions it shows a formal analogy to organic carbenes.

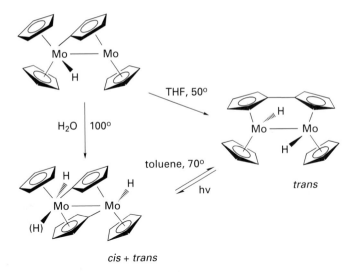

Fig. 2.21 Dimeric isomers of 'molybdocene'.

2.6.3 Dicyclopentadienylmolybdenum dihydride

[Cp$_2$MoH$_2$] can be prepared by reacting MoCl$_5$ with NaC$_5$H$_5$ in the presence of excess NaBH$_4$ or by reaction of dicyclopentadienylmolybdenum halides with NaBH$_4$ (Green *et al.*, 1961). It can be isolated as a yellow, crystalline solid with a melting range of 183–185°C and can be sublimed at 50°C *in vacuo*. It is capable of acting as a Lewis base, being protonated by CF$_3$CO$_2$H to yield [Cp$_2$MoH$_3$]$^+$ and reacting with Lewis acids such as BF$_3$ and AiR$_3$ to form adducts. Neutron diffraction studies show a tetrahedral structure with a dihedral angle between the planes of 34.2°, the Mo—H distances (0.169 nm) and the H—Mo—H angle (75.5°). The d^2 lone pair is located along the y axis and there is good agreement between the structural data and the detailed MO calculations.

2.6.4 Dicyclopentadienyltungsten

The cyclopentadienyl complexes dominate the organometallic chemistry of tungsten, with those containing the [CpW(CO)$_3$] unit being the most common. Like molybdenum, there is no stable monomeric dicyclopentadienyl of tungsten analogous to chromocene. The transient metallocene can be observed via its stabilisation with other reactants and, indeed, the chemistry of [Cp$_2$W] has been crucial in determining the α-elimination processes of metal alkyls and the activation of C—H bonds. There have been many attempted syntheses of the simple mononuclear metallocene [Cp$_2$W] but all preparations have led to isolation of a polynuclear product. For instance, reduction

53

of [Cp$_2$WCl$_2$] by sodium amalgam in THF gave a mixture of [Cp$_2$WH$_2$] and [Cp$_2$W]$_x$ (where $x = 2$). By photolysis of [Cp$_2$WX$_2$] {X$_2$ = H$_2$, D$_2$, H(Me)} or [Cp$_2$WCO] in an argon matrix at 10 K, the metallocene has been observed and characterised by IR spectroscopy. By analogy to the spectra of [Cp$_2$Cr] and [Cp$_2$V], a structure featuring parallel, cyclopentadienyl rings is postulated for [Cp$_2$W]. Conclusive evidence about the structure is lacking and it is also thought that a similar structure to that found for 'titanocene' is likely here. The existence of [Cp$_2$W] has also been postulated after 'trapping' with various reagents in reduction and photochemical reactions, i.e. reduction of [Cp$_2$WCl$_2$] with CO, Me$_2$C$_2$ and C$_2$H$_4$ yields [Cp$_2$WCO], [Cp$_2$W(η^2-C$_2$Me$_2$)] and [Cp$_2$W(η-C$_2$H$_4$)], respectively. Similarly, photolysis of [Cp$_2$WH$_2$] in diethylether gives [Cp$_2$WH(η-C$_2$H$_4$)]$^+$ and a mixture of *cis-* and *trans-*[{Cp(μ-[η^1:η^5-C$_5$H$_4$])WH}$_2$]. The [Cp$_2$W] unit is clearly electron rich, and this is shown by the CO stretching frequency (1684 cm^{-1}) in the infrared spectrum of the compound [Cp$_2$W(CO)]. As a 16-electron unit, it is able to undergo many insertion reactions and acts in a similar fashion to a carbene.

2.6.5 Dicyclopentadienyltungsten dihydride

[Cp$_2$WH$_2$] is synthesised from WCl$_3$, NaCp and NaBH$_4$ in THF. It can be isolated as a yellow, crystalline solid after sublimation with a melting point of 163–165°C. It acts as a Lewis base, reacting with species such as BX$_3$ (X = F, Cl) and AlR$_3$ to form [Cp$_2$(H)W—MX$_3$] but also undergoing protonation to give [Cp$_2$WH$_3$]$^+$. In this species, a typical high field ^{1}H NMR spectrum for metal-bound hydrogens is observed and, in the solid state, the structure **II.8**

M = Nb, Ta, $n = 0$

M = W, $n = 1$

(II.8)

is analogous to those of [Cp$_2$MH$_3$] (M = Nb, Ta) featuring a tilted [Cp$_2$M] unit and three coplanar hydrogen atoms (Wilson *et al.*, 1977).

2.7 Cyclopentadienyl compounds of group 7 elements

2.7.1 Dicyclopentadienylmanganese (manganocene)

Manganocene was first prepared and characterised in the research groups of Wilkinson and Fischer. It is atypical amongst metallocenes, being a high spin 'ionic' compound, and it was many years before complete interpretation of the structural and magnetic data was accomplished. $[(\eta\text{-}C_5H_5)_2Mn]$ was one of the first organometallic complexes of manganese to be characterised and it can be prepared from anhydrous MnX_2 (X = Cl, Br, I) and NaC_5H_5 in THF or from Cp_2Mg and $MnCl_2$. It sublimes as an amber crystalline solid at 100–130°C $(10^{-4}$—10^{-5} mmHg) and undergoes a phase change at 159–160°C, becoming pale pink in colour. It has a melting point of 172–173°C and is thermally stable up to about 350°C. It is soluble in pyridine, THF, benzene and diethylether but insoluble in petroleum ether and various chlorocarbons. The compound is reactive towards air and inflames explosively when finely divided, whilst water reacts instantaneously to produce $Mn(OH)_2$ and cyclopentadiene. It will react with ferrous ions in THF to produce ferrocene and as such resembles an ionic cyclopentadiene, such as $[(C_5H_5)_2Mg]$. Magnetic susceptibility studies on $[(\eta\text{-}C_5H_5)_2Mn]$ indicate a high-spin manganese(II) system, though there is some evidence for a low-spin state. There is very little energy difference between high- and low-spin states, it being around 2.1 kJ mol^{-1}. It is worth noting that $[(\eta\text{-}C_5Me_5)_2Mn]$ has one unpaired electron and it is therefore a low-spin manganese(II) species. It is also interesting that the magnetic moment of manganocene shows an unusual temperature dependence, being high-spin (S = 5/2, μ_{eff} = 5.9 BM) at room temperature but close to the high-spin/low-spin crossover point at lower temperatures.

In the solid state, manganocene has a polymeric structure. Chains are formed via a zig-zag arrangement of $[(\eta\text{-}C_5H_5)Mn]$ units bridged by the second C_5H_5 group (Bünder and Weiss, 1978) (Fig. 2.22). The Mn—C bond distances are long at 0.242 nm whilst the distance between the manganese and carbon atoms in the bridging cyclopentadienyls is between 0.24 and 0.33 nm. However, gaseous $[(\eta\text{-}C_5H_5)_2Mn]$ as determined by electron diffraction experiments has a sandwich structure. The manganese–carbon distance is 0.238 nm and significantly shorter than in the high-spin $[(\eta\text{-}C_5H_4Me)_2Mn]$ analogue. This longer distance in the latter reflects decreased bonding energy and also that two distinct high-spin and low-spin forms of the compound are found in the gas phase (Mn—C distance is 0.2144(12) nm).

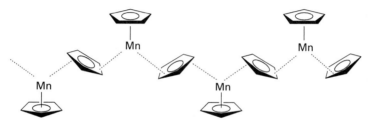

Fig. 2.22 The solid state 'chain' structure of manganocene.

2.7.2 Dicyclopentadienyltechnetium hydride

The organometallic chemistry of technetium is less well-known than that of rhenium due to the lack of availability (and cost) of the materials and its radioactivity. Dicyclopentadienyltechnetium hydride is one of the best known technetium organometallic derivatives but due to air sensitivity and handling problems its chemistry is largely unexplored. It has been obtained from the reaction of $TcCl_4$ with NaCp in tetrahydrofuran and in the presence of sodium tetrahydroborate. After recrystallisation, golden-yellow platelets were obtained (m.p. 150°C). It is very sensitive to air but thermally stable, diamagnetic and soluble in benzene and toluene. A bent sandwich structure is suggested, as in the case of $[(C_5H_5)_2ReH]$, and there are similar spectroscopic observations. IR data show the Tc—H bond to be somewhat weaker than its Re analogue and there is a low field shift of the Tc—H signal compared to that of the Re—H, indicating a weaker shielding in the former. This can be explained by a decreased negative charge on the hydrogen attached to technetium.

2.7.3 Dicyclopentadienylrhenium hydride

Rhenium derivatives containing the (η-Cp) group form a significant part of rhenium organometallic chemistry and the hydrido-species $[(\eta\text{-}C_5H_5)_2ReH]$ is the starting compound for many other derivatives. The species is very air sensitive and, as a result, its chemistry has been largely unexplored. It can be prepared from rhenium pentachloride and sodium cyclopentadienide in THF. Heating the purple reaction mixture to 120–170°C *in vacuo* yields a yellow crystalline sublimate, which melts at 161–162°C (Wilkinson and Birmingham, 1955). The observed compound is diamagnetic and soluble in most organic solvents. In fact, it was discovered to be one of the earliest examples of a hydrido–transition metal complex. It has more recently been synthesised from the co-condensation of rhenium atoms with a mixture of benzene and freshly distilled cyclopentadiene. The hydrogen bonded to the rhenium atom was identified by its low frequency resonance in the ^{1}H NMR (−23.5 ppm). No structural data are currently available but it has been postulated that the structure involves a tilted (bent back) arrangement of the Cp rings with the hydrogen atom located in the electron density surrounding the rhenium atom in the exposed region between the cyclopentadienyl rings **(II.9)**. Interestingly, the compound exhibits no basic properties and, in fact, acts as a proton acceptor. For instance, when treated with mineral acids (i.e. dilute hydrochloric or sulphuric), protonation occurs at the metal centre, showing the same basicity as nitrogen atoms in organic amines. The unipositive cation forms $[(C_5H_5)_2ReH_2]^+$ without the expected evolution of hydrogen. The complex also forms adducts with Lewis acid species such as AlR_3 and BX_3 and it acts as a donor on co-ordination to an acceptor species, $[Mn(Me)(CO)_5]$.

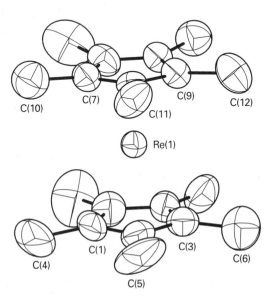

$$\text{(II.9)}$$

Recently, the existence of rhenocene Cp_2Re has been demonstrated, albeit at temperatures of 20 K, from the photolysis of Cp_2ReH in CO or N_2 matrices. It is isoelectronic with the osmocenium cation and thought to have a dimeric structure with a metal–metal bond. From a structural point of view it is worth noting the formation of **decamethylrhenocene**. Cp_2^*ReH can be formed from Re vapour and C_5MeH and will undergo a photochemical conversion into Cp_2^*Re (Bandy *et al.*, 1988) (Fig. 2.23). The X-ray structure of the deep purple, paramagnetic crystalline compound ($\mu_{eff} = 1.70$ BM) shows a parallel ring, eclipsed structure, monomeric in solution and the gas phase with rhenium being 0.1882 and 0.188 nm away from the mean planes of the two rings.

Fig. 2.23 The structure of Cp_2^*Re showing thermal ellipsoids drawn at the 50% probability level. (Bandy *et al.* (1988), with permission.)

2.8 Cyclopentadienyl compounds of group 8 elements

2.8.1 Dicyclopentadienyliron (ferrocene)

There are many possible preparations of ferrocene with the most common being:

1   $[(C_5H_5)_2]$ (M = Li, Na, K) and anhydrous $FeCl_2$.
2   C_5H_6, Et_2NH, anhydrous $FeCl_2$ or $FeCl_3$.
3   C_5H_6, KOH, $FeCl_2 \cdot 4H_2O$.
4   Fe, $[Et_2NH_2]Cl$, C_5H_6.
5   Fe (atoms), C_5H_6.

It is inexpensive and commercially available, and is isolated as an orange, crystalline solid, melting at 173°C and boiling at 249°C. It is soluble in most organic solvents, though insoluble in water, and is stable to air, water, acids and bases. It can exist in ordered and disordered phases and possesses interesting structural variation depending on the phase, i.e. it can crystallise in monoclinic, triclinic or orthorhombic modifications. Original X-ray diffraction studies indicated a staggered configuration of the rings with a molecular centre of symmetry (D_{5d}). However, gas-phase electron diffraction observations projected a D_{5h} geometry (Fig. 2.24) with C—C = 0.1440(2) and Fe—C = 0.2064(3) nm (Haaland, 1979). More recent results for the crystal indicate that below the Λ-point transition at 164 K, the triclinic form persists and the configuration is ordered with a D_5 structure and only 9° rotation from the D_{5h} eclipsed configuration. The room temperature monoclinic crystalline form is a disordered species, indicating a staggered conformation, whilst in the orthorhombic form ($T < 110$ K) the rings are fully eclipsed (D_{5h}). Theoretical studies have shown that, for ferrocene, the eclipsed form is slightly more stable than the staggered form by 2.78 kJ mol^{-1}. Experimental evidence has reinforced this situation with very low rotational barrier values (3.8 kJ mol^{-1}) leading to the possible adoption of either staggered or eclipsed configurations. It is interesting to note that in the

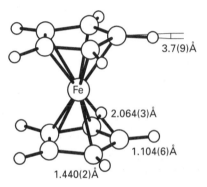

3.7(9)Å

Fe

2.064(3)Å

1.104(6)Å

1.440(2)Å

Fig. 2.24 The gas-phase electron diffraction study of ferrocene. (Haaland (1979), with permission.)

gas phase and in the crystal structure, $[(C_5Me_5)_2Fe]$ adopts a regular D_{5d} geometry, with the staggered form 4.2 kJ mol^{-1} more stable than the eclipsed due to repulsions between the Me groups. There have been extensive studies on the reactivity of ferrocene and these will be covered later, however the exceptional feature is its ability to undergo aromatic-type (electrophilic) substitution reactions in analogy to benzene, hence the name 'ferrocene'. Most metallocenes are destroyed by common electrophiles but acylation of ferrocene was discovered soon after its original synthesis.

Ferrocene is also readily oxidised to the dicyclopentadienyliron(III) cation (called 'ferrocenium'), which is blue or green in dilute solution or blood red when concentrated. X-ray structures of some ferrocenium salts have been determined but oxidation has minimal effect. The C_5H_5 rings remain eclipsed in $[(C_5H_5)_2Fe]^+$ as in ferrocene, but on oxidation of $[(C_5Me_5)_2Fe]$ an internal rotation from staggered to eclipsed forms takes place. In general, upon oxidation there is a slight increase in Fe—C bond lengths and also within the rings, as is to be expected for removal of a bonding electron.

2.8.2 Dicyclopentadienylruthenium (ruthenocene)

Ruthenocene was one of the first organometallic compounds to be formed following the discovery of ferrocene but whilst the chemistry of ferrocene has flourished, much less attention has been paid to the more costly and synthetically more challenging ruthenium (and osmium) analogue and most studies have concerned the relative reactivities of the three 18-electron metallocene systems. Ruthenocene can be formed in a number of ways: (i) $Ru(acac)_3$ and excess $[(C_5H_5)MgBr]$; (ii) $RuCl_3$ and sodium cyclopentadiene; (iii) via the ligand route with ferrocene and anhydrous $RuCl_3$; (iv) the direct reaction between ruthenium trichloride and cyclopentadiene in ethanol in the presence of zinc as a mild reducing system; and recently (v) a high yield synthesis involving $[\{(\eta^4\text{-}C_8H_{12})RuCl_2\}_x]$ and $[(C_5H_5)Sn^nBu_3]$ in ethanol at 80°C. It is a pale yellow crystalline solid, which sublimes *in vacuo* with a melting point of 199–202°C. It is stable in air, soluble in most organic solvents and insoluble in water and is the most thermally stable of the metal dicyclopentadienyls. Its solid state structure shows the same two planar, parallel C_5H_5 rings as with iron. However, it differs from ferrocene (excepting the orthorhombic form) as in the solid state the rings are eclipsed, although there is a low internal barrier to rotation of the rings. This barrier was calculated to favour the eclipsed form relative to the staggered by 2.78 kJ mol^{-1} (ferrocene) and 4.66 kJ mol^{-1} (ruthenocene) and thought to arise from a polarisation of the metal in the field of the rings. Chemically, ruthenocene is not affected by bases or by HCl or H_2SO_4 in the absence of oxygen. Oxidation can be brought about by a number of means, i.e. bromine, iodine, aqueous

Ag$^+$ or electrochemically. Chronopotentiometric studies (potential versus time plots at constant current) indicate that ruthenocene is oxidised irreversibly with the loss of two electrons, undergoing a one-step, two-electron change.

Related spectroscopic investigations (IR, Raman, photoelectron, NMR, electronic) have been undertaken on ruthenocene and in general show expected results adding to the understanding and characterisation of the system. The electronic structure and bonding is similar to that of ferrocene except that now an eclipsed configuration has to be considered. For ruthenocene, the metal orbitals featured are $4d$, $5s$ and $5p$ and incorporation of bigger, more diffuse orbitals on descending this triad enables the heavier metals' valence electrons to approach more closely the ring orbitals than in ferrocene. Thus, this leads to a stronger metal–ring bond from Fe $\rightarrow$ Ru $\rightarrow$ Os.

2.8.3 Dicyclopentadienylosmium (osmocene)

Osmocene [(η-C$_5$H$_5$)$_2$Os] was first prepared by Fischer in 1959. However, its chemistry has not been widely investigated, primarily due to the high costs encountered in the synthesis and the resultant low yields. Osmocene can be prepared by the extended reflux of OsCl$_4$ with excess NaCp in THF or dimethoxyethane, but only affords a yield of 23%. More recently yields of 72% were obtained from the reaction of the polymer $\{[(\eta^4\text{-}C_8H_{12})OsCl_2]_x\}$ with [(C$_5$H$_5$)Sn$''$Bu$_3$] in methanol at 65°C. This has proved to be a reproducible route and successful on a scale of several grams. Other methods include the reactions of C$_5$H$_6$/[(Bu$_4$N)$_2$OsCl$_6$]/Zn/EtOH. It is obtained as a colourless, crystalline solid (m.p. 229–230°C) which undergoes sublimation, is air and moisture stable and soluble in most organic solvents.

From an X-ray structural determination it is isomorphous with ruthenocene, with the cyclopentadienyl rings in an eclipsed (D_{5h}) configuration, **II.10** (Jellinek, 1959). This contrasts with the original ferrocene study that

(II.10)

indicated a staggered conformation although it is similar to the orthorhombic form of ferrocene. Clearly in osmocene there is a greater inter-ring distance than in ferrocene (0.371 cf. 0.332 nm) which thereby reduces the interannular repulsive forces. ^{1}H NMR data show a single resonance at 4.61 ppm (CDCl$_3$) whilst the ^{13}C spectrum {δ,63.9 (C$_6$D$_6$)} indicates that the ^{13}C nuclei of osmocene are amongst the most highly shielded in any metal–Cp compound. Cyclic voltammetry exhibits two irreversible one-electron oxidation peaks whereas controlled-potential electrolysis leads to the osmocenium salt [Cp$_2$Os][BF$_4$]. In general, it is more easily oxidised than ruthenocene. Its chemistry mirrors much of the better-studied ferrocene, but osmocene will only undergo Friedel–Crafts mono-acylation (and that under forcing conditions) and no alkylation. Stronger metal–ring bonding as the iron group is descended is again noted, with IR and Raman spectra producing a series of force constants in the order Cp$_2$Fe < Cp$_2$Ru < Cp$_2$Os.

2.9 Cyclopentadienyl compounds of group 9 elements

2.9.1 Dicyclopentadienylcobalt (cobaltocene)

Cobaltocene was first synthesised by the reaction of NaCp with [Co(NCS)$_2$(NH$_3$)$_4$] in liquid ammonia. Other routes involve using cobalt bromide or [Co(NH$_3$)$_6$]Cl$_2$ giving improved yields (Köhler, 1978).

$$2C_5H_5^-Na^+ [Co(NH_3)_6]Cl_2 \xrightarrow{\text{THF}} [(C_5H_5)_2Co] + 6NH_3 + 2NaCl \qquad (2.22)$$

It can also be formed from cyclopentadiene, CoBr and a base in methanol or by passing cyclopentadiene vapour over cobalt salts or [Co$_2$(CO)$_8$] at 250–300°C. It is an air-sensitive, purple–black pyrophoric crystalline solid which can be sublimed above 40°C *in vacuo* and has a melting point of 173°C. It is soluble in organic solvents and having one more electron than ferrocene is paramagnetic (μ_{eff} = 1.70 BM), thus giving a very broad ^{1}H NMR spectrum. This being said, the most likely reaction of cobaltocene is a one-electron oxidation to give the cobaltocenium cation [Cp$_2$Co]$^+$. Cobaltocene is very rapidly oxidised in air and liberates H$_2$ from water and dilute acids, producing the extremely stable cobaltocenium ion. Cobaltocenium salts are generally prepared via mild oxidation of cobaltocene, e.g. by FeCl$_3$, or oxygen and aqueous acid. The simple salts are yellow, air-stable diamagnetic solids that are soluble in water.

From X-ray crystallography, Cp$_2$Co has been shown to be isostructural with Cp$_2$Fe. Both cyclopentadienyl rings are parallel to each other and in a staggered configuration. The average C—C distance is 0.141 nm and the Co—C distance 0.21 nm (Bünder and Weiss, 1975) (Fig. 2.25). There is no solid state phase transition between 77 and 300 K but electron diffraction studies of the gas phase structure indicate the D_{5h} eclipsed model (Co—C = 0.212 nm), although the staggered ring symmetry (D_{5d}) could not be discounted. The barrier to rotation for the rings is 7.5(4) kJ mol^{-1}. It is interesting to note that crystal structures of some cobaltocenium salts all have

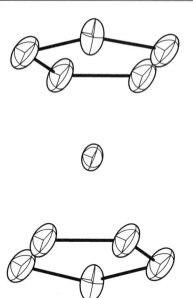

Fig. 2.25 The molecular structure of [(C$_5$H$_5$)$_2$Co].
(Bünder and Weiss (1975), with permission.)

a sandwich structure with the rings staggered but the average Co—C distances are 0.203 nm, which is significantly shorter than in [Cp$_2$Co] (0.21 nm). This is because of the additional antibonding electron in the configuration of cobaltocene.

2.9.2 Dicyclopentadienylrhodium(1+) (rhodocenium)

Stimulated by the discovery and structural characterisation of ferrocene, Wilkinson and co-workers reported the preparation and characterisation of the rhodocenium cation [(η-C$_5$H$_5$)$_2$Rh]$^+$. It was first formed from the reaction of [Rh(acac)$_3$] and a Grignard cyclopentadienyl reagent in benzene.

$$[Rh(acac)_3] + [(C_5H_5)MgBr] \xrightarrow[70°C, 24\ h]{C_6H_6} [(η\text{-}C_5H_5)_2Rh]^+ \qquad (2.23)$$

It is stable in acidic aqueous media but unstable in basic media and can be isolated as a crystalline salt with a number of counteranions, i.e. the tribromide derivative formed as above but followed by treatment with aqueous KBr and Br$_2$ to form a golden-yellow powder. More recent preparations have used anhydrous [RhCl$_3$] instead of the (acac) derivative to give a cleaner synthesis of the colourless cation and it can also be formed from reaction of RhCl$_3$·xH$_2$O and CpH in methanol in a microwave oven for 30 s (Baghurst *et al.*, 1989). A by-product from the reactions has been a neutral air-stable, trinuclear rhodium cyclopentadienyl species, **II.11**. Its structure is composed of a triangle of Rh atoms each bound to a (η-C$_5$H$_5$) ligand. A fourth C$_5$H$_5$

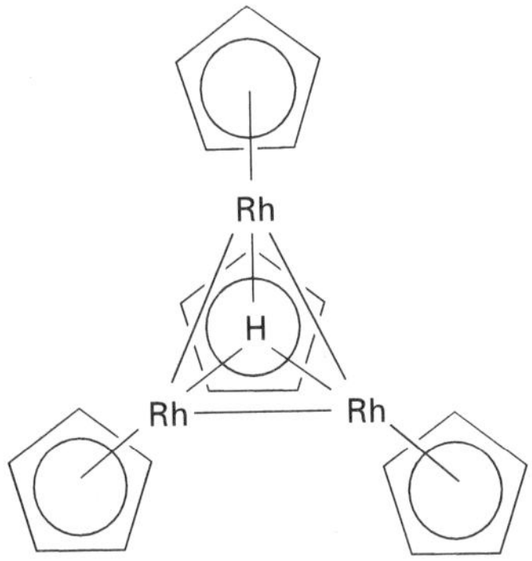

(II.11)

ligand bridges one triangular face and the hydride ligand is assumed to bridge symmetrically on the other face of the triangle.

Polarographic reduction of rhodocenium cation shows a one-electron wave but the expected neutral $[(C_5H_5)_2Rh]$ could not be isolated from aqueous solution. Reduction of the cation has also been carried out using molten Na as the electron source. At $-196\,°C$ black monomeric paramagnetic rhodocene $[(\eta\text{-}C_5H_5)_2Rh]$ is quite stable and can be sublimed onto a cooled probe. At low temperatures it does give an EPR spectrum indicating paramagnetism but on warming it dimerises to form the η^4-diene compound **II.12**.

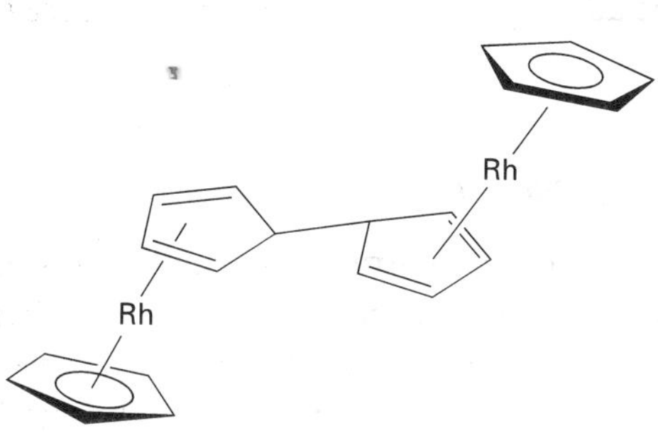

(II.12)

2.9.3 Dicyclopentadienyliridium(1+) (iridocenium)

The reasonably stable iridocenium species can be prepared by treatment of an iridium complex such as $IrCl_3$ or $[Ir(acac)_3]$ with $(C_5H_5)MgBr$, followed by hydrolysis and addition of precipitating counterion. For instance, the tribromide is synthesised from $IrCl_3$ and CpMgBr in THF in an autoclave at $200–220\,°C$ for 5 h followed by treatment with KBr/Br_2 and can be isolated as a golden-yellow powder {m.p. $185\,°C$(dec.)}. The hexafluorophosphate is otherwise synthesised from $[CpIr(\eta^4\text{-}C_5H_5)]$ and $[Ph_3C]PF_6$ in THF in a colourless crystalline form. These formally 18-electron Ir(III) species are diamagnetic but sodium reduction of $[Cp_2Ir]^+$ results in a less stable, uncharged complex Cp_2Ir. In the gas phase and at very low solution temperatures the compound is probably a paramagnetic monomer, but at room temperature it exists as a diamagnetic iridium(I) dimer.

2.10 Cyclopentadienyl compounds of group 10 elements

2.10.1 Dicyclopentadienylnickel (nickelocene)

Since its discovery in 1953, di(η-cyclopentadienyl)nickel has become one of the most thoroughly studied organonickel complexes and the starting point for the formation of many mono(η-cyclopentadienyl)nickel derivatives. It can be prepared in the laboratory by reacting a nickel halide with cyclopentadiene in the presence of KOH or diethylamine (Jolly and Chazan, 1968).

$$NiCl_2 \cdot 6H_2O + 2C_5H_6 + 10KOH \xrightarrow{-8KOH/H_2O} [(\eta\text{-}C_5H_5)_2Ni] + 2KCl \quad (2.24)$$

$$NiBr_2 + 2C_5H_6 + 2Et_2NH \longrightarrow [(\eta\text{-}C_5H_5)_2Ni] + 2Et_2NH_2Br \quad (2.25)$$

It is a dark crystalline solid which can be sublimed *in vacuo* (0.1 mmHg) at 50°C and is paramagnetic ($\mu_{eff} = 2.89 \pm 0.15$ BM). It is reasonably air sensitive, carcinogenic and soluble in most organic solvents. In the solid state and at room temperature, nickelocene is isostructural with ferrocene having planar disordered rings. On cooling to 100 K there is a phase change that forces the system into a centrosymmetric arrangement with staggered rings and a mean Ni—C distance of 0.219 nm. In the gas phase, the rings are free to rotate and electron diffraction gives Ni—C as 0.22 nm. There is a 0.013 nm decrease in metal–carbon bond length on going from nickelocene to ferrocene which reinforces the situation of the two additional electrons in nickelocene occupying anti-bonding orbitals, thus weakening the M—C bond strength. The IR spectrum closely resembles that of ferrocene, as do the ^{13}C and ^{1}H NMR spectra, except that the nickel–ligand vibrations are found at a lower frequency and there is a large chemical shift, respectively, in the techniques. The NMR effect is due to the transfer of positive spin density from the metal to the ring carbon atoms and is accompanied by a change of sign on transfer to the protons.

Nickelocene undergoes many reactions (details in Chapter 4) but, chiefly, ring substitution and displacement of one or both cyclopentadienyl rings is possible and the unusual triple-decker sandwich complex can be formed from the reaction of nickelocene with acids (see Chapter 5).

2.11 Cyclopentadienyl compounds of group 12 elements

The properties of the group 12 dicyclopentadienyls are summarised in Table 2.6. Elucidation of the structure of **dicyclopentadienylzinc (zincocene)**

Table 2.6 Properties of the group 12 dicyclopentadienyls.

Complex	$[(C_5H_5)_2Zn]$	$[(C_5H_5)_2Cd]$	$[(C_5Me_5)_2Cd]$	$[(C_5H_5)_2Hg]$
Synthesis	$[Zn\{N(SiMe_3)_2\}_2] + C_5H_6$	$[Cd\{N(SiMe_3)_2\}_2] + C_5H_6$	$Cd(acac)_2 + Li(C_5Me_5)$	$[Hg\{N(SiMe_3)_2\}_2] + C_5H_6$
Colour	Colourless	Colourless	Red	Yellow
Melting point	Sublimes at 140°C	Decomposition > 250°C		83–85°C
Solubility	Soluble in DMF, diethylether	Soluble in DMSO, pyridine	Soluble in most hydrocarbons	Soluble in most hydrocarbons
Stability	Air and water sensitive	Air and water sensitive	Light sensitive	Light, air, water, heat sensitive
Miscellaneous		Polymeric structure, evidence of η^1-C_5H_5 ligands	Monomeric structure, η^5,η^1 binding of Cp* ligands, fluxional behaviour	

DMSO = dimethyl sulphoxide.
DMF = dimethyl formamide.

has been of particular interest. Zincocene, featuring a borderline element between transition and main group metals, could possess a typical ferrocene-like sandwich structure or form a polymer with bridging cyclopentadienyl groups, as is the trait of electron-deficient cyclopentadienides of main group metals. A solid state X-ray diffraction (Budželaar *et al.*, 1985) shows that the structure consists of infinite chains of zinc atoms with bridging cyclopentadienyl groups. In addition, each zinc carries a terminal cyclopentadienyl group (Fig. 2.26). There is some disorder of the bridging cyclopentadienyl groups and as both σ- and π-type interactions contribute to the Zn—Cp bonds, the bonding situation can best be described as '$\eta^{-2.5}$'. (NB: The structures of

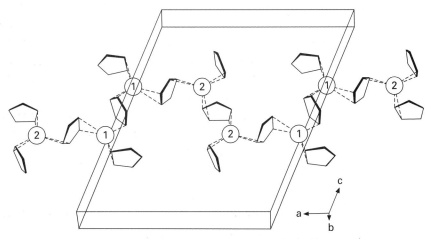

Fig. 2.26 A schematic drawing showing four units of the chain structure of Cp$_2$Zn. (Budzelaar *et al.* (1985), with permission.)

65

many cyclopentadienides can be rationalised on the basis of both the radius of the metal atom and the binding mode of the Cp groups, with more space around the metal being occupied by increasingly delocalised metal—Cp bonds. For instance, the metal ion radius increases in the order Mg < Zn < Mn < Ca = Pb, and in Cp_2Mg the metal is surrounded by only two η^5-bound Cp groups giving a sandwich structure. Zinc can accommodate three Cp groups but none is η^5 bound. In Cp_2Mn, which like Cp_2Zn forms polymeric chains, one η^5-bound Cp group is present but the other two are much less delocalised. This is taken further with polymeric Cp_2Pb where there are three η^5—Cp groups bound to the lead. Cp_2Ca forms a three-dimensional system where each calcium is surrounded by two η^5, one η^3 and one η^1-bound groups and, looking at space requirements, this environment is virtually equivalent to the threefold η^5 co-ordination around lead in Cp_2Pb.

In the solid state, the metal—Cp ring bonding in **dicyclopentadienylmercury (mercurocene)** is η^1 and this is also the case in solution where the σ-bonded monohapto species undergoes an intramolecular rearrangement, sometimes called 'ring whizzing' (see Chapter 4). The vibrational spectrum of Cp_2Hg is too complicated to indicate a centrally bound structure and there is a single resonance in the ^1H NMR spectrum down to $-100°C$ at which some broadening occurs showing a slowing of the fluxional process. $[(C_5Me_5)_2Hg]$, however, shows no fluxionality at room temperature, giving a ^1H NMR spectrum consisting of three methyl peaks in a $2:2:1$ ratio. Structural data for cyclopentadienylmercury derivatives is very limited but a photoelectron spectrum of Cp_2Hg suggested a peripherally bonded structure as indicated by considerable mixing of the Hg—C σ-orbital with the ring π-orbitals. Dicyclopentadienylmercury will undergo Diels–Alder reactions with dienophiles, e.g. maleic anhydride, RCCR (R = $MeCO_2$ or CF_3) and alkenes containing electron-withdrawing groups.

2.12 Cyclopentadienyl compounds of group 13 elements

2.12.1 Dicyclopentadienylaluminium(1+)

One of the last hitherto unknown simple metallocenes, the aluminocenium cation $[\eta\text{-}C_5H_5)_2Al]^+$, has recently been prepared from reaction of $[(C_5H_5)_2AlCH_3]$ with $[B(C_6F_5)_3]$ in dichloromethane at low temperature (Bochmann and Dawson, 1996). The white, crystalline solid $[\eta\text{-}C_5H_5)_2Al]^+[MeB(C_6F_5)_3]^-$ is sparingly soluble in benzene or toluene but dissolves readily in dichloromethane. The compound is highly air and moisture sensitive and, to date, crystals suitable for X-ray diffraction have not been obtained. However, the $[Cp_2Al]^+$ ion is likely to possess a 'normal' metallocene-like structure with η^5-bound Cp ligands (Fig. 2.27). This is

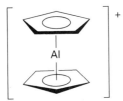

Fig. 2.27 Schematic diagram of the aluminocenium cation.

supported by the structure of $[(\eta^5\text{-}Cp^*)_2Al]^+$, the close agreement of the observed ^{27}Al chemical shift with the one calculated assuming a D_{5d} symmetry, the narrow half-width of the ^{27}Al signal which is typical for a highly symmetric ligand environment for Al as well as the sandwich structures of the isoelectronic magnesocene and related group 2 cyclopentadienyl complexes. This complex has now found use as a highly efficient initiator for the carbocationic polymerisation of isobutene and the industrially important co-polymerisation of isobutene and isoprene.

2.12.2 Tricyclopentadienylgallium

Relatively little is known of the cyclopentadienyl main group compounds and, in particular, the group 13 elements have received little attention. However, $[(C_5H_5)_3Ga]$ has recently been formed and thoroughly characterised, it being the second example of a fully characterised homoleptic cyclopentadienyl group 13 element with the metal in the typical $+3$ oxidation state (the indium derivative $[(C_5H_5)_3In]$ was the first). It can be formed from sublimed $GaCl_3$ and LiC_5H_5 at low temperatures under a scrupulously maintained atmosphere of purified argon.

$$GaCl_3 + 3LiC_5H_5 \xrightarrow{\text{r.t.}} [(C_5H_5)_3Ga] + 3LiCl \qquad (2.26)$$

The product can be purified by careful sublimation at 40–42°C to yield volatile, colourless crystals. It is thermally unstable above 45°C and decomposes without melting. It is extremely sensitive to oxygen and moisture and decomposes slowly at room temperature to yield a yellow solid. $[C_5H_5)_3Ga]$ is soluble in pentane, benzene and cyclohexane and exhibits the properties of a weak Lewis acid, reacting with strong bases NMe_3 and THF to form adducts (and four coordinate complexes). An X-ray structural study indicates that the crystal consists of discrete isolated molecules of $[(C_5H_5)_3Ga]$ separated by normal van der Waals distances and with no abnormally short intermolecular contacts (Beachley et al., 1985) (Fig. 2.28). The cyclopentadienyl rings have the expected η^1 co-ordination to gallium with the average Ga—C bond distance being 0.205(3) nm. Thus, the cyclopentadienyl rings can be classified as an 'allyl' type as opposed to 'vinyl' and it is also interesting to note that the three-carbon atoms of the cyclopentadienyl rings

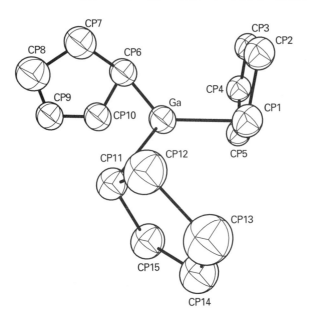

Fig. 2.28 The solid state structure of [(C$_5$H$_5$)$_3$Ga]. (Beachley *et al.* (1985), with permission.)

and gallium are coplanar to within 0.0001 nm. The trigonal–planar group 13 atom configuration represents a departure from the polymeric arrangements found in other structurally characterised group 13 derivatives.

2.12.3 Cyclopentadienylindium

Compounds of indium bound to a cyclopentadienyl group form a transition from low oxidation state species to those in which the element is in the more common +III state. Cyclopentadienylindium(I) and its [C$_5$H$_4$(CH$_3$)] analogue are the only Cp derivatives of In(I). Cyclopentadienylindium(I) was first prepared from InCl$_3$ and a fourfold excess of NaCp. After stirring for 4 h and following sublimation, pale yellow crystalline needles were obtained which possessed a melting point of 169–171 °C. It is interesting to note that the indium(I) species is obtained from an indium(III) starting material with a reaction sequence of

$$InCl_3 + 3MC_5H_5 \longrightarrow [(C_5H_5)_3In] + 3MCl$$
$$[(C_5H_5)_3In] + C_5H_5^- \longrightarrow [C_5H_5In] + 3/2(C_5H_5)_2$$
$$(M = Li \text{ or } Na)$$

(2.27)

It is very air and moisture sensitive in solution but unaffected by water in the solid state and is readily soluble in a wide range of organic solvents. It is particularly useful in synthetic work as it is the only readily available stable and soluble indium(I) compound. Early X-ray studies showed that crystalline

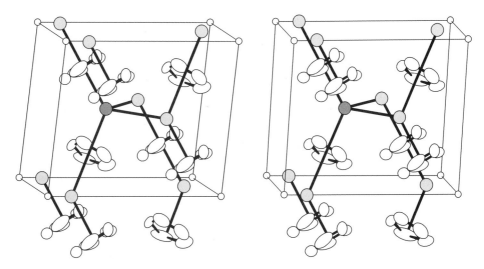

Fig. 2.29 A stereoscopic view of the solid state structure of [(C₅H₅)In] showing formation of polymeric chains (NB: Each In atom is in contact with two other indium atoms – this is shown clearly only for the darkened In atom). (Beachley (1988), with permission.)

CpIn exists as a homopolymeric chain with each indium atom sandwiched between two C_5H_5 rings lying on the C_5 axis of the ring and at a distance of 0.319(10) nm from the centre. Later studies (Beachley *et al.*, 1988) reinforced this, indicating a zig-zag chain structure in which indium atoms interact with each side of the Cp ring and two Cp rings interact with each indium atom. The centroid–In–centroid angle is 128° and the In—C distances range between 0.2853(22) and 0.3091(21) nm (Fig. 2.29). The bonding description can best be described as covalent. There is an interaction of ring carbon p_π orbitals with a hybrid *sp* orbital on indium and with indium p_{xy} orbitals. Additionally, the lone pair on indium points away from the Cp ring.

2.12.4 Tricyclopentadienylindium

As in Section 2.12.3, Cp_3In can be formed from the reaction of $InCl_3$ with an organolithium reagent. It can be isolated as very air-sensitive pale yellow crystals (m.p. 160°C) and proves to be fluxional in solution, with the ^1H NMR spectrum giving a sharp singlet for the Cp resonance at temperatures as low as −90°C. An X-ray crystallographic study at −100°C indicates a structure consisting of infinite polymeric chains with each chain unit comprising an indium atom linked to two terminal and two bridging cyclopentadienyl groups. A slightly distorted tetrahedral environment exists around each indium atom. The indium–carbon bond lengths are 0.224(1) nm for the terminal groups and 0.237(1) and 0.247(1) nm for the two contacted with the bridging groups.

2.12.4 Cyclopentadienyl compounds of thallium

The best known organometallic compound of Tl(I), [(C$_5$H$_5$)Tl], can be prepared in an aqueous medium.

$$Tl_2SO_4 + 2C_5H_6 + 2NaOH \longrightarrow 2[(C_5H_5)Tl] + NaSO_4 + 2H_2O \qquad (2.28)$$

It is sublimable, sparingly soluble in organic solvents and can be handled in air. It is a convenient cyclopentadienyl transfer agent for use with transition metal ions. The structure of [(C$_5$H$_5$)Tl] largely resembles that of [(C$_5$H$_5$)In] and shows an infinite chain structure in which E–(μ-Cp)–E bridges link the metal centres together. The thallocene anion [(C$_5$H$_5$)$_2$Tl]$^-$ is isoelectronic and isostructural with Cp$_2$E (E—Sn, Pb). It can be prepared by stirring a mixture of CpTl and Cp$_2$Mg (1:1) in THF. Addition of N,N,N′,N′,N′′-pentamethyldiethylenetriamine (PMDETA) (1 mol equiv.) produces a colourless precipitate of [(η^5-Cp)$_2$Tl(η^5-Cp)Mg·PMDETA] (Armstrong et al., 1993) (Fig. 2.30). Similarly, an analogous reaction of [CpTl] and [CpLi] (1 : 2) in THF followed by addition of PMDETA (1 mol equiv.) produces [(η^5-Cp)Tl(μ-η^5-Cp)Li·PMDETA] (Fig. 2.31).

Interestingly, addition of CpLi and CpMg to the polymeric structure of (CpTl)$_x$ will 'extrude' the basic 14e$^-$ unit. The structures of ion-separated [Cp$_2$Tl]$^-$[CpMgPMDETA]$^+$ **(II.13**, see Fig. 2.30) and ion-contacted

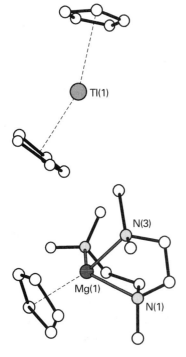

Fig. 2.30 The molecular structure of [(η^5-Cp)$_2$Tl(η^5-Cp)Mg·PMDETA] **(II.13)**.

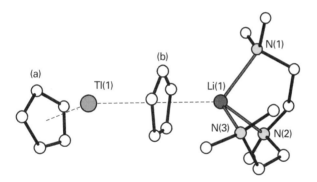

Fig. 2.31 The molecular structure of $[(\eta^5\text{-Cp})\text{Tl}(\mu\text{-}\eta^5\text{-Cp})\text{Li}\cdot\text{PMDETA}]$ **(II.14)**.

$[\text{CpTl}(\mu\text{-Cp})\text{Li}\cdot\text{PMDETA}]$ (**II.14**, see Fig. 2.31) have been determined in the solid state (Figs 2.30 and 2.31, respectively) and contain bent thallocene anions (average $\text{Cp}_{\text{centroid}}\text{-Tl-Cp}_{\text{centroid}}$ 156.7° in **II.13** and 153.3° in **II.14**) which are isoelectronic with stannocene. In **II.13**, the Tl centre of $[(\eta^5\text{-Cp})_2\text{Tl}]^-$ is attached almost equivalently to the η^5-Cp ligands (average Tl–$\text{Cp}_{\text{centroid}}$ 0.272 nm). However, there are further data, most notably from electron density calculations and figures which show that the contacts of the Tl atom with the Cp ligands are not equivalent and that the anion is probably best regarded as a '*close-contact*' complex between CpTl and Cp⁻.

2.13 Cyclopentadienyl compounds of group 14 elements

2.13.1 Dicyclopentadienylgermanium (germanocene)

Dicyclopentadienylgermanium was first prepared by the reaction of sodium cyclopentadienide ($-78°$C, diethylether, 2 h) or thallium(I) cyclopentadienide (20°C, THF, 30 min) with freshly prepared germanium dibromide. A more recent synthesis of germanocene was completed in tetrahydrofuran from sodium cyclopentadienide and the dichlorogermanediyl–dioxane adduct.

$$2\text{NaC}_5\text{H}_5 + \text{GeCl}_2\cdot\text{dioxane} \longrightarrow [(\text{C}_5\text{H}_5)_2\text{Ge}] + 2\text{NaCl} + \text{dioxane} \quad (2.29)$$

Colourless crystals (m.p. 78°C) can be isolated by sublimation that are very air sensitive but can be manipulated under inert gas atmosphere and stored indefinitely at $-30°$C without noticeable polymerisation. There is rapid polymerisation on heating and at room temperature a yellow solid is formed that is insoluble in all common organic solvents, unlike monomeric germanocene which is soluble in THF. The yellow colour suggests the presence of Ge—Ge bonds (similar catenates of silicon are also yellow). It is mono-

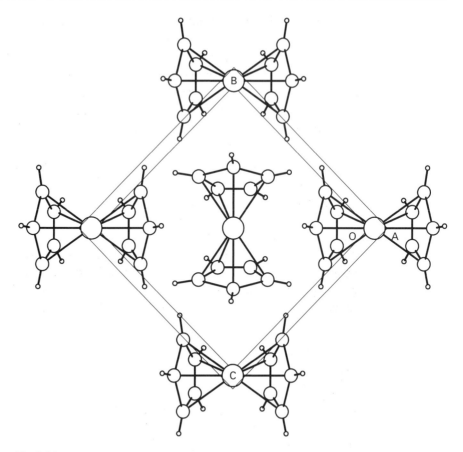

Fig. 2.32 The molecular structure and unit cell packing of germanocene.

meric in benzene and also in the solid state, existing in a bent sandwich structure (Grenz *et al.*, 1984) (Fig. 2.32). The angle between the planes of the cyclopentadienyl rings is 50.4°; a larger deviation from the D_{5d} or D_{5h} symmetry is found in germanocene than in stannocene (45.9 or 48.4°). Both rings are η^5-bound but the Ge—C distances range from 0.235 to 0.273 nm which is quite remarkable and is because of the displacement of the germanium atom from the point of intersection of the fivefold axis towards the apex at which the Cp ring planes touch.

2.13.2 Dicyclopentadienyltin (stannocene)

Cyclopentadienyltin(II) compounds are the only known organotin(II) derivatives which do not clearly owe their stability to steric hindrance. The tin atom can be regarded as sp^2-hybridised, with two of the hybrid orbitals involved in bonding and the third containing an unshared pair of electrons.

Stannocene can be readily synthesised from the reaction between cyclopentadienyllithium or cyclopentadienylsodium and tin(II)chloride in THF. It is isolated as a white crystalline solid melting at 105°C and is soluble in most aprotic solvents. It was assumed to be polymeric (like its lead analogue) but has been shown to be monomeric in the solid state. The structure has been determined crystallographically (Atwood and Hunter, 1981) and by electron diffraction shows that the two rings are not parallel, with the lone pair of electrons controlling the bending back of the rings (Fig. 2.33). The ring centroid–metal–ring centroid angle is large (*c.* 145°) and seems to be governed by electronic rather than steric factors. The rings are best regarded as being pentahapto-bound to the metal. It acts as a reagent for the synthesis of inorganic bivalent–tin compounds; some reactions are detailed in Chapter 4. Stannocene is readily hydrolysed and oxidised and the lone pair of electrons easily coordinates to a Lewis acid such as boron trifluoride or $AlBr_3$. The metallocene reacts with tin(II) bromide or chloride in THF to form corresponding cyclopentadienyltin halides CpSnX. These are white, crystalline solids melting at 130–133°C (X = Br) and 160–162°C (X = Cl) and the

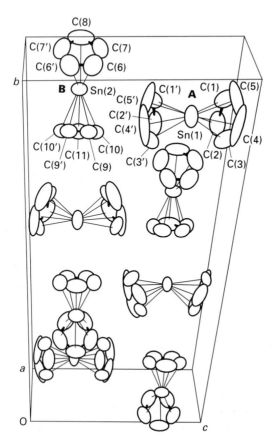

Fig. 2.33 The molecular structure and unit cell packing of $[(C_5H_5)_2Sn]$. (Atwood and Hunter (1981), with permission.)

73

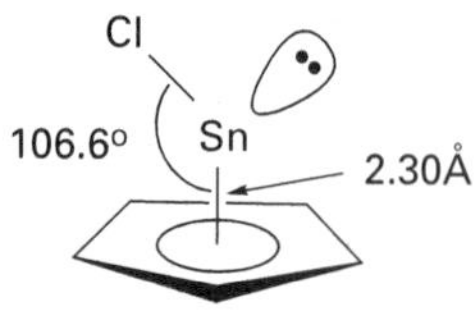

(II.15)

chloride (**II.15**) has a bent, half-sandwich structure which results from π-electron density from the cyclopentadienyl ring being donated into the vacant *p* orbital on tin.

2.13.3 Tetracyclopentadienyltin

This compound and its other cyclopentadienyltin analogues have been of particular interest due to their structural properties. Preparation is via lithium, sodium, potassium or magnesium derivatives of cyclopentadiene with tin tetrachloride in THF or benzene (Scheme 2.3). It is a white crystalline solid with a melting point of 81–82°C. At –60°C, crystalline [(C$_5$H$_5$)$_4$Sn] has a distorted tetrahedral structure at tin with ring C—Sn—ring C angles between 103 and 114°. The average C—Sn bond length is 0.227 nm, which is more than the sum of the covalent radii (0.216 nm). The planar cyclopentadienyl rings are η^1-bound but bent towards the tin atom by an average of 5.1°. The system is fluxional and in solution at room temperature there is only one ^{1}H and one ^{13}C NMR signal. However, at much lower temperatures (–70°C) the signals of the ring protons are broadened and then split into separate signals at –150°C (CF$_2$Cl$_2$ solvent). At room temperature, a time-averaged NMR spectrum is seen with the organotin group undergoing rapid metallotropic migration around the ring (or ring-whizzing), probably involving a series of 1, 2-shifts.

2.13.4 Dicyclopentadienyllead (plumbocene)

There are very few bivalent organolead compounds containing only carbon and hydrogen and the metal; dicyclopentadienyllead [(C$_5$H$_5$)$_2$Pb] is an example of one of them. It was originally prepared from the reaction of lead(II) nitrate and sodium cyclopentadienide. Lead(II) acetate in THF or mono-

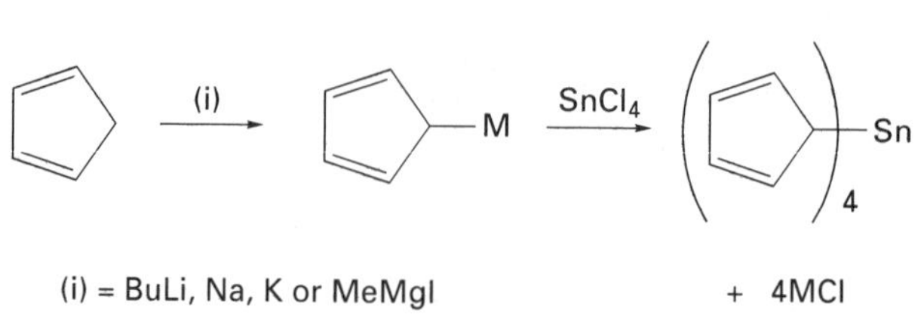

(i) = BuLi, Na, K or MeMgI   + 4MCl

Scheme 2.3 The formation of tetracyclopentadienyltin.

glyme is used though the most successful preparation to date features lead(II) chloride with dicyclopentadienylmagnesium in xylene at 160°C. The compound is an air-sensitive golden-yellow crystalline solid, melting at 140°C and open to sublimation *in vacuo*. It is soluble in most organic solvents and in solution possesses a pentahapto angular sandwich structure **(II.16)** as

(II.16)

determined by electron diffraction in the vapour phase (Almenningen *et al.*, 1967). The ligand rings are not parallel, the angle between the planes being *c.* 45°. However, in the solid state, plumbocene possesses a chain structure (similar to cyclopentadienyl-indium and -thallium) with bridging cyclopenta-dienyl groups (Fig. 2.34) (Panattoni *et al.*, 1966) (cf. dicyclopentadienyltin, which is composed of isolated molecules). The lead—carbon bond distances to the pendant cyclopentadienyl groups are 0.276(1) nm but these become significantly longer in the bridging cyclopentadienyl situation (0.306(1) nm). The lead is sp^2-hybridised and both σ- and π-bonding interactions between the rings and metal are indicated by photoelectron spectra. Dicyclopentadie-nyllead is quite reactive and mirrors the chemistry of the tin analogue. No adduct formation is seen with weak Lewis acids such as BMe_3 or BPh_3 but reaction with boron trifluoride etherate gives a yellow, air-sensitive 1:1

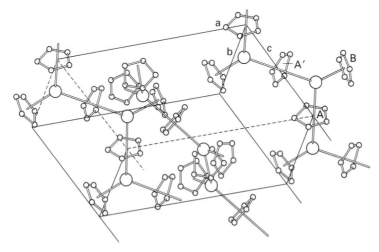

Fig. 2.34 The structure of plumbocene in the solid state with unit cell packing. (Panattoni *et al.* (1968), with permission.)

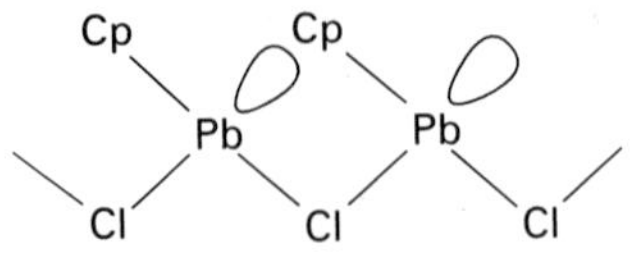

Fig. 2.35 The proposed structure of [(C₅H₅)PbCl].

adduct [(C₅H₅)₂Pb·BF₃] which is totally insoluble and very thermally stable. It was therefore thought to possess a polymeric structure. Hydrogen halides will cleave the cyclopentadienyl group to give cyclopentadienyllead halides [(C₅H₅)PbX] (X = Cl, Br, I) which are insoluble and highly thermally stable. This again suggests a polymeric structure present in this system (Fig. 2.35).

2.14 Cyclopentadienyl compounds of group 15 elements

2.14.1 Tricyclopentadienylbismuth

Two forms of this compound interconvert on heating or cooling. The black form has D_{3h} symmetry and rapid interconversion of σ- and π-bound rings is thought to occur in the orange form. It is monomeric in C_6H_6, synthesised from NaCp and $BiCl_3$ and exists as orange crystals (<15–20°C) and a black insoluble solid (>15–20°C) which decomposes at c. 50°C. Very few structures of the group 15 complexes have been reported and the structures of Cp_3Bi and Cp_3Sb, which have 20-electron metal centres, have not been determined owing to their extreme thermal instabilities.

2.14.2 Tricyclopentadienylantimony

This compound is synthesised from [Sb(NMe₂)₃] and CpH and is a very air-sensitive solid, melting at 56°C. NMR studies once again suggest that there is rapid interconversion between σ- and π-bound forms.

References

Almenningen, A., Haaland, A. and Motzfeldt, T. (1967) *J Organomet Chem*, **7**, 97.

Armstrong, D.R., Herbst-Irmer, R., Kuhn, A. *et al.* (1993) *Angew Chem*, **105**, 1807; *Angew Chem Int Ed Engl*, **32**, 1774.

Atwood, J.L. and Smith, K.D. (1973) *J Am Chem Soc*, **95**, 1488.

Atwood, J.L. and Hunter W.E. (1981) *J Chem Soc Chem Commun*, 925.

Avdeef, A., Raymond, K.N., Hodgson K.O. and Zalkin, A. (1972) *Inorg Chem*, **11**, 1083.

Baghurst, D.R., Mingos, D.M.P. and Watson, M.J. (1989) *J Organomet Chem*, **368**, C43.

Bandy, J.A., Cloke, F.G.N., Cooper, G. *et al.* (1988) *J Am Chem Soc*, **110**, 5039.

Beachley, O.T., Jr, Getman, T.D., Kirss, R.U., Hallock, R.B. Hunter W.E. and Atwood, J.L. (1985) *Organometallics*, **4**, 751.

Beachley, O.T., Jr, Pazik, J.C., Glassman, T.E., Churchill, M.R., Fettinger, J.C. and Blom, R. (1988) *Organometallics*, **7**, 1051.

Beattie, J.K. and Nugent, K.W. (1992) *Inorg Chim Acta*, **198–200**, 309.

Blackborow, J.R. and Young, D. (1974) *Metal Vapour Synthesis in Organometallic Chemistry*. Springer-Verlag, Berlin.

Bochmann, M. and Dawson, D.M. (1996) *Angew Chem*, **108**, 2371; *Angew Chem Int Ed Engl*, **35**, 2226.

Budzelaar, P.H.M., Boersma, J., Van der Kerk, G.J.M., Spek, A.L. and Duisenberg, A.J.M. (1985) *J Organomet Chem*, **281**, 123.

Bünder, W. and Weiss, E. (1975) *J Organomet Chem*, **92**, 65.

Bünder, W. and Weiss, E. (1978) *Z Naturforsch*, **B33**, 1235.

Burns, J.H. (1974) *J Organomet Chem*, **69**, 225.

Calderon, J.F., Cotton, F.A., DeBoer, B.G. and Takats, J. (1971) *J Am Chem Soc*, **93**, 3592.

Eggers, S.H., Kopf, J. and Fischer, R.D. (1986a) *Organometallics*, **5**, 383.

Eggers, S.H., Schulltze, H., Kopf, J. and Fischer, R.D. (1986b) *Angew Chem*, **98**, 631.

Eggers, S.H., Schulltze, H., Kopf, J. and Fischer, R.D. (1986c) *Angew Chem Int Ed Engl*, **25**, 656.

Gard, E., Haaland, A., Novak, D.P. and Seip, R. (1975) *J Organomet Chem*, **88**, 181.

Green, M.L.H., McCleverty, J.A., Pratt, L. and Wilkinson, G. (1961) *J Chem Soc*, 4854.

Grenz, M., Huhn, E., du Mont, W.W. and Pickardt, J. (1984a) *Angew Chem*, **98**, 69.

Grenz, M., Huhn, E., du Mont, W.W. and Pickardt, J. (1984b) *Angew Chem Int Ed Engl*, **23**, 61.

Guggenberger, L.J. (1973) *Inorg Chem*, **12**, 294.

Haaland, A. (1979) *Acc Chem Res*, **12**, 415.

Hodgson, K.O. and Raymond, K.N. (1972) *Inorg Chem*, **11**, 3030.

Jellinek, F. (1959) *Z Naturforsch*, **B14**, 737.

Jolly, W.L. and Chazan, D.J. (1968) *Inorg Synth*, **11**, 122.

Köhler, F.H. (1978) *J Organomet Chem*, **160**, 299.

Labinger, J.A. (1982) *Comprehensive Organometallic Chemistry*. (eds. E.W. Abel, F.G.A. Stone and G. Wilkinson) vol. 3, ch. 25, pp. 706–80. Pergamon Press, Oxford.

Panattoni, C., Bombieri, G. and Croatto, U. (1966) *Act Crystallogr*, **21**, 823.

Rogers, R.D., Bynum, R.V, and Atwood, J.L. (1978) *J Am Chem Soc*, **100**, 5238.

Rogers, R.D., Bynum, R.V, and Atwood, J.L. (1980) *J Organomet Chem*, **192**, 65.

Timms, P.L. and Turney, T.W. (1977) *Adv Organomet Chem*, **15**, 53.

Wasserman, H.J., Zozulin, A.J., Moody, D.C., Bryan, R.R. and Salazar, K.V. (1983) *J Organomet Chem*, **254**, 305.

Wilkinson, G. and Birmingham, J. M. (1955) *J Am Chem Soc*, **77**, 3421.

Wilson, R.D., Koetzle, T.F., Hart, D.W., Kuick, A., Tipton, D.L. and Bau, R. (1977) *J Am Chem Soc*, **99**, 1775.

Zinnen, H.A., Pluth, J.J. and Evans, W.J. (1980) *J Chem Soc Chem Commun*, 810.

3 Electronic Structure and Bonding of Metallocenes

Simple bonding theories such as the '18-electron rule' are useful starting points for discussion of structure and bonding (see Chapter 1) but for a rigorous study of these aspects of metallocenes, application of molecular orbital theory is a prerequisite.

3.1 Molecular orbital calculations on parallel, planar 'ferrocene-like' sandwich compounds

3.1.1 Di(cyclopentadienyl)metal compounds

Over recent years, there have been many theoretical investigations into the bonding of sandwich compounds and nowadays the essential features are well understood (Mingos, 1982). It is essential to formulate a model for the electronic structure of sandwich molecules and relate the model to known compounds and structures to aid the understanding of the chemistry of these molecules. The fundamental questions of structure and bonding can be approached by examining some representative examples especially with regard to the normal, parallel, planar sandwich structure of ferrocene-like systems. Most studies have been concerned with the overlap of the metal orbitals with matching symmetry ligand orbitals, derived by taking in-phase and out-of-phase linear combinations of the ligand π-molecular orbitals (MO).

Construction of MO diagrams for ferrocene-type systems

The best starting point is to consider the π-molecular orbitals formed from the set of p_π oribitals of the C_5H_5 fragment. The five p orbitals on the planar C_5H_5 group can be combined to produce five molecular orbitals (Fig. 3.1) of which one is strongly bonding (a), a degenerate pair are weakly bonding (e_1) and another degenerate pair are weakly anti-bonding (e_2). For the metallocene molecular orbital description, there are two cyclopentadienyl groups so the sum and difference of each molecular orbital has to be considered. The five ligand π orbitals are combined pairwise in a $(+)(-)$ antisymmetric to inversion combination ($= u$, from the German *ungerade*, meaning odd) and a $(-)(+)$ symmetric to inversion combination ($= g$, *gerade*, even). There are now three sets of π orbitals: a low-lying filled pair of a_{1g} and a_{2u} symmetry; a

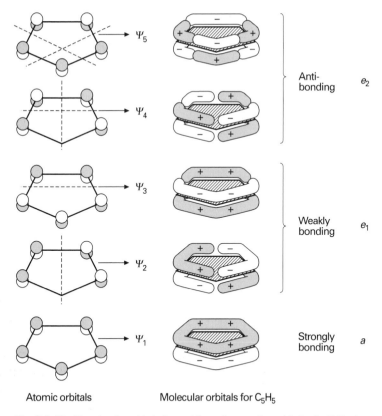

Fig. 3.1 The Π molecular orbitals formed from the set of p_π orbitals of a C_5H_5 ring.

filled set of e_{1g} and e_{1u} symmetry; and an unfilled set of anti-bonding orbitals of e_{2g} and e_{2u} symmetry at higher energy. This is the situation for a staggered D_{5d} symmetry molecule (i.e. when there is a centre of symmetry within the metallocene) (Fig. 3.2). Then by considering these ligand orbitals and how overlap with metal orbitals of appropriate symmetry can be effected, the molecular orbital bonding picture for a metallocene can be constructed. Each of the combinations (ligand orbitals + metal orbitals) leads, in principle, to a bonding molecular orbital of the molecule, assuming that the energy of the two components is not very different. There are, of course, an equal number of anti-bonding combinations derived from (ligand orbitals – metal orbitals). For example, a combination of the ψ_1's of both rings, $(\psi_1 + \psi_1)$, gives a_{1g} which interacts with the d_{z^2} orbital on the metal (also a_{1g}). With the opposite combination of $\psi_1 - \psi_1$ (a_{2u}), the metal p_z is the interactive orbital. Similarly, ψ_2 and ψ_3 combinations are strongly stabilised by interactions with the d_{xz}, d_{yz}, p_x and p_y orbitals. These combinations are shown in Fig. 3.3.

Symmetry considerations allow determination of which molecular orbitals are possible but knowledge of overlap integrals and relative energies is

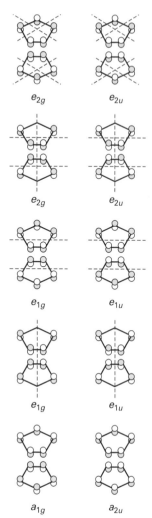

e_{2g} e_{2u}

e_{2g} e_{2u}

e_{1g} e_{1u}

e_{1g} e_{1u}

a_{1g} a_{2u}

Fig. 3.2 Examples of the overlap of the sum and difference ligand orbital–metal orbital combinations.

necessary to allow estimation of the resulting energy levels. There has been a certain amount of uncertainty and controversy regarding several of these values and their respective energy levels due to slightly differing assumptions made in the calculations and the type of computational method adopted (Lauher and Hoffmann, 1976). A qualitative molecular orbital diagram for a normal, planar metallocene in its staggered conformation (D_{5d}) is shown in Fig. 3.4. The a_{1g} (σ–completely cylindrical symmetry–no nodes) bonding MO is mainly ligand-based as it is so stable relative to the metal orbitals (for Fe, $4s$ and $3d_{z^2}$) that they interact very little. Similarly, the a_{2u} level has virtually no interaction with the higher-lying Fe $4p_z$ orbital which formally has the correct symmetry to combine. The most well-matched orbitals are those of the ligand e_{1g} orbitals and the Fe $3d_{xz}$ and $3d_{yz}$ and, consequently,

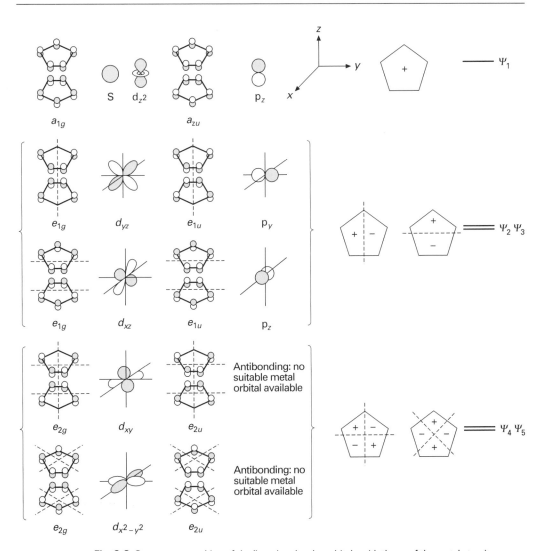

Fig. 3.3 Symmetry matching of the ligand molecular orbitals with those of the metal atomic orbitals.

the two strong π bonds created form most of the strength of the metallocene molecule. The corresponding antibonding e_{1g}^* set are unoccupied in the ground state, but have an impact in optical transitions. The e_{1u} bonding molecular orbitals are mainly ligand-based with some contribution from the Fe $4p$ orbitals, although these are at a high energy so the e_{1u} set do not contribute much to the bonding. The remaining three d orbitals of the metal, the a_{1g}' (d_{z^2}) and the e_{2g} ($d_{x^2-y^2}$, d_{xy}) set, remain essentially non-bonding. This is because for the former, the ligand π orbitals of a_{1g}' symmetry point towards the nodal cone of the metal $3d_{z^2}$ orbital (Fig. 3.5), whilst for the latter the

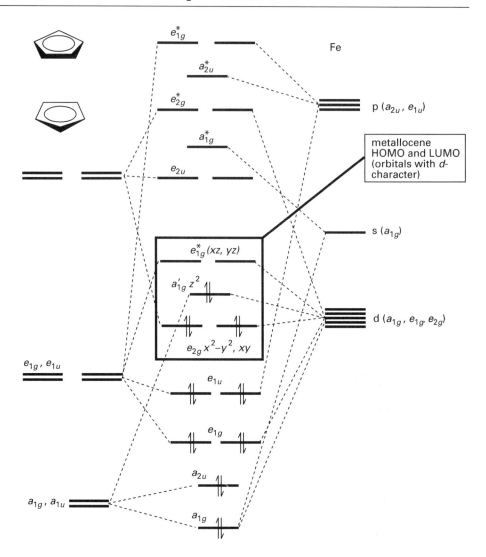

Fig. 3.4 A qualitative molecular orbital diagram for ferrocene.

δ-type overlap (two nodes) with the ligand e_{2g} set is poor (for ferrocene in particular).

When dealing with larger carbocyclic ligands, the metal–ligand e_{2g} overlap is more significant with the energies of these orbitals far more favourably matched. The e_{2g} molecular orbital therefore becomes more bonding in character. As with the bonding situation in CO and C_2H_4 compounds, electron donation takes place from the filled ligand orbitals to the metal whilst back donation from occupied metal d orbitals fills π^* levels of the ligand. If the metallocene is denoted as featuring M^{2+} and $2Cp^-$ fragments, synergistic theories may be employed (Scheme 3.1).

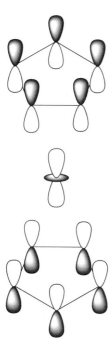

Fig. 3.5 Illustration of the interaction of the ligand a'_{1g} symmetry π orbitals with the metal d_{z^2} orbital.

Examination of the MO diagrams in relation to the 18-electron rule

From the simple molecular orbital picture, it is clear why ferrocene is the most stable of all the metallocenes; it has an ideal number of electrons for Cp_2M complexes. If each $C_5H_5^-$ ligand is considered as a six electron donor and this is coupled with the six d electrons of Fe(II) then the molecule has 18 valence electrons. These nine pairs are accommodated precisely by filling all the bonding and non-bonding molecular orbitals and none of the anti-bonding ones. The *frontier orbitals* can be regarded as bonding (e_2), non-bonding (a_1) and anti-bonding (e_1^*). The scheme is, of course, applicable to other metallocenes of D_{5d} symmetry. The molecular orbital energy level diagrams indicate that the chemically relevant frontier orbitals are neither strongly bonding nor anti-bonding and this characteristic permits the possibility of the existence of metallocenes that diverge from the 18-electron rule. Deviations from the rule do lead to significant changes in M—C bond lengths

Donor Bond
$$\left\{\begin{array}{l} \sigma \quad Cp(a_1) \ \text{-----> } M\ (4s,\ 3p_z) \\[2em] \pi \quad Cp(e_1) \ \text{-----> } M\ (d_{xz,\ yz}\ ;\ p_{x,\ y}) \end{array}\right.$$

Acceptor Bond
(Backbonding)

$\delta \quad Cp(e_2) \ \text{<----- } M\ (d_{x^2-y^2,\ xy})$

Scheme 3.1 Synergism concepts in metallocenes Cp_2M.

83

Table 3.1 Electron configurations and M—C bond length in $[(C_5H_5)_2M]$ complexes.

Complex	Valence electrons	Electron configuration	R (M—C) (nm)
$[(C_5H_5)_2V]$	15	$(e_{2g})^2(a_{1g}')^1$	0.228
$[(C_5H_5)_2Cr]$	16	$(e_{2g})^3(a_{1g}')^1$	0.217
$[(C_5H_4CH_3)_2Mn]$	17	$(e_{2g})^3(a_{1g}')^2$	0.211
$[(C_5H_5)_2Fe]$	18	$(e_{2g})^4(a_{1g}')^2$	0.206
$[(C_5H_5)_2Co]$	19	$(e_{2g})^4(e_{1g}^*)^1(a_{1g}')^2$	0.212
$[(C_5H_5)_2Ni]$	20	$(e_{2g})^4(e_{1g}^*)^2(a_{1g}')^2$	0.220

that correlate quite well with the molecular orbital scheme. For instance, the short M—C distance in ferrocene results from the a_{1g}' frontier orbital and all lower orbitals being full and the e_{1g}^* frontier orbital and all higher orbitals being empty. Similarly, the redox properties of the complexes can also be understood in terms of electronic structure (Table 3.1). Due to the non-bonding nature of the metal a_{1g}' and e_{2g} orbitals, transition metal metal-locenes with 15, 16 and 17 valence electrons are well known, e.g. vanadocene (15 valence electrons, d^3, V^{2+}) and chromocene (16 valence electrons, d^4, Cr^{2+}) are electron deficient, paramagnetic and highly reactive with the three non-bonding orbitals being only partially filled (Fig. 3.6). It is probable that there is some change in relative energies of the molecular orbitals on going across the Periodic Table and, indeed, it seems that the a_{1g} and e_{2g} energy levels converge and possibly cross on moving from ferrocene to the titanium analogue. The Cp_2V system has three unpaired electrons with a probable electronic configuration of $(e_{2g})^2(a_{1g}')^1$. Conversely, cobaltocene (a 19-electron system featuring d^7, Co^{2+}) and nickelocene (20-electron system, d^8, Ni^{2+}) feature electrons in the degenerate anti-bonding e_{1g}^* (d_{xz}, d_{yz}) orbitals. The species are paramagnetic and much more reactive than ferrocene with both being very easily oxidised (Green *et al.*, 1968).

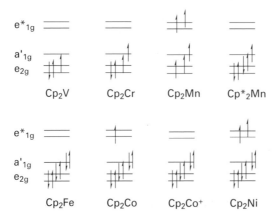

Fig. 3.6 The orbital occupancy for some first-row metallocenes.

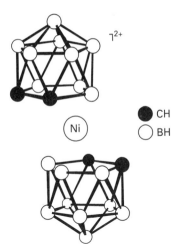

● CH
○ BH

Ni

Fig. **3.7** The 'slipped' sandwich structure of [(η-C$_2$B$_9$H$_{11}$)$_2$Ni].

Distortion in sandwich compounds not following the 18-electron rule

The interesting situation of stable compounds featuring at least partially filled anti-bonding orbitals can also be examined. For example, in metallo-carborane sandwich compounds that have too many *d* electrons, i.e. d^8 Ni(II) in [(η-C$_2$B$_9$H$_{11}$)$_2$Ni] (Fig. 3.7), a distortion from the symmetrical sandwich geometry is apparent (Hawthorne, 1975). Normally, there is equal bonding to the two carbon and three boron atoms, but in the 'slipped' case the metal is positioned closer to the boron atoms. It can be visualised as a *sideways* (lateral) displacement of the ligands resulting in pseudo π-allylic bonding to the metal (Fig. 3.8). To maintain the C_{2h} symmetry of the molecule, the 'slip' distortion can be accompanied by a slight folding of the C$_2$B$_3$ face away from the metal atom. For 18-electron dicarbollide sandwich complexes, the 'slip' distortion is unfavourable as the bonding nature of the molecule orbitals would be lessened by a reduction of the metal–ligand overlap integrals which result from the distortion. With higher electron complexes (19–22), the additional electrons are contained in the anti-bonding $5e_{1g}^*$ set and the components are considerably stabilised by '*slippage*'. Returning to the 19- and 20-electron metallocenes [(C$_5$H$_5$)$_2$Co] and [(C$_5$H$_5$)$_2$Ni], by analogy it

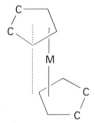

Fig. **3.8** The lateral displacement of the dicarbollide ligands.

85

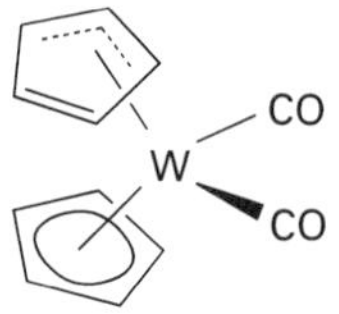

Fig. 3.9 The slipped allylic structure of $[(\eta^5\text{-}C_5H_5)(\eta^3\text{-}C_5H_5)W(CO)_2]$.

could be supposed that these systems may also undergo a 'slip' distortion. Longer metal–ring distances and smaller metal–ligand bond dissociation energies in these compounds with respect to $[(C_5H_5)_2Fe]$ indicate that the extra electrons of the Co and Ni metallocenes occupy the anti-bonding $5e_{1g}^*$ (d_{xz}, d_{yz}) orbitals. However, there is a less effective driving force for the distortion due to the smaller amount of covalency in the $5e_{1g}^*$ orbitals of the metallocenes (because carbon is of a higher electronegativity than boron). This is illustrated by the fact that in the 20-electron compound $[(\eta\text{-}C_5H_5)_2W(CO)_2]$ the metal–cyclopentadienyl bond features greater covalent character and a 'slip' distortion giving an 18-electron allylic-like structure (Fig. 3.9).

The ordering of the energy levels and electronic and magnetic behaviour from the general MO schemes

As mentioned previously there has been some disagreement as to the relative ordering of the energy levels, in particular the a'_{1g} and e_{2g} levels in ferrocene (and, indeed, dibenzenechromium) (Cowley, 1979; Green, 1981). Early theoretical calculations suggested an ordering of $a'_{1g} < e_{2g} < e_{1g}^*$, although the optical absorption spectrum of ferrocene plus further calculations indicated the sequence $e_{2g} < a'_{1g} < e_{1g}^*$. Additionally, UV photoelectron studies suggested that the first two bands in the spectrum are due to $^2E_{2g}$ (6.88 eV) and $^2A_{1g}$ (7.23 eV) ion states, thereby agreeing with the ordering $a'_{1g} < e_{2g} < e_{1g}^*$. The exact detail in the MO diagram, such as the energetic separation or sequence of the molecular orbitals, can vary depending on the quantum chemical method used. Upon ionisation of the molecule, there may be an inversion of certain molecular orbitals indicating that the scheme can be subject to modification with changing complex charge. As the occupied orbitals are either of 'a' type (symmetric around the fivefold molecular axis) or pairs of 'e$_1$' or 'e$_2$' type which are also symmetrical (in pairs) about the axis, no intrinsic barrier to internal rotation is predicted. The low barriers actually observed can be attributed to van der Waals forces directly between the rings, packing forces or repulsive effects of peripheral substituents.

For most planar, parallel metallocenes, the general molecular orbital scheme can predict electronic configurations and magnetic properties (Table 3.2) (Green, 1981). In other situations, a more exact approach is

Table 3.2 Electron configurations, ionisation potentials and magnetism of metallocenes.

Compound	Electronic ground state configuration*	Electron count	Number of unpaired electrons	Spin-only value $[n(n+2)]^{1/2}$	Magnetic moment (Bohr magnetons) (Experimental)	(Found)	First ionisation potential (eV)
Cp_2Ti^+	$(e_{2g})^1$	13	1	1.73	>1.73	2.29 ± 0.05	—
Cp_2V^{2+}	$(e_{2g})^1$	13	1	1.73	>1.73	1.90 ± 0.05	—
Cp_2V^+	$(e_{2g})^2$	14	2	2.83	2.83	2.86 ± 0.06	—
Cp_2V	$(e_{2g})^2(a_{1g}')^1$	15	3	3.87	3.87	3.84 ± 0.04	6.78
Cp_2Cr^+	$(e_{2g})^2(a_{1g}')^1$	15	3	3.87	3.87	3.73 ± 0.08	—
Cp_2Cr	$(e_{2g})^3(a_{1g}')^1$	16	2	2.83	>2.83	3.20 ± 0.16	5.71
Cp_2Mn	$(e_{2g})^2(a_{1g}')^1(e_{1g}^*)^2$	17	5	5.92	5.92	5.81	6.91
Cp_2^*Mn	$(e_{2g})^3(a_{1g}')^2$	17	1	—	—	—	—
Cp_2Fe^+	$(e_{2g})^3(a_{1g}')^2$	17	1	1.73	>1.73	2.34 ± 0.12	—
Cp_2Fe	$(e_{2g})^4(a_{1g}')^2$	18	0	0	0	0	6.88
Cp_2^*Fe	$(e_{2g})^4(a_{1g}')^2$	18	0	—	—	—	5.88
Cp_2Co^+	$(e_{2g})^4(a_{1g}')^2$	18	0	0	0	0	—
Cp_2Co	$(e_{2g})^4(a_{1g}')^2(e_{1g}^*)^1$	19	1	1.73	>1.73	1.76 ± 0.07	5.56
Cp_2^*Co	$(e_{2g})^4(a_{1g}')^2(e_{1g}^*)^1$	19	1	—	—	—	4.71
Cp_2Ni^+	$(e_{2g})^4(a_{1g}')^2(e_{1g}^*)^1$	19	1	1.73	>1.73	1.82 ± 0.09	—
Cp_2Ni	$(e_{2g})^4(a_{1g}')^2(e_{1g}^*)^2$	20	2	2.83	2.83	2.86 ± 0.11	6.51

* $[(a_{1g})^2(a_{2u})^2(e_{1g})^4(e_{1u})^4] + \cdots$

87

needed to predict a high-spin or low-spin form. For example, $[(\eta\text{-}C_5H_5)_2Mn]$ has a high-spin $^6A_{1g}$ $[(e_{2g})^2(a'_{1g})^1(e^*_{1g})^2]$ configuration whereas the analogous pentamethylated derivative $[(\eta\text{-}C_5Me_5)_2Mn]$ has a low-spin $^2E_{2g}$ $[(e_{2g})^3(a'_{1g})^2]$ configuration. This spin change has been shown to be coupled with a considerable shortening (0.03 nm) of the metal—carbon bond length. Another example of the limitation in predictions is that the magnetic moment of the Cp_2Fe^+ cation is much higher than the '*spin-only*' value and is not directly deduced from a rigid molecular orbital scheme. This suggests a ground state $^2A_{1g}$ state and not the experimentally found $^2E_{2g}$ ground state.

Reports (Cowley, 1979; Green, 1981) have been made on noting trends in the ionisation energies of transition metal sandwich compounds and certain conclusions reached: (i) ionisation potentials for related first-, second- and third-row transition metal sandwich compounds follow the order 3rd row $\simeq$ 2nd row > 1st row, and is probably due to the ionisation potentials of the isolated metal atoms; (ii) for $[(\eta\text{-}C_5H_5)_2M]$ compounds, the $^2E_{2g}$ ---- $^2A_{1g}$ energy separation decreases with the atomic weight of the central metal atom and is regarded as an effect of the increased covalent bonding in compounds of the second- and third-row transition elements; (iii) ionisation energies for ionisations from the a'_{1g} and e_{2g} molecular orbitals are found to increase with atomic number across the Periodic Table from vanadium to nickel and is a reflection of the increased nuclear charge on the central metal atom; (iv) substitution of methyl groups for hydrogen on the cyclopentadienyl rings results in a decrease in ionisation potential, the size of which depends on the amount of substitution.

Of the other types of symmetrical, planar, parallel metal sandwich compounds with larger (than cyclopentadienyl) polyene rings, **di(benzene)chromium** and **di(cyclooctatetraene)uranium** are the best known examples and their electronic structures and bonding are worth considering as representative examples.

Di(benzene)chromium $[(\eta^6\text{-}C_6H_6)_2Cr]$. Following the synthesis of di(benzene)chromium (Fischer and Hafner, 1955), the structure and symmetry (D_{3d} or D_{6h}) was a controversial subject. Infrared spectra and neutron diffraction studies indicated 'localised' electron pairs and alternating C—C bond lengths as in a cyclohexatriene but low temperature X-ray diffraction confirmed a geometry with equal bond lengths (D_{6h}) (Fig. 3.10). Using this symmetry, it is possible to construct a bonding model for dibenzenechromium in a similar fashion to that for ferrocene. The MO treatment initiates from a consideration of the ligand π-MOs and again one ligand ring can be placed on top of the other thereby generating *gerade* and *ungerade* combinations (Fig. 3.11). Combination of the symmetry-adapted linear combinations of the C_6H_6 ligand π-MOs and the appropriate metal orbitals can then be

Fig. 3.10 Diagram showing the orientation of the aromatic rings in di(benzene)chromium.

effected, i.e. the *ungerade* combinations may combine with the metal p orbitals whilst the *gerade* combinations interact with the metal s and d orbitals (Fig. 3.12). From this, a qualitative MO diagram for di(benzene)chromium can be constructed (Fig. 3.13).

The situation is very similar to that in ferrocene and any shifts in the energy levels of the diagram result from differences in the energies of the respective basis orbitals. The basis orbitals of symmetry e_{2g} are closer in energy for the combination Cr^0/C_6H_6 than for $Fe^{2+}/C_5H_5^-$ so there is a slightly larger metal e_{2g} ($3d_{x^2-y^2}$, $3d_{xy}$)–ligand e_{2g} interaction resulting in a stabilisation of these orbitals relative to a_{1g} ($3d_{z^2}$). Therefore, the δ bond (backbonding, $C_6H_6 \leftarrow M$) contributes more significantly in di(benzene)chromium, as compared to ferrocene. Following this, photoelectron spectroscopy experiments have shown that the first ionisation process in $[(\eta\text{-}C_6H_6)_2Cr]$ generates the $^2A_{1g}$ ion state $[(e_{2g})^4(a'_{1g})^2]$. It is interesting to plot the ionisation energies for a series of isoelectronic metallocenes $[(\eta\text{-}C_6H_6)_2Cr]$, $[(\eta\text{-}C_5H_5)Mn(\eta\text{-}C_6H_6)]$ and $[(\eta\text{-}C_5H_5)_2Fe]$ (Fig. 3.14) as this indicates the cross-over from $(e_{2g})^4(a'_{1g})^2$ to $(a'_{1g})^2(e_{2g})^4$ configurations (Evans *et al.*, 1972). Note also that the $^2A_{1g}$ ionisation energies closely parallel those of the isolated metal atoms in these complexes and this is consistent with the non-bonding nature of the a'_{1g} ($3d_{z^2}$) orbital (Cowley, 1979).

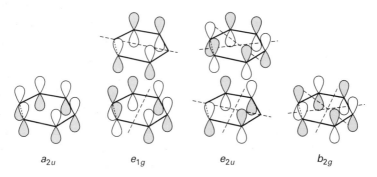

Fig. 3.11 The benzene ligand p_π molecular orbitals.

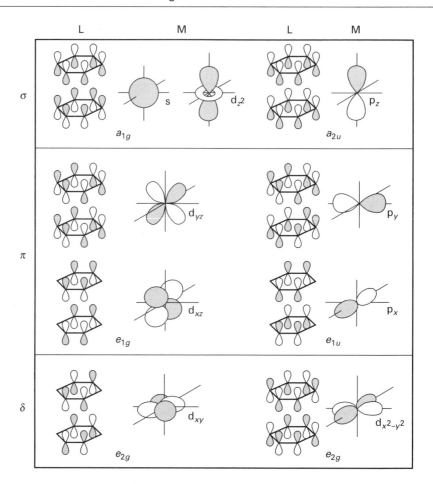

Fig. 3.12 Interaction of the symmetry-adapted linear combinations of the π MOs of two C_6H_6 ligands with matching metal orbitals in di(benzene)chromium.

Di(cyclooctatetraenyl)uranium ('uranocene'). This compound was synthesised in 1968 and isolated as a green, crystalline species being pyrophoric and paramagnetic but stable in water (Streitwieser and Muller-Westerhof, 1968). X-ray crystallography showed that the molecule was a sandwich structure with the metal located equidistantly between two, uniform, planar cyclo-octatetraene (COT) rings oriented in an eclipsed configuration (D_{8h}) (Fig. 3.15). The planarity of the COT^{2-} ligands, the equality of their C—C bond lengths (mean 0.139 nm) and the symmetrical positioning of the U^{4+} ion suggested significant covalent bonding between the metal and the aromatic system. From this information, a qualitative molecular orbital scheme can once again be produced. For the aromatic cyclooctatetraene dianion $C_8H_8^{2-}$, the 10 π-electrons occupy the molecular orbitals $(a_{1g})^2(e_{1u})^4(e_{2g})^4$ and when two COT^{2-} ions are placed parallel to one another (Fig. 3.16) with

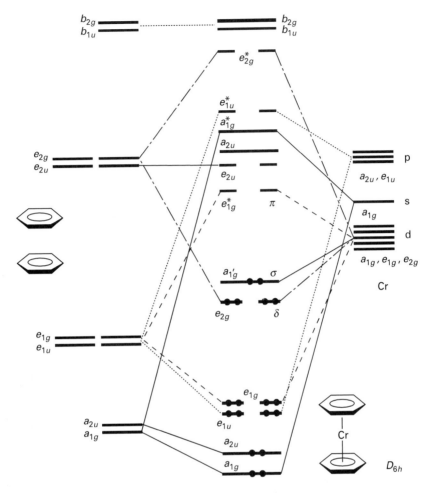

Fig. 3.13 A qualitative MO diagram for di(benzene)chromium.

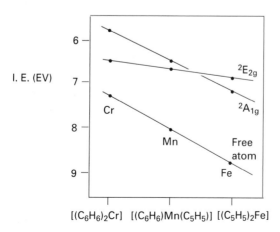

Fig. 3.14 A plot of the energies of the $^2A_{1g}$ amd $^2E_{2g}$ ion states for isoelectronic metallocenes derived from UV photoelectron spectral studies.

91

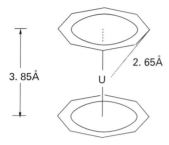

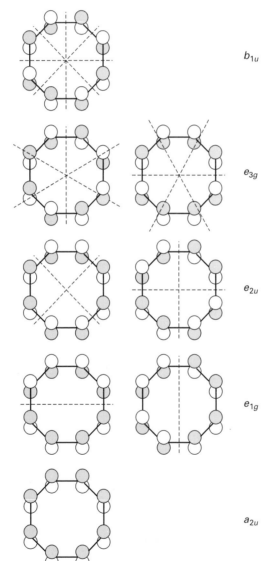

Fig. 3.15 The 'eclipsed' configuration of the cyclooctatetraene rings in uranocene.

b_{1u}

e_{3g}

e_{2u}

e_{1g}

a_{2u}

Fig. 3.16 The symmetry-adapted linear combinations of the π-MOs of two parallel cyclooctatetraene $[(C_8H_8)^{2-}]$ ligands.

eclipsed C atoms, '*symmetry-adapted linear combinations*' of the π-MOs are obtained. Then after matching with the appropriate metal orbitals, the molecular orbital scheme can be initiated (Fig. 3.17) (Rosch and

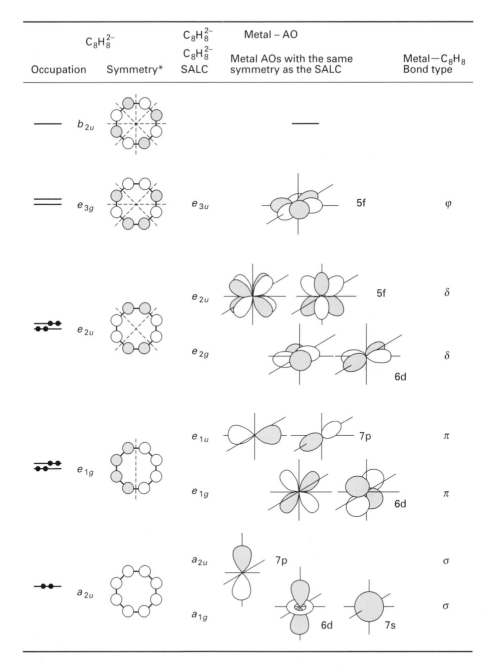

Fig. 3.17 Interaction of the ligand π-MOs with the appropriate metal orbitals.

Streitwieser, 1978). So for uranocene, the filled ligand π orbitals are a_{1g}, a_{2u}, e_{1u}, e_{1g}, E_{2g}, and e_{2u}.

The interactions as found for the d-block metallocenes are relevant here but now there are also bonding interactions between the metal f orbitals (transforming as a_{2u}, e_{1u}, e_{2u} and e_{3u} in the D_{8h} point group) and the symmetry matching ligand π orbitals. The energetic sequence for uranocene in a qualitative sense resembles that for ferrocene up to level e_{2g}. Then instead of the non-bonding level a'_{1g} (Fe, $3d_{z^2}$) in ferrocene, uranocene features molecular orbitals derived from C_8H_8–U($5f$) interactions, namely the frontier orbitals e_{2u} and e_{3u} (Fig. 3.18). Interactions of symmetry e_{2u} cannot occur in d-block sandwich complexes but it is thought that they are particularly important in stabilising this class of f-metal species in that they allow the covalent bonding of large, electron-rich π-perimeters. Further studies have demonstrated that contributions of U($5f$) orbitals to bonding in uranocene are, however, dominated by those from U($6d$) orbitals, in that they have the effect of depressing the e_{2g} level below that of the e_{2u} (Rosch and Streitwieser, 1978; Bursten, 1989). Generally in actinoid–C_8H_8 complexes, a considerable contribution from covalent interactions may be assumed as $5f$ orbitals are less shielded than $4f$ orbitals and, consequently, can overlap more extensively with the ligand orbitals. This is different from the

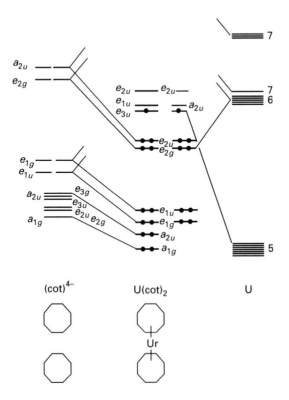

Fig. 3.18 A schematic illustration of the energetic ordering of the molecular orbitals of [(η-C_8H_8)$_2$U].

lanthanide species where the $4f$ orbitals have no radial nodes and are deeply buried in the atomic core, leaving negligible overlap with the ligand orbitals. The metal–ligand bond in lanthanide–C_8H_8 complexes is, therefore, considered to be predominantly ionic.

3.2 Molecular orbital calculations on bent di(cyclopentadienyl) metal complexes

Returning to the di(cyclopentadienyl) systems, similar metal–ligand interactions of different group theoretical designation are applicable to eclipsed metallocene structures (D_{5h}) or other rotamers (D_5) with the bond energy not affected by rotation (as opposed to the staggered D_{5d} conformations detailed previously). The preference for a particular conformation is governed by packing forces or repulsive effects of peripheral substituents. However, incorporation of additional groups or ligands onto the metal (normally to relieve electron deficiency and achieve the desired 18-electron configuration) will destroy the D_{5d} or D_{5h} symmetry of the simple metallocene, i.e. [(η^5-$C_5H_5)_2Cr$] ($16e^-$) and [(η^5-$C_5H_5)_2W$] ($15e^-$) have unfilled bonding MOs and are highly reactive. This destruction of the D_{5d} (or D_{5h}) symmetry of the simple metallocene alters the molecular orbital picture and so the situation has to be reconsidered.

In bent di(π-cyclopentadienyl) transition metal complexes, the rings are not parallel, i.e. the angle between the normals to the planes of the cyclopentadienyl ligands is less than 180° and there is binding to one, two or three ligands in addition to the two cyclopentadienyl groups. Examples are shown in Scheme 3.2. If the cyclopentadienyl ligands are in an eclipsed configuration, a bent Cp_2M fragment will have C_{2v} symmetry but if the rings are staggered it will be C_s. In these compounds, the firmly bound Cp ligands can

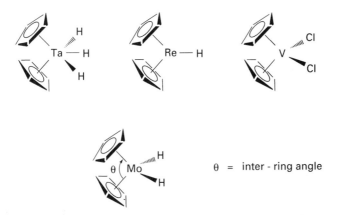

Scheme 3.2 Examples of some bent di(cyclopentadienyl) metal complexes.

95

be regarded as 'protecting groups' and chemical transformations involve only the frontier orbitals of the molecule, i.e. those oriented towards the non-Cp ligands. From crystallographic evidence, the inter-ring angle (θ) can range from 148° in [(η-C$_5$H$_5$)$_2$MoH$_2$] to 126° in [(η-C$_5$H$_5$)$_2$ZrI$_2$]. The low symmetry in these systems allows a wide variety of overlap and the molecular orbital picture is more complicated than the planar, parallel species. However, empirically, the cyclopentadienyl rings in these compounds are not bonded that differently from those in ferrocene.

The bonding situation has been the subject of considerable discussion but models have now been established that can rationalise the bonding in molecules of the general formula [η-C$_5$H$_5$)$_2$ML$_x$] (Lauher and Hoffmann, 1976; Peterson *et al.*, 1975). These compounds are usually electron deficient, so to attain the favourable 18-electron configuration, additional ligands contributing extra electrons are incorporated. In doing this, the steric requirements of these additional ligands mean that the Cp ligands bend back. This bonding results in a mixing or rehybridisation of the three previously non-bonding or weakly bonding a_{1g} and e_{2g} orbitals to form three new, highly directed orbitals that point out of the open side of the metallocene away from the rings and towards the additional ligands (Fig. 3.19).

All the bent metallocenes have these three directed orbitals but their occupancy depends on the *d*-electron count of the metal, i.e. how many electrons the metal provides (Scheme 3.3). For instance, (a) for [Cp$_2$TiCl$_2$], the centre orbital is left unfilled and this can act as a Lewis acid (as shown in (c) where the Ti interacts with a lone pair of π-basic ligand). For Mo, this orbital is filled (b) and so can act as a Lewis base, i.e. in (c) it is shown in back

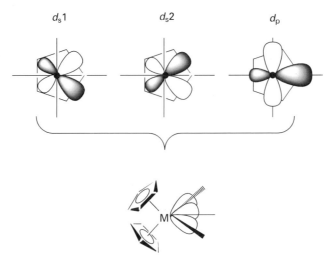

Fig. 3.19 Orientation of the 'directed' frontier orbitals of a bent Cp$_2$M fragment.

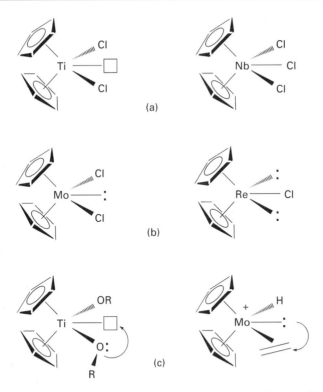

(a)

(b)

(c)

Scheme 3.3 Examples of the various occupancy of the directed orbitals of some bent metallocenes.

donation to a π-acid ligand such as ethylene. Complexes of d^0 configuration, such as $[Cp_2M]^{2+}$ (M = Ti, Zr or Hf), can co-ordinate up to three donor ligands (Ti, having a small ionic radius, is normally restricted to two), while $[Cp_2^*ZrH_2]$ accepts only one. In contrast, the d^2 complex $[Cp_2MoH_2]$ clearly possesses an electron pair and acts as a donor to BF_3 (Fig. 3.20). The fragments 'Cp$_2$Mo' and 'Cp$_2$W' have two fewer electrons than ferrocene and can therefore bind two $1e^-$ ligands (e.g. Cp_2MH_2) or one $2e^-$ ligand (e.g. $Cp_2M(CO)$) to reach 18 electrons. Only two of the available three rehybridised orbitals are used to bind the H atoms in the group 6 metallocene dihydrides and the third orbital is a lone pair that points between the two

Fig. 3.20 Molybdocene dihydride donating electrons to BF_3.

substituents but can be protonated to give the trihydride cations $[Cp_2MH_3]^+$. The group 5 metals can bind three X ligands (e.g. Cp_2NbCl_3) whilst the group 4 metals only bind two X ligands (e.g. Cp_2TiCl_2) and, having only four valence electrons, the maximum oxidation state is therefore M(IV). Thus, an empty orbital is left in the $16e^-$ titanocene dihalide rather than a filled one in molybdocene dihalides. This explains the differences in their chemistry: group 4 metallocene complexes act as Lewis acids and bind π-basic ligands (e.g. ^-OR) whilst group 6 metallocene species act as Lewis bases and bind π-acceptor species such as ethylene (Scheme 3.3).

Examination of the frontier orbitals of a bent Cp_2M fragment (e.g. Cp_2Ti) enables an understanding of how the additional ligands bind, and the way in which the d-type MOs of the $[(\eta\text{-}C_5H_5)_2M]$ fragment change as a function of the bending angle (θ) can be seen in Fig. 3.21. The trend shows that the orbitals derived from the e_{1g}^* set are stabilised with bending, while those from the a_{1g} and e_{2g} sets are destabilised. Increased σ-anti-bonding on departure from the θ = 180° geometry affects the lower orbitals $1a_1$, b_2 and $2a_1$ but most noticeable is the effect on a_{1g}–$2a_1$ which was initially non-bonding. Another factor in the steep rise in energy of $2a_1$ with smaller θ is its interaction with $1a_1$. In D_{5d} or D_{5h} configurations the orbitals possess different symmetries, but in C_{2v} they are able to mix and repel each other. Therefore, the upward slope of $2a_1$ is increased whilst $1a_1$ is approximately constant in energy. The a_2^* and b_1^* orbitals arise at a lower energy from their original e_{1g}^* set due to a decrease in the overlap of the metal d orbitals with the filled ligand orbitals. Additionally in the C_{2v} geometry, the d_{xy} and d_{xz} orbitals are able to interact with two of the empty anti-bonding orbitals of

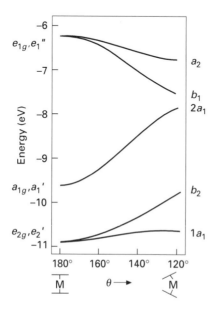

Fig. 3.21 Molecular orbitals of $[(\eta\text{-}C_5H_5)_2M]$ as a function of the bending angle θ. (Hoffmann (1976), with permission.)

the ligands, which is a net stabilising effect and lowers the energy of these orbitals.

Considering complexes with d^0–d^{10} configurations, there is little driving force for the sandwich complex to take up a bent configuration and the extra ligands co-ordinated provide a crucial stabilising interaction for the formation of the bent sandwich structure. Usually the metals have four or fewer valence electrons and so in the formation of co-ordinate bonds to the additional ligands it is the three low-lying levels (Fig. 3.21) that are most important. Contour diagrams for these three new bonding orbitals (b_2, $1a_1$ and $2a_1$) obtained as θ gets smaller and as the cyclopentadienyl rings are tilted back have been calculated and are shown in Fig. 3.22. The b_2 molecular orbital is mainly d_{yz} in character whilst the two a_1 orbitals are formed

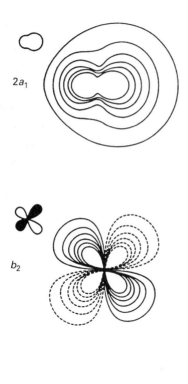

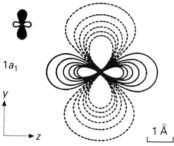

Fig. 3.22 Electron density contour diagrams of the three new bonding orbitals of the $[(\eta\text{-}C_5H_5)_2M]$ angular fragment. (Lauher and Hoffmann (1976), with permission.)

Fig. 3.23 Orientation of the hybridised orbitals of a bent Cp$_2$M fragment.

from the metal s, p_z, d_{z^2} and $d_{x^2-y^2}$ atomic orbitals. The $1a_1$ molecular orbital is described primarily along the y axis and it remains essentially non-bonding or weakly bonding with respect to ligands L located in the zy plane of the molecule. The $2a_1$ orbital is hybridised away from the C$_5$H$_5$ ligands and ensures a very good overlap with the donor orbitals of the extra ligands (Fig. 3.23) (which shows the orientation of the hybridised orbitals).

The simplest, bent di(cyclopentadienyl) transition metal complexes are the monohydrides [Cp$_2$MH] where the simple hydride ligand can only bind in a σ-fashion. Well-known examples are [Cp$_2$ReH] and [Cp$_2$FeH$^+$] and the most symmetrical and sterically favourable structure has the H$^-$ ligand situated along the z axis (Fig. 3.24). Here the hydride ligand overlaps well with the molecular orbital $2a_1$, slightly with $1a_1$ but not at all with b_2. This leads to a situation where there is strong bonding interaction between the $2a_1$ orbital and the σ-orbital of the H$^-$ ligand, the b_1 orbital is unaffected and the $1a_1$ orbital is slightly destabilised. The b_2 and $1a_1$ orbitals can accommodate four electrons, therefore d^4 complexes (e.g. [Cp$_2$ReH] and [Cp$_2$FeH$^+$]) are favoured (Fig. 3.25).

Two of the three low-lying orbitals of the bent Cp$_2$M fragment will be utilised when two σ-bonding ligands are incorporated. There are many examples of this type of compound and they all generally possess the same geometry (**III.1**) where the angle (φ) between the two X ligands depends on

(**III.1**)

the number of d electrons the metal possesses. This can be seen from the molecular orbital scheme for [(η-C$_5$H$_5$)$_2$ML$_2$] species (Fig. 3.26) where the important features are the strong interactions between $2a_1$ and the symmetric

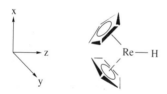

Fig. 3.24 Axial orientation of the ligands in [(η-C$_5$H$_5$)$_2$ReH].

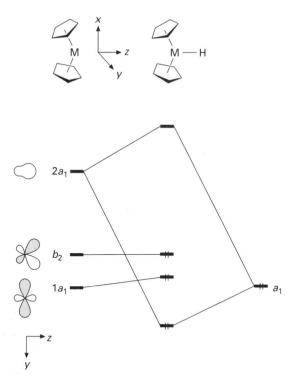

Fig. 3.25 A qualitative MO diagram involving the three new MOs for a bent metallocene. (Lauher and Hoffmann (1976), with permission.)

ligand combination, and b_2 and the anti-symmetric ligand combination. Modelling the simple hydride system, an 18-electron d^2 complex features the highest occupied molecular orbital (HOMO) $1a_1$, which retains a non-bonding nature if the ligands are located along the nodes of the orbital. These nodal lines are around 78° and so this is the expected X—M—X angle. In d^0

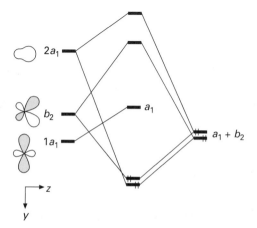

Fig. 3.26 A qualitative MO diagram for $[(\eta\text{-}C_5H_5)_2ML_2]$ formed by the combination of $[(\eta\text{-}C_5H_5)_2M]$ (d^0) and the σ-donating two-electron ligands L_2. (Lauher and Hoffmann (1976), with permission.)

(16-electron) systems, the maximum metal–ligand bonding is observed when the ligand donor orbitals overlap effectively with both $2a_1$ and $1a_1$ which raises the energy of the empty $1a_1$ orbital; φ here is roughly $110°$ while in d^1 complexes an intermediate situation is reached with a φ angle of $85°$ (Lauher and Hoffmann, 1976), i.e. the angle φ decreases as d electrons are added. Structures of the $[Cp_2MX_2]$ compounds have been experimentally studied and the angles determined for d^0 complexes are $94–97°$, d^1 $85–88°$ and d^2 $76–82°$, a trend that reinforces the above model dihydride calculations. Electron spin resonance (ESR) studies (Peterson *et al.*, 1975) on d^1 species have also demonstrated that the non-bonding electron resides in an a_1 orbital which is localised on the metal and made up mainly of d_{z^2} character. In $[Cp_2MX_3]$ complexes, the metal is generally a d^0 species because all three valence orbitals of the fragment are utilised to bond the three σ-ligands (usually hydride species). All the ligands lie in the yz plane as this orientation maximises the overlap between the metal $1a_1$, $2a_1$ and b_2 orbitals and the hydrogen $1s$ orbitals (Bau *et al.*, 1979). Ultra-violet photoelectron studies have also reinforced the bonding pictures described above. They have demonstrated the presence of two non-bonding molecular orbitals for $[(\eta\text{-}C_5H_5)_2MX_2]$ and none for $[(\eta\text{-}C_5H_5)_2MH_3]$, all arising from various utilisations of the $1a_1$, $2a_1$ and b_2 molecular orbitals in the bent sandwich compounds.

In this section, we have only dealt with additional ligands that interact with metal through σ-donor orbitals. There are, of course, bent Cp_2M species with π-bonded ligands or a mixture of π- and σ-bound species and detailed analysis is given in Lauher and Hoffmann (1976), the most comprehensive treatise on the bonding in bent di(cyclopentadienyl) metal complexes.

3.3 Molecular orbital calculations on tri(cyclopentadienyl) metal complexes

These complexes are common in lanthanide and actinide chemistry and possess the general formulae Cp_3M, Cp_3MX and Cp_3MB (X = halogen, B = base) (**III.2**) but there are also some yttrium and zirconium transition metal species, e.g. Cp_4M (M = Ti, Zr or Hf). They are fluxional molecules on the NMR timescale with a probable intermediate for the interchange of the ligands (from $3\eta^5$-Cp ligands and $1\eta^1$-Cp (Zr) to two η^5-Cp and two η^1-Cp (Ti)) being a molecule with three η^5-C_5H_5 rings.

(**III.2**)

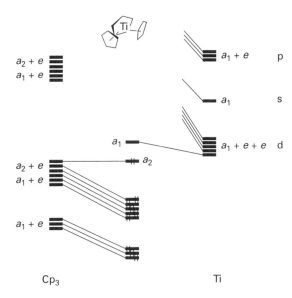

$a_2 + e$
$a_1 + e$

a_1

$a_2 + e$
$a_1 + e$

$a_1 + e$

Cp$_3$

a_2

$a_1 + e$ p

a_1 s

$a_1 + e + e$ d

Ti

Fig. 3.27 Interaction diagram and qualitative MO diagram for a {Cp$_3$Ti$^+$} fragment. (Lauher and Hoffmann (1976), with permission.)

A molecular scheme for the hypothetical {Cp$_3$Ti$^+$} fragment can be drawn, illustrating C_{3v} symmetry (Fig. 3.27). Although the three ligands can donate 18 electrons in total, one of the donor orbitals is of a$_2$ symmetry and as there is no a$_2$ metal orbital in the C_{3v} point group it cannot act as a donor orbital. This results in three ligands donating 16 electrons to the metal, leaving one orbital on the metal fragment empty. This lone orbital is of a$_1$ symmetry and mainly d_{z^2} in character. Now d^2 complexes of Cp$_3$MR are theoretically possible with the lone a$_1$ orbital being used to form a σ-bond to the alkyl group or the base.

The scarcity of Cp$_3$M complexes within the transition metal series is due to poor orbital overlap and unfavourable sterics. The lanthanides and actinides have larger radii thus alleviating steric problems and of course they possess f orbitals. There is an f orbital of the required a$_2$ symmetry in C_{3v}, thereby making Cp$_3$M complexes stable and easier to obtain within the lanthanide and actinide series. The bending back of the cyclopentadienyl rings can also be effected by the lone pair steric requirements in species such as Cp$_2$Sn and Cp$_2$Pb which leads us to main group cyclopentadienyl complex bonding (Jutzi, 1986).

3.4 Molecular orbital calculations on main group cyclopentadienyl complexes

The bonding in π-complexes of the main group elements differs quite significantly from nearly all of the transition metal π-species. Main group fragments do not possess empty or partially filled d subshells and this leads to (i) only a small donation of π-electron density from the ligand and (ii) only donation into s- or p-type orbitals. As a result, π-electrons normally remain in non-bonding orbitals centred at the ligands. There is also a very high energy difference between filled d subshells (if present) and the ligand π^* orbitals so that electron transfer (or back donation) is not significant. Synergistic effects, so typical within transition metal π-complexes, therefore do not feature in π-complexes of the main group elements.

3.4.1 The bonding in group 4 metallocenes

Germanocene

Monomeric germanocene [$(C_5H_5)_2Ge$] (**III.3**) (like its congeners, stannocene and plumbocene) has a tilted sandwich structure. This allows the presence of

Germanocene

14 - electron compound

(III.3)

a stereochemically active lone pair on the metal. The distortion can be thought of as a small bending of the $Cp_{centroid}$—Ge—$Cp_{centroid}$ angle from linearity and as small tilts of the cyclopentadienyl ring planes relative to the germanium–centroid vectors ($\alpha = 50.4°$) (Fig. 3.28). A qualitative comparison of the values of α for germanocene derivatives indicates that bulky substituents at the cyclopentadienyl rings force the two rings into a more parallel conformation (Jutzi, 1986). The bonding in germanocenes and, in general, all group 4 metallocenes can be rationalised qualitatively by construction of molecular orbital schemes for species with parallel (D_{5d} symmetry) and with bent (C_{2v} symmetry) cyclopentadienyl rings (Fig. 3.29). It should be noted that with D_{5d} symmetry, interaction between the e_{1g} set in the Cp–Cp unit and a germanium–centred orbital is symmetry forbidden. This means that four of the overall 14 electrons in the germanocene are

Fig. 3.28 The distortion arising from a small bending of the $Cp_{centroid}$–Ge–$Cp_{centroid}$ angle from linearity and small tilts of the Cp ring planes relative to the germanium–centroid vectors.

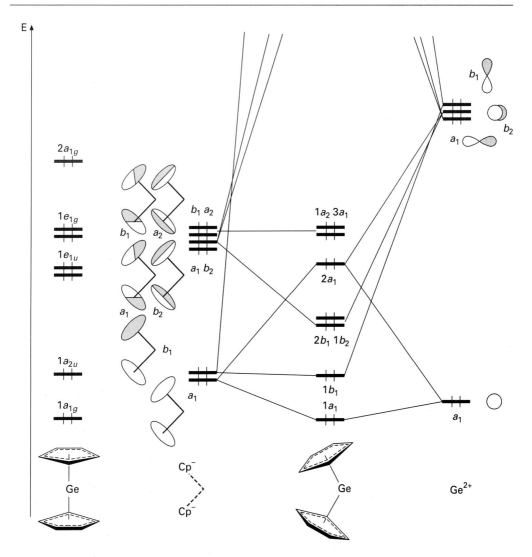

Fig. 3.29 Qualitative molecular orbital diagrams for germanocenes with D_{5d} and C_{2v} symmetry.

situated in non-bonding molecular orbitals ($1e_{1g}$). Qualitatively, this does not change on moving to the C_{2v} symmetry species. Also in the parallel species, the HOMO ($2a_{1g}$) is highly destabilised and anti-bonding with respect to the germanium cyclopentadienyl ring interactions. Bending of the rings greatly reduces the energy of the $2a_{1g}$ orbital but leaves the other molecular orbitals relatively untouched. There is a reduction in germaniun–cyclopentadienyl ring anti-bonding through admixing of p character from germanium; the resulting orbital $2a_1$ basically represents the non-bonding lone pair on the germanium atom, thus leading to significantly greater stability of the bent structure (Fig. 3.30). For bent germanocene, the three highest orbitals are

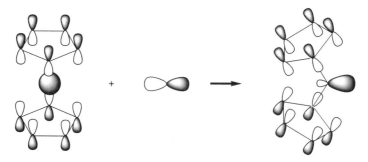

Fig. 3.30 The structure of germanocene resulting from the admixing of *p* character from the metal.

close in energy and the HOMO will vary according to a particular situation, i.e. the energy of the lone pair orbital $2a_1$ is very dependent on the interplanar angle α whilst the energies of the non-bonding orbitals $1a_2$ and $3a_1$ and the bonding orbitals $2b_1$ and $1b_2$ vary greatly according to the energies of the original cyclopentadienyl orbitals.

Stannocene

Stannocene (Cp_2Sn) also has a tilted sandwich structure in the gas and solid states. The bending depends on (i) the balance between the steric bulk of the substituents attached to Cp, where the tilted structure approaches an axial structure with increasing steric demand from the peripheral substituents, and

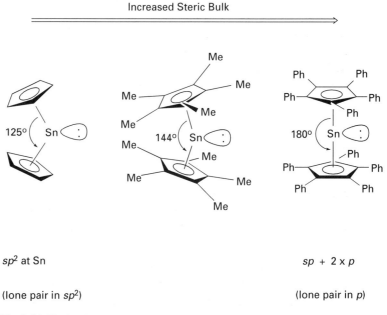

Fig. 3.31 The bending in related stannocene derivatives.

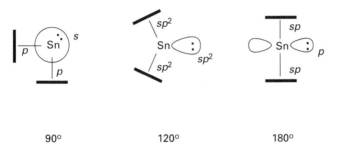

90°   120°   180°

Fig. 3.32 Examples of various cases of hybridisation resulting from σ-interaction between the atomic orbitals of tin and the a_{1g} molecular orbitals of the ligands C_5H_5.

(ii) the desire to retain the lone pair in an orbital with as much s (Sn $5s$) character as possible (Fig. 3.31). The situation below only considers the limiting hybridisation and σ-interaction between the tin atomic orbitals and the a_{1g} molecular orbitals of the C_5H_5 ligands (Fig. 3.32). In this simple picture, stannocene features Sn sp^2 hybridisation and the sp^2 hybrid orbital not required for meta-ring bonding is occupied by a non-bonding pair of electrons. Additional contributions to metal–ring bonding can result from π-interactions between filled MOs e_1 (C_5H_5) and empty atomic orbitals (AOs) of Sn which possess the correct symmetry. More exact calculations have shown that the highest occupied MOs ($3a_1$ and $1a_2$) are of π-character and highly localised on the cyclopentadienyl rings. In decreasing energy, the MOs arising from the π-bound Cp system to the tin are $2b_1$, $1b_2$, $1b_1$ and $1a_1$, with the largest lone pair character being in $2a_1$. The $2b_1$ orbital features the weakest ring–tin bonding but a stronger ring–tin interaction takes place in the $1b_2$ MO which contains a substantial amount of Sn p_x character. The strongest bonding between a tin $5p$ orbital and the Cp rings occurs in the $1b_1$ MO. The lowest energy MO involved in ring–tin bonding is $1a_1$, which is derived from the interaction of the Sn $5s$ orbital with an a_1 combination of cyclopentadienyl ring π-MOs.

Plumbocene

Plumbocene has been known since the 1950s and in the solid state it exists in a polymeric form consisting of zig-zag chains of lead atoms separated by bridging cyclopentadienyl ligands. However, from dipole moment measurements in solution as well as from electron diffraction experiments in the gas phase, the compound features a monomeric bent-sandwich structure. The bonding in these monomeric plumbocene moieties can be explained in the same way as the situation in other monomeric group 4 metallocenes (Baxter *et al.*, 1982). Other studies have shown that the polymeric structure minimises the anti-bonding interaction between occupied lead and bridging cyclopentadienyl orbitals. The group 4 metallocenes have also been the feature of UV and photoelectron (PE) experiments along with theoretical ionisation energy computations (Table 3.3).

As would be expected, the energies of the HOMOs in the unsubstituted and permethylated metallocenes are virtually identical and independent of the central atom, thus showing the non-bonding character of these orbitals. It is also worth noting that the 'lone pair' MO energies are lower in the lead species than the tin analogues. This can be explained on the basis of the larger ring–metal–ring angles in the lead complexes compared to the tin species, thereby indicating that in the former the lone pair possesses more s character (Baxter, 1982). In general, we can expect complexes involving the larger elements at the bottom of the group, where the s and p orbitals are more separated in energy, to be more linear, i.e. featuring less effective admixing of orbitals (Scheme 3.4).

Table 3.3 Ionisation energies and orbital assignments for group 4 metallocenes.

	Calculated	Experimental				
	$(H_5C_5)_2Ge$	$(Me_5C_5)_2Ge$	$(H_5C_5)_2Sn$	$(Me_5C_5)_2Sn$	$(H_5C_5)_2Pb$	$(Me_5C_5)_2Pb$
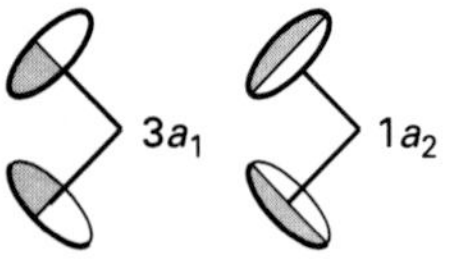 $3a_1$ $1a_2$ =	6.60* 6.61	6.60 6.75	7.57 7.91	6.60 6.60	7.55 7.85	6.33 6.88
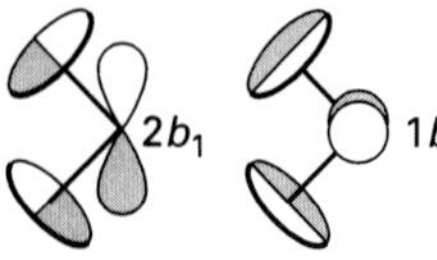 $2b_1$ $1b_2$ =	7.31 7.64	7.91 8.05	8.85 8.85	7.64 7.64	8.54 8.88	7.38 7.38
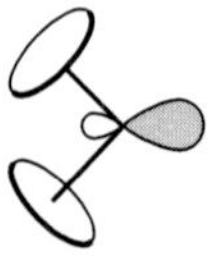 $2a_1$ —	8.74	8.36	9.58	8.40	10.10	8.93
 $1b_1$ —	11.25	–	10.5	9.4	10.6	9.38
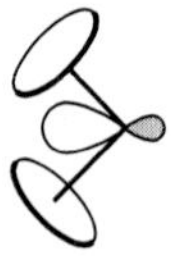 $1a_1$ —	~13.40	–	–	–	–	–

* All values in eV.

108

s and p orbitals relatively matched

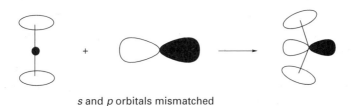

s and p orbitals mismatched

Scheme 3.4 Relative admixing of orbitals in bent metallocene structures.

3.4.2 The bonding in thallium and indium cyclopentadienyl species

Studies on the di(cyclopentadienyl)thallate(I) anion (Armstrong *et al.*, 1993) (which is isoelectronic with stannocene) have revealed a slightly different situtation than for the previously mentioned bent main group systems. The ion-separated $[\eta^5\text{-Cp}_2\text{Tl}]^-.[\eta^5\text{-CpMgPMDETA}]^+$ structure has been determined in the solid state and contains bent thallocene anions (with the average $\text{Cp}_{\text{centroid}}$—Tl—$\text{Cp}_{\text{centroid}}$ angle being $156.7°$ and the π-centre being attached almost equivalently to the η^5-Cp ligands (average Tl—$\text{Cp}_{\text{centroid}}$ 0.272 nm) (see Chapters 2 and 4). *Ab initio* MO calculations on various isomeric models of the discrete $[(\eta^5\text{-Cp})_2\text{Tl}]^-$ anion show that the association of Cp$^-$ and CpTl is reasonably favourable and, furthermore, that the most stable bent structure (with C_s symmetry) is only slightly more favourable than the most stable linear structure (with D_{5d} symmetry) (Fig. 3.33). These results suggest that orbital admixing is ineffective in the discrete $[\eta^5\text{-Cp})_2\text{Tl}]^-$ anion where an essentially spherical (s-type) lone pair is present on the Tl centre of the bent $[(\eta^5\text{-Cp})_2\text{Tl}]^-$ anion. The findings are contrary to those calculated for $[\text{Cp}_2\text{Ge}]$ and similar theories concerning $[\text{Cp}_2\text{Sn}]$ where the bent structures are supposedly preferred on electronic grounds; this is because bonding stabilises the lone pair in the anti-bonding a^*_{1g} molecular orbital as a consequence of the admixing of this orbital with the metal p_x atomic orbital (Jutzi, 1986).

Whilst indium(I) and thallium(I) cyclopentadienyls are not true metallocene sandwich species (they have been described as having '*open-faced*' sandwich structures), the bonding situation can serve as a model for many main group element cyclopentadienyls so is worth mentioning here (Canadell

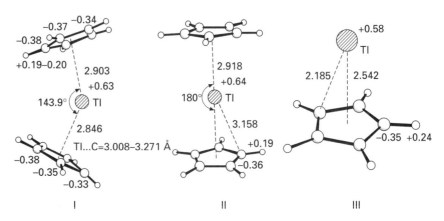

Fig. 3.33 Optimised geometry of the [CpTl]⁻ ion – the bent structure with C_s symmetry (I) and the linear structure with D_{5d} symmetry (II) and that of monomeric [CpTl] (III).

et al., 1984). In the gas phase, CpIn and CpTl both possess a half-sandwich η^5 structure (Fig. 3.34). In CpIn, the cyclopentadienyl hydrogens are bent away from the cyclopentadienyl plane by approximately 4° and in the solid state both CpIn and CpTl adopt a zig-zag chain structure of alternating metal atoms and planar cyclopentadienyl rings. The M—Cp distances in the polymer are equal and considerably larger than those in the monomer (0.319 versus 0.232 nm for CpIn) (Fig. 3.35). There has been much controversy over the bonding in CpIn but generally it has been suggested that the bonding between the indium atom and the cyclopentadienyl unit has a large covalent component (Canadell *et al.*, 1984). The small difference in electronegativity between carbon and indium does not imply high ionic character, nor does the noticeably short 0.262 nm In—C distance in the gaseous species, when the 0.13 nm ionic radius of In⁺ and the 0.17 nm van der Waals radius of carbon would indicate an ionic bond length of around 0.3 nm. The long distances in the polymeric structures do indicate an enhancement of ionic character in the solid state but the zig-zag orientation of the chain also suggests some covalent metal–ligand interaction.

The bonding situation is characterised by three bonding molecular orbitals, arising from interactions of a_1- and e_1-type orbitals on the π-cyclopentadienyl fragment and on the indium atom and by one further occupied

(C_5H_5)In, in the gas phase, a monomeric half - sandwich complex

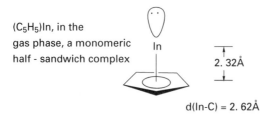

In

2. 32Å

d(In-C) = 2. 62Å

Fig. 3.34 The monomeric, half-sandwich [(C_5H_5)In] in the gas phase.

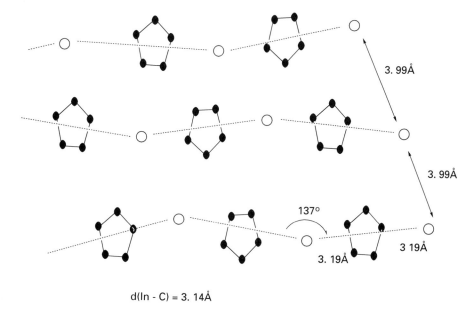

d(In - C) = 3. 14Å

Fig. 3.35 [(C₅H₅)In] in the solid state: polymer chains of [(C₅H₅)In]ₓ.

orbital which represents the 'lone pair' at indium. With the metal–ligand bond in isolated [(C₅H₅)In] molecules being predominantly covalent, the bonding situation can be constructed from the fragments In⁺ (sp hybrid) and C₅H₅⁻ (6π-e) (Scheme 3.5). These three metal–ligand interactions each generate an L → M charge transfer. The extent of the charge transfer and the ensuing covalent nature of the bond are governed by the energies of the basis orbitals, which in individual cases have to be established by means of

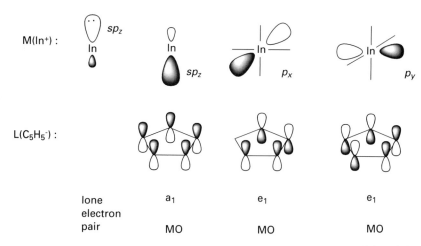

Scheme 3.5 The possible metal–ligand interactions formed from the fragments In⁺ (sp hybrid) and C₅H₅⁻ (6π-e).

Fig. 3.36 The solid state, staggered and the gas phase, eclipsed conformations of magnesocene, $[(C_5H_5)_2Mg]$.

staggered   eclipsed

quantum chemical calculations. These considerations can similarly be applied to the isolobal fragments $R-Be^+$, $R-Mg^+$, $R-B^{2+}$, $R-Al^{2+}$ and also Ge^{2+}, Sn^{2+} and Pb^{2+} (Canadell *et al.*, 1984).

3.4.3 The bonding in alkaline earth metal sandwich complexes

Considering cyclopentadienyl species of earlier main group elements, e.g. alkaline earth metals, the example of magnesocene is a good one. It was easily prepared in 1954 and is a useful reagent for the introduction of C_5H_5 groups. In the crystalline state X-ray diffraction shows a typical sandwich structure with the two parallel rings in a staggered conformation, whereas the electron scattering pattern in the gas phase shows the eclipsed conformation (Fig. 3.36).

The bonding situation is still a matter of debate – ionic or covalent? *Ab initio* MO–linear combination of atomic orbitals (LCAO) investigations predict the charge separation in magnesocene to be slightly larger than in ferrocene, though not enough to justify the classification of one as ionic and the other as covalent. The metal–ring bond in magnesocene is much weaker than in ferrocene, as indicated by the lower force constant of the metal–ring stretch and by the larger metal–ring distance. Structural similarities to ferrocene and the absence of a dipole moment do not conclusively indicate covalency, as with purely ionic bonding an axially symmetric sandwich structure would also be the (electrostatically) favoured configuration. Additionally, a molecular crystal lattice does not necessarily mean covalent bonding within the $[(C_5H_5)_2Mg]$ units because, due to the disparate sizes of Mg^{2+} and $C_5H_5^-$, an arrangement in which the lattice points are occupied by triple ions $\{Mg^{2+}(C_5H_5^-)_2\}$ is more favourable than a genuine ionic lattice. A high degree of polarity in $[(C_5H_5)_2Mg]$ bonding is suggested by the electric conductivity in solutions of NH_3 or THF, rapid hydrolysis to $Mg(OH)_2$ and C_5H_6 and the similarity of ^{13}C NMR shifts to those of alkali metal cyclopentadienyls (Table 3.4). However, from chemical shifts and linewidths observed in ^{25}Mg NMR spectra, conclusions have been reached that mainly covalent interactions determine the bonding, a theory also shared after electron diffraction and bonding models.

Magnesocene is an interesting compound as it is situated at the lighter

Table 3.4 ^{13}C NMR shifts of 'ionic' and 'covalent' cyclopentadienyl compounds.

Dominant bond type	Ionic			Covalent
Compound	[(C$_5$H$_5$)Li]	[(C$_5$H$_5$)Na]	[(C$_5$H$_5$)Mg]	[(C$_5$H$_5$)$_2$Fe]
^{13}C NMR (δ/ppm)	103.6	103.4	108.9	68.2

end of the series of dicyclopentadienyl metal sandwich compounds and is notable for the fact that the central metal atom has no *d* electrons available for bonding. A qualitative molecular orbital diagram for [(C$_5$H$_5$)$_2$Mg] (assuming staggered D_{5d} symmetry) is shown in Fig. 3.37 (including only the

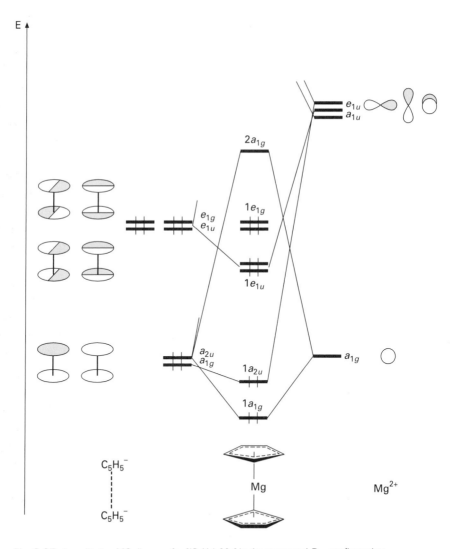

Fig. 3.37 A qualitative MO diagram for [(C$_5$H$_5$)$_2$Mg] in the staggered D_{5d} configuration.

113

Table 3.5 Vertical ionisation energies and their assignment for some magnesocenes.*

Compound	e_{1g} (π)	e_{1u} (π)	a_{1g}, a_{2u}, σ C—H
	8.11	9.03	12.2
[(C$_5$H$_5$)$_2$Mg]	8.23	9.26	12.5
	8.44		13.5
	7.78	8.62	11.7
[(C$_5$H$_4$Me)$_2$Mg]	7.90	8.86	12.4
	8.10		13.0
			11.09
[(C$_5$Me$_5$)$_2$Mg]	7.06	7.75	12.38
			13.98

* All values are in eV.

interactions between the metal valence AOs and the ligand π orbitals). The highest occupied MOs (e_{1g}) are essentially non-bonding orbitals belonging to the ligand π-framework. The following MOs (e_{1u}) then feature the most significant bonding character (some data on vertical ionisation energies obtained from the photoelectron spectra on some magnesocenes have been elucidated and are shown in Table 3.5).

It is interesting to compare the molecular orbital picture of magnesocene with that of calciocene where low-lying d orbitals *are* available for bonding. The MO scheme for both metallocenes and the (C$_5$H$_5$)$_2$ skeleton and the free metal atom is shown in Fig. 3.38 (Blom *et al.*, 1990). The main difference is that the calcium analogue has empty low-lying d orbitals that can be used for bonding whereas magnesium does not. For both compounds, the highest metal s orbital combines with an a_{1g} orbital of the (C$_5$H$_5$)$_2$ skeleton. For [(C$_5$H$_5$)$_2$Mg] there is additional bonding between the p_x, p_y and p_z orbitals on magnesium and the e_{1u} and a_{2u} orbitals of the rings. However, for [(C$_5$H$_5$)$_2$Ca], the p_z orbital on the calcium atom is involved in metal–ring bonding to a far lesser extent and the highest e_{1g} orbital of the (C$_5$H$_5$)$_2$ skeleton is stabilised by interaction with the d_{xz} and d_{yz} orbitals of the calcium atom. Overall for [(C$_5$H$_5$)$_2$Mg] the molecular orbitals that are most destabilised on bending are the bonding $3a_{2u}$ orbital and the C—H orbitals $2e_{2g}$, $2e_{2u}$, $2e_{1g}$ and $2e_{1u}$. With bending of 20°, these orbitals are all noticeably destabilised via inter-ring anti-bonding interactions which helps to explain why magnesocene possesses a regular sandwich structure. [(C$_5$H$_5$)$_2$Ca], because of the 0.06 nm longer inter-ring distance, does not experience the destabilisations via inter-ring anti-bonding interactions. This is part of the explanation of why calciocenes have bent sandwich structures, but more weight is placed on the effect of the low-lying empty d orbitals that are available for bonding (Blom *et al.*, 1990). There is a covalent contribution to the bonding in [(C$_5$H$_5$)$_2$Ca],

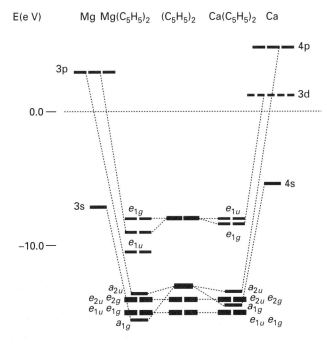

Fig. 3.38 Qualitative diagram of the highest MOs of $[(C_5H_5)_2Mg]$ and $[(C_5H_5)_2Ca]$ along with those of the $(C_5H_5)_2$ skeleton and the atomic orbitals of Mg and Ca.

similar to that proposed for rare earth cyclopentadienides. The possible valence configurations which can form covalent bonds in a trigonal configuration are sp^2, dp^2, sd^2 and d^3 hybridisations. The $4s$, $4p$ and $3d$ orbitals of calcium are of suitable size and energy to form covalent metal–ligand bonds, where two of the three trigonal orbitals are directed towards the centres of the $\eta^5\text{-}C_5H_5$ groups with the third orbital pointing toward the allylic protons of the $\eta^3\text{-}C_5H_5$ species. Trigonal hybridisation of the orbitals in the x and y planes allows the orbitals in the z direction to interact with the fourth ring in the calcium co-ordination sphere.

This confusion over covalent or ionic bonding in certain metallocenes (detailed for magnesocene) leads us into the final section of this chapter.

3.5 Metallocene bonding–ionic or covalent?

As a conjugated saturated hydrocarbon the cyclopentadienyl ligand possesses π and π^* orbitals and acts as both a π-donor and π-acceptor. As a consequence, most Cp complexes are essentially covalent with little charge separation between the metal and ring ligand. Metallocenes such as Cp_2Cr, Cp_2Fe and Cp_2Co are considered to feature strong covalent bonding and although not all are stable with respect to oxidation, they all have strong bonding with

Table 3.6 Dissociation energies of metallocene cations.

Reaction		$\Delta H\,(\text{kJ mol}^{-1})$
Cp_2Mg^+	$\rightarrow$ $CpMg^+ + Cp$	310
Cp_2V^+	$\rightarrow$ $CpV^+ + Cp$	515
Cp_2Cr^+	$\rightarrow$ $CpCr^+ + Cp$	633
Cp_2Mn^+	$\rightarrow$ $CpMn^+ + Cp$	364
Cp_2Fe^+	$\rightarrow$ $CpFe^+ + Cp$	641
Cp_2Co^+	$\rightarrow$ $CpCo^+ + Cp$	754
Cp_2Ni^+	$\rightarrow$ $CpNi^+ + Cp$	524

respect to dissociation of the rings from the metal atom (Table 3.6). There is of course, some polarity in the bonds between the metals and rings but the compounds do not react like polar organometallic compounds (cf. a Grignard reagent).

$$RMgX + H_2O \longrightarrow RH + MgXOH \qquad (3.1)$$

$$Cp_2Fe + H_2O \longrightarrow \text{No reaction} \qquad (3.2)$$

There are some compounds that are essentially ionic systems and these normally involve the very electropositive metals. An indication of this character is the presence of a very reactive C_5H_5 group, e.g.

$$Cp^-Na^+ + H_2O \longrightarrow C_5H_6 + Na^+OH^- \qquad (3.3)$$

$$2Cp^-Mg^{2+} + 2H_2O \longrightarrow 2C_5H_6 + Mg(OH)_2 \qquad (3.4)$$

$$3Cp^-Ln^{3+} + 3H_2O \longrightarrow 3C_5H_6 + Ln(OH)_3 \qquad (3.5)$$

and reactions with ferrous chloride to give ferrocene, e.g.

$$Cp_2Mn + FeCl_2 \longrightarrow Cp_2Fe + MnCl_2 \qquad (3.6)$$

They are considered to have a salt-like nature and are often known as '*metal cyclopentadienides*'. (NB: Cyclopentadienides is applied to compounds M^+Cp^- and cyclopentadienyl to Cp_2M, but this is open to interpretation and the -yl nomenclature is often used for all of these compounds, so caution needs to be exercised in reading too much into the bonding via the name.) As is usual with polar bonds, there is no sharp distinction between covalent and ionic bonding.

3.5.1 Ionic bonding in the *d*-block sandwich compounds

Of the *d*-block species, only manganocene (Cp_2Mn) is thought to possess largely ionic character. This arises from manganese(II) being a high-spin d^5 ion with zero crystal field stabilisation energy and behaving as a 'hard' ion in

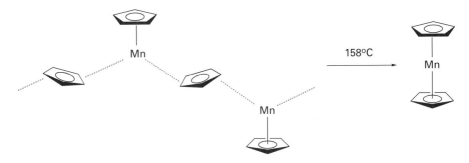

Fig. 3.39 The two phases (polymeric and monomeric) of $[(C_5H_5)_2Mn]$.

a similar fashion to Zn^{2+} or Mg^{2+}. It exists in two phases, with one form consisting of infinite chains of Cp_2Mn fragments bridged by cyclopentadienyl rings in the solid. Then, on warming to 158°C, the colour changes from brown to orange/pink and the structure reverts to a 'normal', monomeric metallocene one (Fig. 3.39). The evidence for ionic bonding is from three main sources: (i) manganocene reacts instantaneously with ferrous chloride in tetrahydrofuran to form ferrocene and is hydrolysed immediately by water; (ii) the dissociation energy ($\Delta H = 364$ kJ mol^{-1}) is closer to that of magnesium cyclopentadienide ($\Delta H = 310$ kJ mol^{-1}) than to those of other transition metal metallocenes; and (iii) the magnetic moment of manganocene is 5.86 Bohr magnetons, corresponding to five unpaired electrons, and is consistent with the presence of a d^5 Mn^{2+} ion. However, this evidence cannot be called unequivocal. Other metallocenes will react with iron(II) chloride to form ferrocene although not so rapidly. The presence of high-spin manganese and a low dissociation energy for the compound suggest the absence of strong, covalent bonding but do not exclude the possibility of *some* covalent bonding. The stability of the half-filled subshell can be a reason for much of the anomalous behaviour of manganocene.

3.5.2 Ionicity or covalency in main group cyclopentadienyl sandwich species

The situation is more complicated in cyclopentadienyl complexes of the main group elements (Jutzi, 1990). In general, a high degree of ionic character can be assumed for the three alkali metal and heavier earth alkali metal (Ca, Sr, Ba) compounds. The environment of the corresponding cations is governed by the steric requirements of the cyclopentadienide anions and additional coordinating species. Electron-counting rules cannot be applied to the structures of these compounds. Borderline bonding situations are found in magnesiun(II), indium(I), thallium(I), lead (II) and bismuth(III) complexes. Conclusive information on the bonding modes in these species is as yet unavailable and postulations presently rely on NMR and structural parameters. In the other complexes, the bonding between the main group element and the cyclopentadienyl ligand can be regarded as predominantly covalent. The structural types within these main group systems have also been generally classified according to bonding types and these are expanded upon elsewhere in this book and by Jutzi (1990) (Fig. 3.40).

Beryllium and magnesium are particularly interesting with regard to covalent versus ionic bonding. Although magnesium cyclopentadienide is structurally almost identical to ferrocene, it is thought to be essentially ionic. The sandwich structure is most stable, not only for covalent complexes utilising d orbitals, but also from an electrostatic situation for a cation and two negatively charged rings. The structure of the beryllium compound is unusual and still quite uncertain. From electron diffraction in the gas phase, the beryllium atom is closer to one ring than the other. The distance between the rings (0.337 nm) is governed by non-bonded repulsions between the rings and can be expected from the van der Waals radii of carbon ($r_{vdw} = 0.165$–0.17 nm). Thus, the small Be^{2+} ion polarises the π-cloud of one ring and an

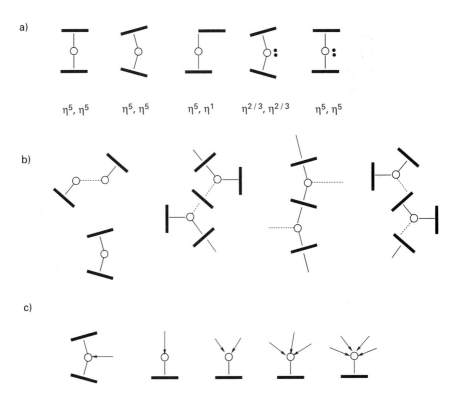

Fig. 3.40 (a) Structure types for sandwich compounds with predominantly covalent Cp-element bonding; (b) structure types for non-donor-stabilised cyclopentadienyl complexes with predominantly ionic Cp-element bonding; (c) structure types for donor-stabilised cyclopentadienyl complexes with ionic Cp-element bonding.

energetically more favourable situation is obtained from a strong, short 'covalent' bond and a longer weaker ionic bond than from two bonds of intermediate length. An X-ray diffraction study of solid beryllocene indicates a '*slipped sandwich*' molecule with a σ- or monohapto bond to one ring and a π- or pentahapto linkage to the other (see Chapter 2).

3.5.3 Ionicity in lanthanide and actinide cyclopentadienyls

The tri(cyclopentadienides) of the lanthanides and actinides are generally thought to feature ionic bonding (Nugent *et al.*, 1971) but more covalent character is observed through the actinide series. The considerable degree of ionic character associated with the metal–carbon bonds in these compounds makes them extremely air and moisture sensitive. However, the possibility of bonding interactions between the metal f orbitals and the symmetry matching ligand π-orbitals 'opens up other bonding scenarios'. For example, in the cyclooctatetraene complexes of the lanthanoids and actinoids, there is a marked difference in the extent to which f orbitals participate in metal–ligand bonding. The $4f$ orbitals in the lanthanide species have no radial nodes, being deeply buried in the atomic core, and this results in negligible overlap with the ligand orbitals. Therefore, the metal–ligand bond in lanthanoid–C_8H_8 complexes is predominantly ionic. In actinoid–C_8H_8 complexes, a considerable contribution from covalent interactions may be assumed because the $5f$ orbitals are subject to less shielding than $4f$ orbitals and can, therefore, overlap more extensively with the ligand orbitals.

In general, it can be stated that the degree of covalency of the metal–carbon bond distances leads to the sequence transition metal–R > actinide–R > lanthanide–R where the organolanthanoids are predominantly ionic. The trend is backed up by chemical, spectroscopic and theoretical results and the main reason has been suggested as being due to the nature of the valence orbitals. They change from $3d$, $4d$, $5d$ (transition metals) to $4f$ (lanthanoids) whilst the actinoids fit somewhere in the middle, featuring $5f$ and $6d$ orbital bonding. With lanthanoid elements, the highly contracted nature of the $4f$ orbitals results in efficient shielding by filled $5s$ and $5p$ orbitals, rendering overlap of Ln ($4f$) with ligand orbitals vanishingly small. Therefore, Ln^{2+}, Ln^{3+} ions resemble the main group metal ions M^{2+}, M^{3+} as electrostatic factors and steric considerations, as opposed to metal–ligand orbital interactions, will govern structure and chemical reactivity.

There are a number of points of evidence for lanthanoid species: (i) organolanthanoid compounds are extremely air and moisture sensitive, which is due to high anionic character of the organic ligand and the oxophilicity of the Ln^{2+}, Ln^{3+} ions. The Cp_3Ln species can behave as Cp-transfer agents (similar to Cp_2Mg) which is not a property possessed by the cyclopentadienyl compounds of the actinoids and the transition metals; (ii) there is a distinct lack of extensive metal ($4f$)–ligand orbital overlap preventing the lanthanoid cations from acting as π-bases. This being the case, π-backbonding effects are virtually non-existent, i.e. to date, no compounds containing Ln—CO bonds that are stable at ambient temperature have been prepared; (iii) finally, the dominance of ionic character in the lanthanoid–ligand bond is also shown by the variable hapticity that the rings display in binary lanthanide cyclopentadienyls (Cp_3Ln). For example, in [$(MeC_5H_4)_3Nd$], each Nd atom is η^5-bonded to three cyclopentadienide rings and η^1-bonded to another ring of an adjacent (MeC_5H_4)$_3$ unit. In doing this, the central metal maximises the number of cation–anion contacts compatible with its size. This is also often observed in ionic cyclopentadienide complexes of the group I and II metals. It is also interesting to note the bent structure of metallocenes of ionic alkaline earth – and lanthanoid – cyclopentadienides, e.g. [$(Me_5C_5)_2M$] (M = Ca, Sr, Ba and Eu, Sm and Yb) (Scheme 3.6). This bent geometry does not follow traditional steric, electrostatic or covalency principles. Reasoning has been put forward using a '*polarizable ion model*'. This suggests that the total electrostatic bonding is maximised in the bent structure because then the dipole moment induced on the central cation interacts favourably with the two adjacent anions. As some of the anion–anion repulsion has to be offset for this effect to operate

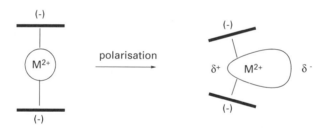

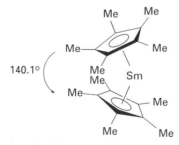

Scheme 3.6 The bent structure of metallocenes of ionic alkaline earth and lanthanoid cyclopentadienides.

it is perhaps unsurprising that only ionic metallocenes of relatively large M^{2+} ions are bent, whereas Cp_2Mg is axially symmetric.

Photoelectron spectral analysis has led to the suggestion that *all* systems with centrally bonded rings have a significant degree of π-bonding, with the exception of purely ionic compounds. For the transition metals, the d subshell is clearly involved and the additional stability and electroneutralisation is gained by π-back donation from filled d orbitals to the antibonding π orbitals of the rings. Main group elements must be assumed to use their vacant valence shell p orbitals for π-bonding interactions to supplement the σ-overlap normally involved. There does not, though, seem a clear distinction between σ-bonded and π-bonded schemes. There is certainly a greater proportion of π-bond character in Cp_2Fe than in Cp_2Mg or Cp_2Pb but there seems little doubt that both σ and π interactions occur in all three compounds. There seem to be two primary factors that contribute to the geometry around the metal atom in CpM complexes: (i) steric interactions which are directly dependent on the size of the metal atom radius and (ii) maximum interaction between the metal orbital–ligand orbital overlap, electrostatic interaction or a combination of both (Zerger and Stucky, 1974).

The electronic structure of the metallocenes and other parallel sandwich complexes is well-described by the ligand field model (Warren, 1976; Green, 1981; Robbins *et al.*, 1982). This suggests that the d orbitals are split into three sets, in order of decreasing binding energy: e^2 ($xy, x^2 - y^2$), a_1 (z^2) and e_1 (xz, yz). When the complexes have a d^5 configuration, the choice of electronic ground state is very sensitive to the particular combination of metal and ligand. There are three possible configurations (Fig. 3.41) with the $^2A_1$ state being adopted by all the di(arene) metal complexes, whilst Cp_2Fe^+, $Cp_2^*Fe^+$ and Cp_2^*Mn adopt $^2E_2$ states ($Cp^* = \eta^5$-C_5Me_5). In contrast, Cp_2Mn and $[(\eta^5$-$C_5H_4R)_2Mn]$ both may be switched between $^2E_2$ and $^6A_1$ states by a change of host material or exhibit an equilibrium between them. In general, the bonding and characterisation of various metal cyclopentadienyl species can be summarised in Table 3.7.

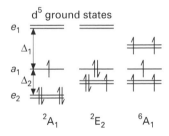

Fig. 3.41 The three possible electron configurations and terms available to d^5 parallel sandwich complexes in their ground states.

Table 3.7 The character and bonding of metal cyclopentadienyl [$(C_5H_5)_nM$] compounds.

Character	Bonding	Properties	Examples
Ionic	Ionic lattice M^{n+} $[C_5H_5]_n^-$, as in halides MX_n	Highly reactive towards air, water and other compounds with active hydrogen, not sublimable	$n = 1$: alkali metals, $n = 2$: heavy alkaline earth metals, $n = 2, 3$: lanthanides
Intermediate		Partially sensitive to hydrolysis, sublimable	$n = 1$: In, Tl, $n = 2$: Be, Mg, Sn, Pb, Mn, Zn, Cd, Hg
Covalent	Molecular lattice	Only partially air sensitive, usually stable to hydrolysis, sublimable	$n = 2$: (Ti), V, Cr, (Re), Fe, Co, Ni, Ru, Os, (Rh), (Ir), $n = 3$: Ti, $n = 4$: Ti, Zr, Nb, Ta, Mo, U, Th

References

Armstong, D.R., Herbst-Irmer, R., Kuhn, A. *et al.* (1993) *Angew Chem*, **105**, 1807; *Angew Chem Int Ed Engl*, **32**, 1774.

Bau, R., Teller, R.G., Kirtley, S.W. and Koetzle, T.F. (1979) *Acc Chem Res*, **12**, 176.

Baxter, S.G., Cowley, A.H., Lasch, J.G., Lattman, M., Sharum, W.P. and Steart, C.A. (1982) *J Am Chem Soc*, **104**, 4064.

Blom, R., Faegri, K., Jr and Volden, H.V. (1990) *Organometallics*, **9**, 372.

Bursten, B.E. (1989) *Comments Inorg Chem*, **9**, 61.

Canadell, E., Eisenstein O. and Rubio, J. (1984) *Organometallics*, **3**, 759.

Cowley, A.H. (1979) *Prog Inorg Chem*, **26**, 46.

Evans, S., Green, J.C. and Jackson, S.E. (1972) *J Chem Soc, Faraday Trans II*, **68**, 249.

Fischer, E.O. and Hafner, W. (1955) *Z Naturforsch*, **10b**, 665.

Green, J.C. (1981) *Struct Bonding (Berlin)*, **43**, 37.

Green, M.L.H., Coates, G.E. and Wade, K. (1968) *Organometallic Compounds*, vol. 2, 3rd edn, pp. 90–164. Methuen and Co., London.

Hawthorne, M.F. (1975) *J Organomet Chem*, **100**, 97.

Jutzi, P. (1986) *Adv Organomet Chem*, **26**, 217.

Jutzi, P. (1990) *J Organomet Chem*, **400**, 1.

Lauher, J. W. and Hoffmann, R. (1976) *J Am Chem Soc*, **98**, 1729, and references therein.

Mingos, D.M.P. (1982) in *Comprehensive Organometallic Chemistry*, (eds. E.W. Abel, F.G.A. Stone and G. Wilkinson) vol. 3, ch. 19, p. 28. Pergamon Press, Oxford.

Nugent, L.G., Laubereau, P.G., Werner, G.K. and Vander Sluis, K.L. (1971) *J Organomet Chem*, **27**, 365.

Peterson, J.L., Lichtenberger, D.L., Fenske, R.F. and Dahl, L.F. (1975) *J Am Chem Soc*, **97**, 6433.

Robbins, J.L., Edelstein, N., Spencer, B. and Smart, J.C. (1982) *J. Am Chem Soc*, **104**, 1882.

Rosch, N. and Streitwieser, A. Jr. (1978) *J Organomet Chem*, **145**, 195.

Streitwieser, A. Jr and Muller-Westerhoff, U. (1968) *J Am Chem Soc*, **90**, 7364.

Warren, K.D. (1976) *Struct Bonding (Berlin)*, **27**, 45.

Zerger, R. and Stucky, G. (1974) *J Organomet Chem*, **80**, 7.

4 Chemical and Spectroscopic Properties of Metallocenes

Part A: Chemical properties

4.1 The organic chemistry of ferrocene

There is now a huge area of study involving the chemistry of metallocenes with that of ferrocene predictably being the most well documented. The cyclopentadienyl rings are aromatic and, in general, most of the chemistry of ferrocene and its derivatives may be predicted on this basis. Due to its great stability and the ability to maintain the ligand–metal bonding under harsh conditions, there is a vast range of organic chemistry and it is possible to carry out a variety of transformations on the cyclopentadienyl ligands. An outline of this chemistry is shown in Scheme 4.1.

4.1.1 General electrophilic substitution

In general, metallocenes are more reactive towards electrophilic substitution than benzene indicating that more electrons are readily available. In fact compared to benzene, ferrocene reacts 3×10^6 faster. It is thought that the electrophilic substituents (E^+) interact first with the weakly bonding electrons of the iron atom and then transfer to the C_5H_5 ring with proton elimination (Route A) (Scheme 4.2). The intermediate cation with the electrophile (E) bound to the metal rearranges to a cyclopentadiene complex with E in the *endo* (i.e. metal side) position which then loses a proton to give the substituted ferrocene. (NB: The electrophile should not be an oxidising agent as substitution would be suppressed by oxidation to the ferrocenium ion $[Cp_2Fe]^+$ which is inert to attack by electrophiles.) Alternatively, it has also been proposed that attack takes places on the ring (as in benzene and other aromatic species) and does not involve direct participation of the metal (Route B). This route involves direct addition of the electrophile to the less hindered *exo* face of the ligand to give an intermediate which then loses a proton to give the product (Scheme 4.3). Kinetic evidence has been found to support (B) using the fact that intramolecular acylation of two isomeric ferrocene carboxylic acids can occur from the *exo* direction (B) or the *endo* direction (A). It was shown that the formation of the *exo* isomer, where the acylium ion cannot interact directly with the metal, proceeds preferentially and is the rate-determining step ($k_{exo}/k_{endo} = 4$) (Scheme 4.4) (Rosenblum

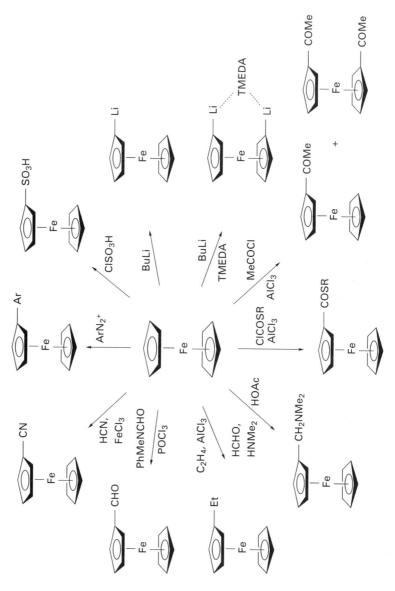

Scheme 4.1 Some organic reactions of ferrocene.

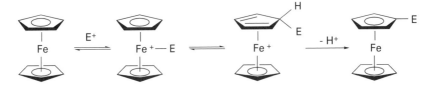

ROUTE A

Scheme 4.2 A mechanism for electrophilic substitution of ferrocene.

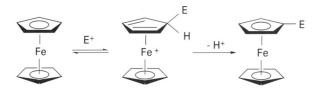

ROUTE B

Scheme 4.3 A second mechanism for electrophilic substitution of ferrocene.

and Abbate, 1966). However, there is also plenty of evidence for the metal participation route (A). Examples are known of intramolecular migration of a group from a metal atom to the *endo* face of an attached π-hydrocarbon ligand and there are a wealth of studies showing protonation of the metal atom in ferrocene and a host of other organometallic complexes.

There is a third possible mechanism which involves addition of the electrophile to the *endo* face of the ligand but not via any metal interaction

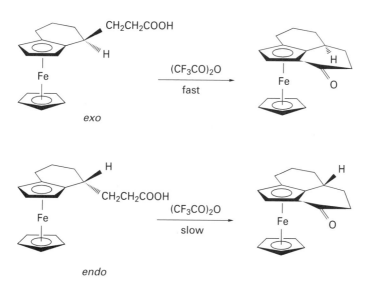

Scheme 4.4 The intramolecular acylation of two isomeric ferrocene carboxylic acids.

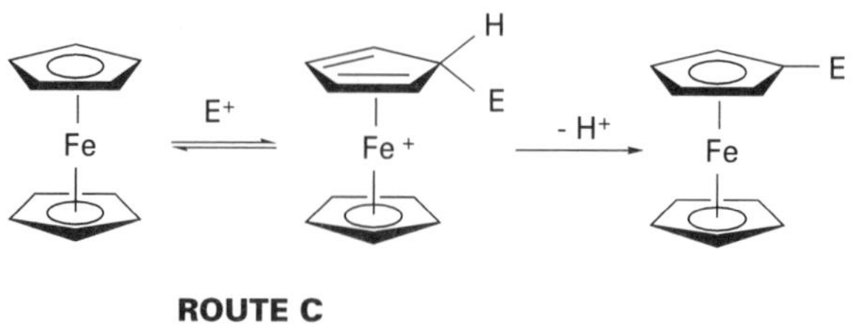

H
E
E+
Fe
Fe +
- H+
Fe
ROUTE C

Scheme 4.5 A third mechanism for electrophilic substitution of ferrocene.

(Route C) (Scheme 4.5). This is attractive because the electronic charge concentration is unsymmetrical with respect to the ligand ring plane with a higher concentration on the *endo* side adjacent to the metal. In truth, each route is plausible and probably all do occur, but it has been suggested that the stereochemistry of electrophilic substitution and the kinetic features of the reaction are governed by the nature of the electrophile. For instance, where reagents are more electrophilic than the proton (e.g. RC^+O) the rate-determining step is addition of the electrophile which favours attack from the *exo* direction (B), but for reagents less electrophilic than the proton (e.g. $AcOHg^+$), the *endo* pathway is favoured (C) and deprotonation is rate limiting. In the proton-exchange reaction, it seems that the *exo* and *endo* pathways are equally likely.

Some examples of the major reaction types observed with ferrocene are now detailed.

4.1.2 Friedel–Crafts acylation

This reaction is of historical importance as it established the aromatic character of ferrocene (Scheme 4.6). The metallocene undergoes Friedel–Crafts catalysed acylation very readily on one ring and less readily on both rings. If the two rings are free to rotate only one 1,1′-disubstituted compound is isolated, whereas three 1,1′-disubstituted isomers could be formed in the

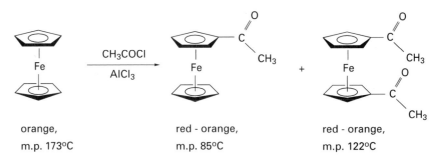

Fe
CH3COCl
AlCl3
Fe
O
C
CH3
+
Fe
O
C
CH3
O
C
CH3
orange,
m.p. 173°C
red - orange,
m.p. 85°C
red - orange,
m.p. 122°C

Scheme 4.6 The Friedel–Crafts acylation of ferrocene.

absence of rotation. As expected, only one compound is isolated in practice. The reaction can be catalysed by any Lewis acid, most commonly $AlCl_3$ but the use of H_3PO_4 as catalyst can be effective as it limits the amount of disubstituted product formed. The disubstituted 1,2-isomer is always only a minor by-product. Deactivation by acyl groups is expected by analogy with other organic ring systems. It is interesting to note that as more rigorous conditions are necessary to effect substitution even on the second ring some electronic effects must be passed from ring to ring through the iron atom.

Ferrocene also undergoes alkylation on treatment with alkyl halides or alkenes under Friedel–Crafts conditions. However, such reactions are not synthetically useful because of poor yields and side reactions; mixtures of polyalkylated products generally result because the first alkyl substituent activates the molecule towards further substitution. The effects of a ring substituent on ferrocene in this context have been reviewed by Slocum and Ernst (1972). Introduction of an electron-donating substituent activates the molecule towards electrophilic substitution whereas an electron-withdrawing substituent has the opposite effect. The unsubstituted ring of an alkylferrocene is about twice as reactive towards acylation as a ring in ferrocene itself whilst the acyl substituent in FcCOMe reduces the reactivity towards acylation of the unsubstituted ring by a factor of $c.\ 2 \times 10^4$. It seems that, in general, electrophilic substitution at the 2-position of a ferrocene ring bearing either an activating or deactivating group is favoured on electronic grounds (although this can be outweighed by adverse steric factors).

4.1.3 Mannich reaction (aminomethylation)

This reaction involves the condensation of ferrocene with formaldehyde and amines. Using dimethylamine gives dimethylaminomethylferrocene, a compound useful in the preparation of many other derivatives (Scheme 4.7). As such, ferrocene demonstrates the reactivity of its rings and resembles the reactive thiophene and phenol species rather than benzene which does not undergo Mannich condensations.

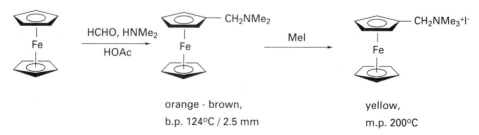

orange - brown,
b.p. 124°C / 2.5 mm

yellow,
m.p. 200°C

Scheme 4.7 Aminomethylation of ferrocene.

4.1.4 Metallation

Another reaction typical of aromatic systems is metallation and a wide range of reactions of metallocenes have been found in which a ring hydrogen atom is replaced by a metal atom. Alkali metal derivatives have found extensive application as intermediates in the synthesis of other ring-substituted species and lithium, sodium, mercury and boron derivatives can be usefully employed (Bublitz and Rinehart, 1969) (Scheme 4.8). Clearly, with a view to the formation of substituted ferrocenes, metallation is a synthetically very useful complement to electrophilic substitution. Lithiation reactions are thought to involve nucleophilic attack of the hydrocarbon portion of the Li-containing reagent on a hydrogen atom of the compound undergoing metallation and this proton must be relatively acidic. Ring hydrogen atoms in metallocenes are indeed weakly acidic. The extent of lithiation can be controlled by careful choice of reaction conditions. For example, mono-lithiated ferrocene can be prepared by treating ferrocene with stoichiometric quantities of n-BuLi or t-BuLi in hexane/ether or by substitution of HgCl by Li. However, the 1,1′-dilithio species can be formed exclusively by using n-BuLi with TMEDA (N,N,N',N'-tetramethylethylenediamine) or with PMDETA (1,1,4,7,7-pentamethyldiethylenetriamine). The effectiveness of the amines comes from the fact that the lithium atom is strongly complexed by the diamine, thus rendering the organic group even more carbanionic, and that the amine breaks down the less reactive RLi tetramer. The structure of 1,1′-dilithioferrocene· 2PMDETA shows a dinuclear arrangement, with two ferrocenylene groups linked by two bridging Li atoms.

Competition experiments have shown that the acidity of cyclopentadienyl ring protons in ferrocene is reduced when an alkyl substituent is introduced. Alkylferrocenes undergo homo- (1,3) and hetero-annular lithiation with the ring positions adjacent to the alkyl group remaining unreactive. However, lithiation of ferrocene species containing a ring substituent with an accessible non-bonding electron pair occurs with predominant or exclusive 2-lithiation. This is attributed to an intramolecular co-ordination between the lithium atom and the lone pair of electrons in the substituent and forms a '*directive mechanism*', driving the second substitution to the 2 or *ortho* position on the rings (Watts, 1980).

Sodation (using pentylsodium) and potassiation (with n-BuK) of ferrocene gives predominantly the 1,1′-dimetallated products but they are of limited synthetic use as methods specific to mono- and di-substituted forms are unavailable. Mercuration reactions are usually carried out using mercuric acetate and this reaction can occur up to 10^9 times faster than for benzene. It has been concluded that mercurations are electrophilic substitutions that follow a classical mechanistic pattern where there is rapid complex formation followed by rate-determining formation of products.

4.1.5 Nitration and halogenation

There are reactions such as nitration and halogenation that although typical of aromatic systems are not feasible with metallocenes due to their sensitivity to oxidation. For example, using NO_2^+ leads to the ferrocenium ion $[(C_5H_5)_2Fe]^+$ and other degradation products. Sulphonation cannot be carried out directly using concentrated sulphuric acid as this again leads to oxidation but use of chlorosulphonic acid in acetic anhydride or with the SO_3–dioxane complex gives good yields of $[(C_5H_5)Fe(C_5H_4SO_3H)]$. The great propensity for ferrocene to undergo electrophilic substitution is due to the electron-donating or -releasing abilities. In comparison to the butyl group, ferrocenyl is amongst the strongest inductive electron-releasing agents known (Slocum and Ernst, 1972). This is emphasised by aminoferrocene being found to be a stronger base than aniline by a factor of 20 and ferrocene carboxylic acid being a weaker acid that benzoic acid. α-Ferrocenylcarbenyl acetates also undergo solvolysis more rapidly than the phenyl analogues and the α-ferrocenylcarbenium ions are stable enough to be isolated, using BF_4^- and PF_6^- as counterions (Watts, 1979) (Scheme 4.9). The carbenium ions feature α-carbon–metal bonding with the iron atom being displaced from its position below the centre of the substituted ring towards the substituent.

4.2 The organic chemistry of ruthenocene and osmocene

The organic chemistry of ruthenocene and osmocene closely resembles that of ferrocene in many ways, i.e. undergoing reactions characteristic of an aromatic system; however, there are some notable differences as the group is descended. Ruthenocene and osmocene are thermally more stable than ferrocene but in general the latter undergoes aromatic substitution reactions more easily. The general reactivity is shown in Schemes 4.10 and 4.11. Friedel–Crafts acylation, metallation, arylation, formylation and sulphonation reactions are all possible but the degree of aromatic reactivity has been shown to be markedly different in ferrocene, ruthenocene and osmocene. From an exhaustive series of competitive acylation reactions, the following order of reactivity was observed:

$[(C_5H_5)_2Fe] > [(C_5H_5)_2Ru] \geqslant [(C_5H_5)Mn(CO)_3] > [(C_5H_5)_2Os] > [(C_5H_5)Cr(CO)_2(NO)] \geqslant C_6H_6$.

The electron-withdrawing properties of the CO and NO ligands will destabilise the transition states for addition of E^+ to $[(C_5H_5)M(CO)_m(NO)_n]$ relative

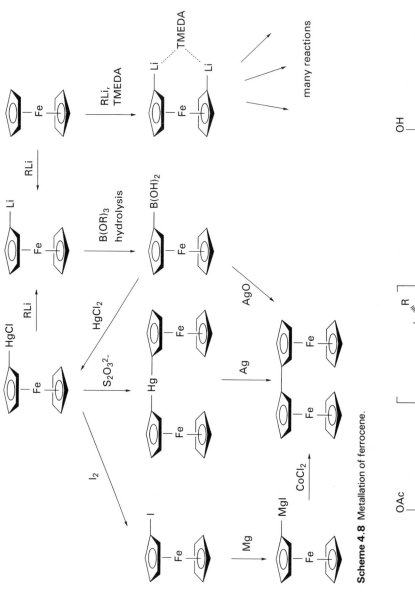

Scheme 4.8 Metallation of ferrocene.

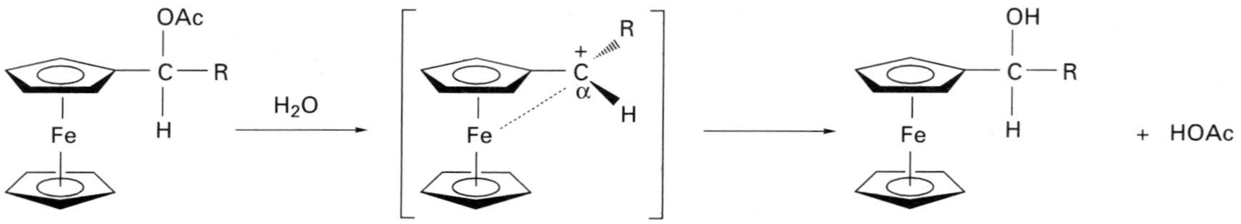

Scheme 4.9 The formation of α-ferrocenyl carbenium ions.

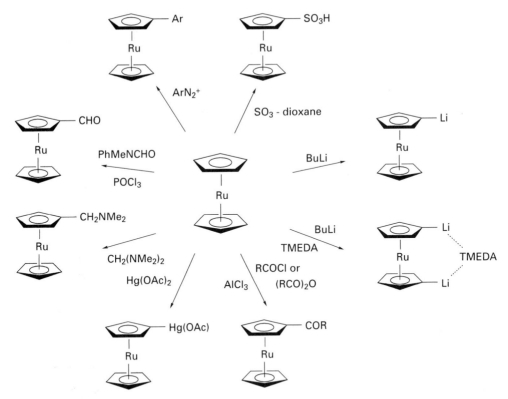

Scheme 4.10 The organic chemistry of ruthenocene.

to those for Cp$_2$M, resulting in a generally higher reactivity of the metal-locenes. The order of reactivity of the metallocenes is in agreement with the relative availability of metal electrons, or basicity, and gives: ferrocene > ruthenocene > osmocene, i.e. osmocene experiences only mono-acylation even under very forcing conditions while ruthenocene gives a mixture of mono- and di-substitution products and ferrocene exclusively undergoes di-substitution under these conditions. This behaviour can be explained by the electronic structures and the characteristics of the different metals and therefore the effective electronegativity differences at the ring carbon atoms. As mentioned previously, there is increased ring–metal bond strength on descent of the group. This results in a lower π-electron density around the rings and so accounts for the decreased electrophilic reactivity.

The powerful electron-releasing nature of the metallocenyl group is again an important feature and results in exceptional stabilisation of the α-metallocenylcarbenium ions (Fig. 4.1). In contrast to the electrophilic sub-stitution reactivity, α-carbenium ions are stabilised by the metals in the order osmocene > ruthenocene > ferrocene. Stabilisation results from electron do-nation from metal to ligand through overlap of a filled metal d orbital of

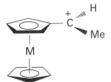

Fig. 4.1 An α-metallocenylcarbonium ion.

appropriate symmetry with the lowest unoccupied molecular orbital (LUMO) of the ligand. Therefore, there must be better overlap of the Ru 4d and Os 5d orbitals with the exocyclic carbon p-orbital than in the case of iron.

4.2.1 Friedel–Crafts acylation

Acylation of ruthenocene with acid chlorides or anhydrides under Friedel–Crafts conditions gives both mono- and 1,1′-di-substituted derivatives (the latter being favoured by excess catalyst such as $AlCl_3$). Examples of derivatives formed are RcCONHPh and ·RcCOSMe. Subsequent reactions of acylruthenocenes result in a variety of compounds (Scheme 4.12). Osmocene will undergo Friedel–Crafts mono-acylation but not alkylation (see Scheme 4.11). Acylosmocenes can be reduced to the corresponding hydroxyalkylosmocenes or alkylosmocenes and the acyl derivatives will also react with

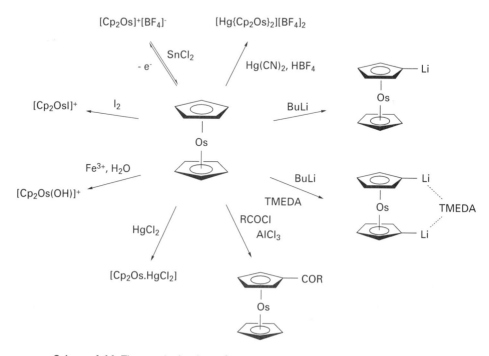

Scheme 4.11 The organic chemistry of osmocene.

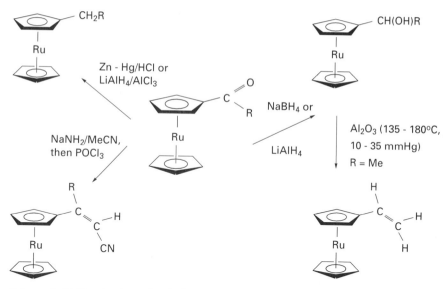

Scheme 4.12 The chemistry of acylruthenocene.

MeCN/MeNH$_2$ to give β-hydroxynitriles which in turn can be dehydrated to the α, β-unsaturated nitriles.

4.2.2 Mannich reactions

Ruthenocene and CH$_2$(NMe$_2$)$_2$ react to form RcCH$_2$NMe$_2$, the ferrocene analogue of which is a key synthetic intermediate. Metallation of RcCH$_2$NMe$_2$ with Li$_2$PdCl$_4$/NaOAc produces a substituted ruthenocenyl complex **IV.1a** which is more reactive than the ferrocene derivative, giving good yields of the olefins **IV.1b** (R = COMe, COPh or Ph) following reaction with RCH=CH$_2$.

(IV.1a)

(IV.1b)

4.2.3 Metallation reactions

Ruthenocene and osmocene have been successfully metallated. Reaction with n-BuLi results in lithio intermediates which, when condensed with carbon dioxide, yield a mixture of mono- and di-carboxylic acids. When using a 6 : 1 excess of n-BuLi, an 86% yield of 1,1'-ruthenocene dicarboxylic acid is obtained with only 1% of the mono-species being isolated. Osmocene is also lithiated under similar conditions but separation of mono- and di-acids is very difficult. With ferrocene and reactions under similar conditions, the yields of mono- and di-carboxylic acids were 35 and 39%, respectively. Ferrocene is less readily metallated due to the higher acidity of the ring protons in the ruthenium species. The lithioruthenocenes are the source of many otherwise inaccessible derivatives, some of which are shown in Scheme 4.13. Reaction of ruthenocene with $Hg(OAc)_2$ gives a complex mixture of products which with KCl yields $RcH : HgCl_2$. This compound is also obtainable from RcH and $HgCl_2$. Ruthenoceneboronic acid reacts with $HgCl_2$ to form RcHgCl, which on reaction with sodium thiosulphate yields Rc_2Hg.

4.3 The inorganic chemistry of ferrocene, ruthenocene and osmocene

This has again been well documented (Deeming et al., 1982) and there are some notable differences between the congeners.

4.3.1 Oxidation

Ferrocene is readily oxidised to dicyclopentadienyliron(III) cation (known as ferrocinium, ferrocenium or ferricenium), it being blue or green in dilute solution and blood-red in concentrated solutions. It can be obtained in a number of ways: electrochemically, photochemically or by various oxidising agents (e.g. HNO_3, $FeCl_3$, I_2, Ag^+, quinone, N-bromosuccinamide).

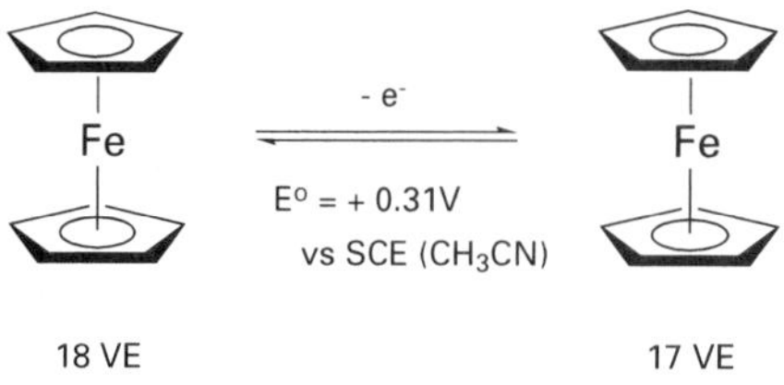

$$\text{(4.1)}$$

The oxidation (4.1) is chemically and electrochemically reversible and the potential shows a sensitivity to the nature of the substituents. Alkyl substituents increase the tendency towards oxidation whilst this trait is reduced in the presence of phenyl groups. For instance $E_{1/2}$ decreases by 0.047 V upon

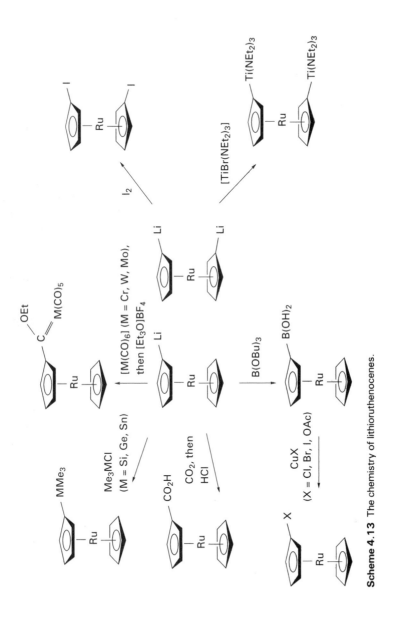

Scheme 4.13 The chemistry of lithioruthenocenes.

the introduction of each alkyl group, wherever the group is introduced. Conversely, Ph and CO_2H groups do not favour oxidation and $E_{1/2}$ is increased by 0.023 and 0.280 V by the introduction of each of these groups, respectively. The ferrocenium cation can be isolated using anions such as BF_4^- and $FeCl_4^-$ and X-ray structures of various salts are known. There are no gross structural changes on oxidation. The C_5H_5 rings are eclipsed in $[(C_5H_5)_2Fe]^+$ as in ferrocene but there are the expected increases in Fe—C and inter-ring bond lengths on oxidation owing to the removal of a bonding electron. Ferrocenium salts are widely used as one-electron oxidising agents in many solvents and the ferrocene/ferrocenium couple is often used as a secondary standard in electrochemical studies, being insensitive to solvent variations. This redox activity has resulted in the incorporation of metallocenes into systems to act as molecular switches or receptors. Selective binding of the systems can affect the redox potential of the metallocene unit and this offers application in the fields of chemical sensors and molecular electronics (see Chapter 6).

Ferrocene is much less readily oxidised than both ruthenocene and osmocene which is to be expected within the general differences between first-, second- and third-row transition metals, i.e. although ruthenocene and osmocene can be oxidised with the loss of two electrons, $[(C_5H_5)_2Fe]^+$ cannot be oxidised any further even at extremely high potentials. Comparative studies on these metallocenes showed the ease of electrochemical oxidation to be Fe < Os < Ru, i.e. in the presence of the non-co-ordinating anion $\{B[C_6H_3(CF_3)_2-2,6]_4\}^-$, $E°$ values for Cp_2M were found to be +0.11 V, +0.55 V and +0.46 V for iron, ruthenium and osmium, respectively. The chemical and electrochemical oxidation of osmocene and ruthenocene is still the subject of controversy. Initial reports of the oxidation of ruthenocene involving electrochemical experiments or use of reagents such as halogens, aqueous Ag_2SO_4, p-quinone and $FeCl_3$ suggested a one-electron oxidation process. However, chronopotentiometric studies indicated that a two-electron oxidation (one step) occurs which is irreversible at a platinum electrode. In molten $AlCl_3$/[1-Bu-py]Cl, a two-electron irreversible oxidation of Cp_2Ru occurs at +0.68 V and a quasi-reversible one-electron reduction at −0.65 V (both versus standard calomel electrode (SCE)). For the mixed metallocene ferrocenylruthenocenylmethane, the irreversible two-electron process is observed for the Ru species along with a reversible one-electron oxidation of the ferrocene centre. To date, preparations of $[(C_5H_5)_2Ru]^+$ have been rather anomalous and the formation of this cation has yet to be substantiated. Chemical oxidation of ruthenocene by halogens gives the bent compound **IV.2** and a dication is formed from the direct reaction of ruthenocene with $Hg(CN)_2$ in perchloric acid.

(IV.2)

X = Cl, Br, I

Oxidation of [(C$_5$H$_5$)$_2$Os] by aqueous iron(III) yields diamagnetic Os(IV) salts of the [(C$_5$H$_5$)$_2$Os(OH)]$^+$ cation whilst using iodine gives [(C$_5$H$_5$)$_2$OsI]$^+$ salts. Electrochemical oxidation gives a reversible one-electron process ($E_{1/2}$ = – 0.15 V) but upon oxidation at a platinum electrode osmocene undergoes a two-step, one-electron each, reversible change ($E_{1/2}$ = 0.48, 1.52 V). Dimeric structures can arise from Os—Os bond homolysis followed by hydrogen abstraction from the [Cp$_2$Os]$^{+\bullet}$ radical or the [Cp$_2$OsH]$^+$ cation. The bridged dimer results from one of the C$_5$ rings on each osmium being metallated by the other osmium atom and is very stable and does not react further. It is difficult to estimate the relative susceptibility towards oxidation of the metallocenes because of the different mechanisms of the products generated. However, since oxidation removes one electron from the highest molecular orbital of the complex, the different behaviour within the group must be because of changes in electronic structures.

4.3.2 Protonation

In strong acids such as H$_2$SO$_4$/CF$_3$CO$_2$H, BF$_3$·H$_2$O/CF$_3$CO$_2$H, the weakly basic metallocenes can be protonated at the metal atom which is illustrated by a low frequency ^{1}H NMR hydride signal at around –2 ppm and a C$_5$H$_5$ doublet. It has been suggested that the metal-protonated intermediate possesses tilted, no longer parallel cyclopentadienyl rings (IV.3) and that

(IV.3)

the angle (α) can vary from 135 to 180° without any significant loss in metal–ring bond strength. Evidence for this ring-tilting was provided by a ^{1}H NMR study of protonated alkyl ferrocenes which showed that free rotation of the substituted rings was not possible. The non-ring hydrogen in [(C$_5$H$_5$)$_2$FeH]$^+$ [BCl$_4^-$] has been identified as a hydride by the v(Fe—H) absorption at 1630 cm^{-1}. Alkyl groups increase and chloro-substituents decrease the metal basicity, and the acyl group in MeCOFc is much more readily protonated than the metal atom to give the ferrocenyl-stabilised carbocation **IV.4**.

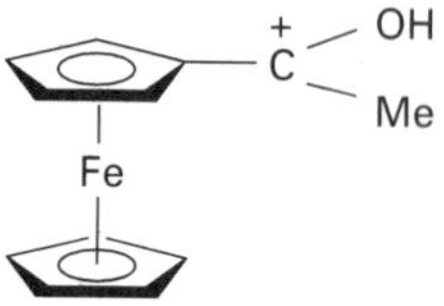

$$(\text{IV.4})$$

Two mechanisms for exchange of ring protons of ferrocene under acidic conditions have been suggested. The direct attack of a proton at a ring would give the 16-electron complex $[(\eta^5\text{-}C_5H_5)Fe(\eta^4\text{-}C_5H_6)]^+$ (with *exo* attack) whereas the hydride ligand of $[(C_5H_5)_2FeH]^+$ might be transferred to give **IV.5** causing the attack to be *endo*. The direction of attack (*endo* or *exo*)

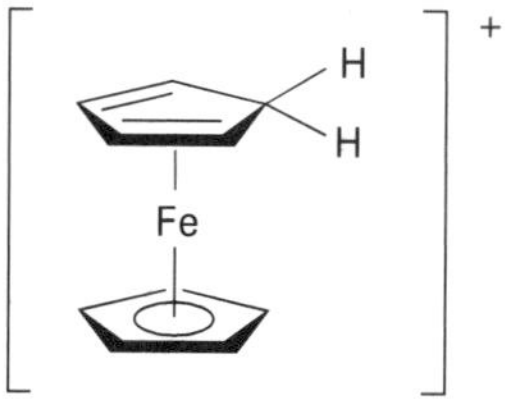

$$(\text{IV.5})$$

or whether protonated ferrocenes can lead to exchange is unknown. There is evidence that substitution at one ring has a marked effect on the exchange rate at the unsubstituted ring. In $[(C_5H_5)Fe(C_5H_4X)]$, the relative rates of H—D exchange at the C_5H_5 ring are 1 (X = CN), 5 (X = CO_2H), 12 (X = Br) and 260 (X = H). The formation of these protonated species has been investigated and the ease of protonation of the metallocene decreases in the order ferrocene > ruthenocene > osmocene, i.e. ferrocene is the strongest (though still very weak) metallocene base in $BF_3 \cdot H_2O$. More recent studies have found that the metal basicity of ruthenocene is slightly higher than that of ferrocene. This is explained by the lower repulsion of the proton by the tilted C_5H_5 rings due to the higher inter-annular distance and by the greater extension of the 4*d* orbitals. As protonation of the metal becomes more extensive in more acidic solvents, the rate of ring H/D exchange decreases. There is no evidence for the intermediacy of any metal-protonated species in the exchange reaction. The protonated metal atom is electron withdrawing, and thus has a deactivating effect.

4.3.3 Reactions with Lewis acids

Another interesting difference in iron group metallocene behaviour is the formation of donor–acceptor complexes with weak Lewis acids. Ruthenocene and osmocene prefer to form adducts with species that accept an electron pair whilst ferrocene will release just one electron for adduct formation. For example with HgX_2, ruthenocene and osmocene will produce stable complexes of definite composition, i.e. 1:1 or 3 : 1 with excess $HgCl_2$. The 1:1 adducts (**IV.6**) are simple halide-bridged compounds with a Ru—Hg bond whilst the extra $HgCl_2$ molecules in the 3 : 1 adduct (**IV.7**) are incorporated

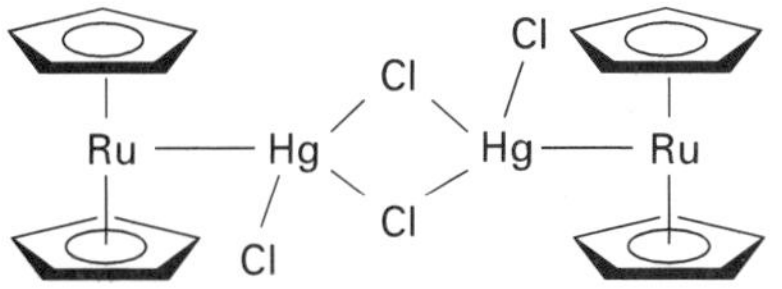

(**IV.6**)

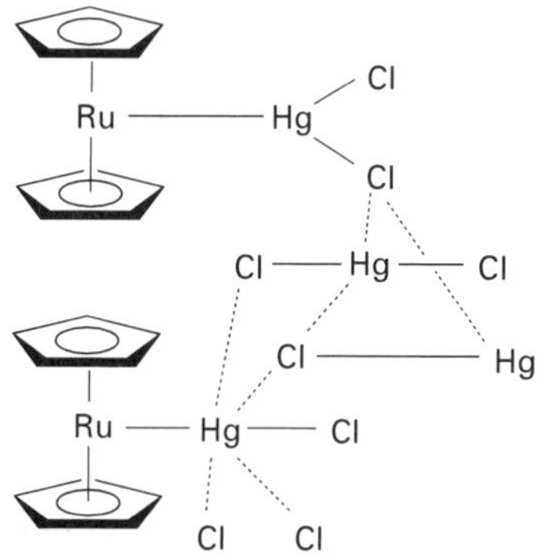

(**IV.7**)

via Cl—Hg—Cl bridging bonds. Ferrocene forms unstable adducts which readily undergo redox reaction with formation of the ferrocenium cation. Generally, ferrocene forms far more adducts with π-acids and Lewis acids than the other metallocenes but it is clear that there is significant enhancement of donor properties as one descends the group.

4.4 The chemistry of first-row d-block metallocenes

Although the chemistry of the iron metallocenes is extensive, many other cyclopentadienyl compounds do not survive the reaction conditions involved in aromatic substitution and similar reactions. The chemical reactivity of these metallocenes varies widely across the series and is dominated by the number of valence electrons each compound possesses due to not having the stable 18-electron structure of the iron group metallocenes.

4.4.1 Vanadocene

This compound has a high reactivity, acting like a carbene in many ways, and the reactions of the 15-electron metallocene demonstrate electron-deficient character and the tendency to reach higher oxidation states. Electrochemical studies show that in THF Cp_2V undergoes reversible one-electron reduction

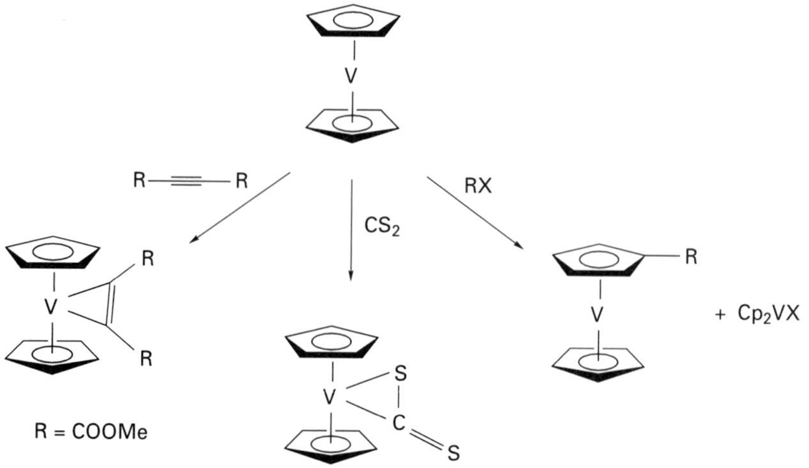

Scheme 4.14 Examples of the reactivity of vanadocene.

at a very negative potential (-3.0 V versus SCE) and indicates formation of a stable anion $[Cp_2V]^-$. The oxidation of Cp_2V involves two one-electron transfer reactions. The first step is reversible and occurs at $c.$ -0.7 V and the second is irreversible, showing that there is almost certainly a severe structural change with the formation of $[(C_5H_5)_2V(THF)_2]^{2+}$. As mentioned before, Cp_2V shows high bond dissociation energies so it is not surprising to observe slow ring-exchange reactions. There is no exchange with $[(\eta\text{-}C_5D_5)_2Ni]$ in heptane at $51\,°C$, very little formation of ferrocene with $FeCl_2$, and thermal equilibrium with $[(\eta\text{-}C_5D_5)_2V]$ takes one week. Some examples of typical reactions are detailed in Scheme 4.14 and adduct formation with Cp_2V and neutral ligands such as alkynes can be thought of as giving oxidative addition products due to extensive back bonding.

4.4.2 Chromocene

With only 16 electrons in its valence shell this compound is unstable and air sensitive but undergoes a typical range of reactions (Davis and Kane-Maguire, 1982). Chemical and electrochemical oxidation of chromocene is possible, forming $[Cp_2Cr]^+$ at a half-wave potential of -0.67 V (versus SCE), and is reversible. The ion can be stabilised in solution by addition of a suitable counterion. Oxidation of Cp_2Cr using $CH_2{=}CHCH_2I$ yields $[Cp_2Cr]I$ in high yield, although the product is pyrophoric. It has three unpaired electrons but is more stable as the tetraphenylborate. Chromocene also undergoes reversible one-electron reduction to form $[Cp_2Cr]^-$ ($E_{1/2} = -2.30$ V versus SCE). The extremely reactive anion cannot be isolated because it reacts with the solvent.

4.4.3 Manganocene

This compound possesses significant ionic character and as such is quite reactive towards atmospheric oxygen, sometimes being pyrophoric. Protonic reagents react instantaneously to form cyclopentadien and $Mn(OH)_2$ (if water is used) or from acids, salts of manganese(II). Additionally, ferrocene is liberated from a reaction of ferrous chloride in THF with manganocenes. As a further expression of its ionic character the compound forms 19- to 21-electron adducts with donor ligands, i.e. $[Cp_2Mn(THF)_2]$, $[Cp_2Mn(PMe_3)]$ and $[Cp_2Mn(dmpe)]$. Manganocene does not possess significant redox chemistry (unlike $[(\eta^5\text{-}C_5Me_5)_2Mn]$) but carbon monoxide converts the metallocene to η^5-cyclopentadienylmanganese tricarbonyl, a species that has a noble gas configuration for its metal and this undergoes aromatic substitution reactions.

4.4.4 Cobaltocene

The chemistry of the metallocenes of cobalt (and nickel) is influenced by their tendency to attain an 18 VE configuration.

$$[(C_5H_5)_2Co] \underset{\substack{E^0 = -0.90 \text{ V} \\ \text{versus SCE}}}{\overset{-e^-}{\rightleftharpoons}} [(C_5H_5)_2Co]^+ \qquad (4.2)$$

$$\text{19 VE} \qquad\qquad\qquad \text{18 VE}$$

The loss of this anti-bonding, unpaired electron enables the compound to act as a powerful reducing agent. Cobaltocenium salts can be obtained by mild oxidation of cobaltocenes, e.g. by $FeCl_3$, oxygen or aqueous acid. The salts are yellow, diamagnetic solids and are very soluble in water, although anions such as $[PF_6]^-$, $[I_3]^-$ and $[BPh_4]^-$ precipitate cobaltocenium ions from aqueous solution. $[Cp_2Co]^+$ is more stable than ferrocene and unaffected by strong aqueous bases and acids, and incorporation of electron-withdrawing substituents lower the reduction potentials of $[Cp_2Co]^+$, while they are raised by introduction of electron-releasing substitutents. Due to the inductive effects of alkyl substituents, the pentamethylcyclopentadienyl (Cp*) ligand is a better donor than the cyclopentadienyl group; thus, the reduction potential of $[Cp^*_2Co]^+$ (-1.48 V versus SCE in CH_2Cl_2) is 0.62 V more cathodic than that of $[Cp_2Co]^+$ (-0.86 V).

Some typical reactions of cobaltocene are shown in Scheme 4.15. Generally, electrophilic substitution reactions on $[Cp_2Co]$ lead to oxidation to cobaltocenium salts, but milder electrophiles such as alkyl halides can lead to the formation of substituted cyclopentadiene complexes. From this reaction, equimolar amounts of cobaltocenium halide and *exo* substituted η^4-cyclopentadiene compounds can be obtained. The reaction rates rise in the

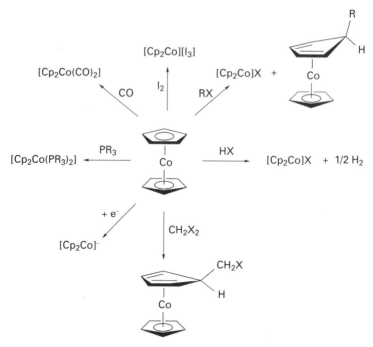

Scheme 4.15 Examples of the reactivity of cobaltocene.

order Cl < Br < I and extending the degree of halogenation will also increase the rate of reaction, e.g. $CH_3I < CH_2I_2 < CHI_3$. Two mechanisms have been postulated to account for these reactions (Scheme 4.16). Evidence for mechanism (a) comes from: (i) a charge-transfer complex is formed between cobaltocene and trichloroacetic acid; (ii) reaction rates are faster in polar aprotic solvents due to formation of a charged species; and (iii) the rate of reaction of benzyl bromide with cobaltocene is faster when electron-withdrawing substituents such as p-CN or p-CO$_2$Me are present on the benzene ring, as would be expected in the formation of a charge-transfer intermediate. It seems clear that organic radicals add to cobaltocene and are accommodated in the antibonding e_{1g}^* level. The η^5-cyclopentadienyl ligand is converted to a η^4-cyclopentadiene with change from 19 VE to 18 VE configuration and formation of a σ-bond on the cyclopentadienyl ring (Scheme 4.17).

One of the simplest radical additions is the reaction of cobaltocene with the diradical oxygen. An unusual orange dioxygen adduct can be isolated at low temperatures, **IV.8**. Reaction of cobaltocene with phosphines or carbon

(IV.8)

Mechanism (a)

$[Cp_2Co]$ + RX $\longrightarrow$ $[CpCo(C_5H_5X)]$ + R$^\bullet$

$\downarrow$

$[Cp_2Co]^+X^-$

$[Cp_2Co]$ + R$^\bullet$ $\longrightarrow$ $[CpCo(\eta^4\text{-}C_5H_5R)]$

Mechanism (b)

$[Cp_2Co]$ + RX $\longrightarrow$ $[Cp_2Co]^+[RX]^{-\bullet}$

$\downarrow$

$[Cp_2Co]^+X^-$ + R$^\bullet$

$[Cp_2Co]$ + R$^\bullet$ $\longrightarrow$ $[CpCo(\eta^4\text{-}C_5H_5R)]$

(R = CCl$_3$, Bz, CH$_2$I,
CH = CHPh, CH$_2$C $\equiv$ CH,
CH$_2$CO$_2$R, CCl$_2$CO$_2$R;
X = Cl, Br, I)

Scheme 4.16 Mechanisms for the reactions of organic halides with cobaltocene.

monoxide under pressure causes the loss of one Cp ring and formation of $[CpCo(PR_3)_2]$ or $[CpCo(CO)_2]$. Cobaltocene is also known to intercalate into inorganic solids with layer structures, i.e. SnS$_2$ or TaS$_2$ to form compounds such as $[SnS_2(Cp_2Co)_{0.31}]$. These interesting materials have beneficial magnetic and electronic properties, e.g. intercalation raises the temperature at which superconductivity is observed (O'Hare, 1996). X-ray diffraction shows

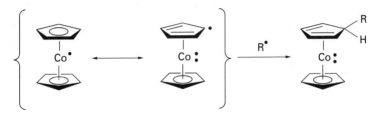

Scheme 4.17 Organic radicals adding to cobaltocene.

the cobaltocene molecules to be orientated parallel to the layers of the host lattice (Fig. 4.2).

Nucleophilic addition reactions to cobaltocenium cations have been well documented (Davies *et al.*, 1978), and simple rules have been proposed that enable the prediction of the most favourable position of nucleophilic attack on 18-electron organotransition metal cations containing unsaturated hydrocarbon ligands. Two-electron electrochemical reduction of the $[CpCo]^+$ cation in glyme with exclusion of oxygen gives the unstable orange-brown $[Cp_2Co]^-$. Protonation of this anion leads to decomposition but with alkyl halides substituted cyclopentadiene complexes may be obtained. The anion is one of the most strongly nucleophilic organometallic complexes known and the nucleophilic displacement of halide from alkyl or even aryl halides can be effected to give neutral 5-*exo*-substituted η^4-cyclopentadiene products. These species are easily prepared by nucleophilic additions to $[Cp_2Co]^+$ but $[Cp_2Co]^-$ is synthetically useful via the quantitative electrophilic addition of CO_2 to give a carboxylate from which the free acid and methylester can be readily obtained.

4.4.5 Rhodocene and iridocene

These species are much more reactive than cobaltocene and can be formed and isolated only by reduction in the absence of solvents and in matrix isolation. At room temperature, they behave as free radicals and dimerise spontaneously (Fig. 4.3). From electron paramagnetic resonance (EPR) measurements and molecular orbital (MO) calculations, it can be shown that the unpaired electron is located mainly on the C_5H_5 rings. As with $[Cp_2Co]^+$, the 18 VE ions $[Cp_2Rh]^+$ and $[Cp_2Ir]^+$ are very stable as opposed to the respective neutral monomers. Controlled one-electron electrochemical reduction of $[Cp_2Rh]^+$ in MeCN gives a dimerised dihydrofulvalene species.

4.4.6 Nickelocene

Nickelocene is the only metallocene with 20 valence electrons. It is paramagnetic and is easily oxidised to the nickelocenium ion (19 VE) which is quite unstable. Careful oxidation to the dication has been observed at low temperatures and very pure solvents but usually removal of the second electron leads to decomposition rather than formation of an 18-electron dication. An irreversible reduction to a 21-electron anion has been reported and the formation of the anion is unusually slow, suggesting a structural distortion. The electrochemical and $E_{1/2}$ values are summarised in Scheme 4.18. The monocation can be generated chemically by oxidation of nickelocene with halogens or nitric acid or by irradiation of nickelocene in the presence of

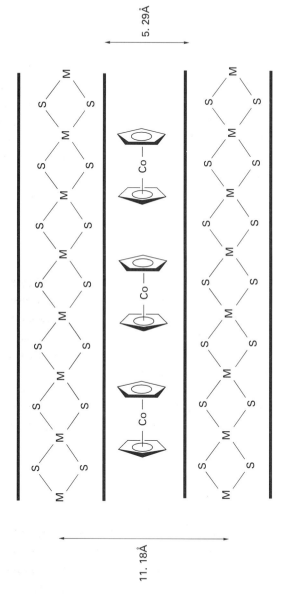

Fig. **4.2** The intercalation of cobaltocene.

5.29Å

11.18Å

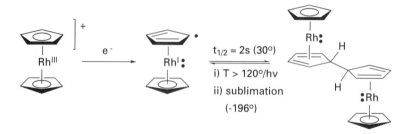

Fig. 4.3 The dimerisation of rhodocene.

$$[(C_5H_5)_2Ni]^- \underset{-1.6V}{\overset{-e^-}{\rightleftharpoons}} [(C_5H_5)_2Ni] \underset{0.1V}{\overset{-e^-}{\rightleftharpoons}} [(C_5H_5)_2Ni]^+ \underset{0.74V}{\overset{-e^-}{\rightleftharpoons}} [(C_5H_5)_2Ni]^{2+}$$

| 21 VE | -60°C 20 VE | 19 VE | 18 VE |

Scheme 4.18 The electrochemical oxidation and reduction of nickelocene.

CCl_4. Mono- and di-cationic species have been detected in the reversible electrochemical oxidation of $[(C_5H_5)Ni(C_5Me_5)]$ and $[(C_5Me_5)_2Ni]$; the latter complex can also be oxidised chemically.

The reactivity of nickelocene generally reflects the tendency of the Ni atom to obtain an 18-electron configuration (Scheme 4.19). As nickelocene is easily oxidised, there is not the extensive aromatic substitution chemistry compared to ferrocene. However, a number of ring-substituted species have been made by different methods, generally by the reaction of nickel salts with metallated substituted cyclopentadiene derivatives. Ring exchange reactions have been found with $LiC_5H_4CHPhNMe_2$ and LiC_5D_5. One or both of the cyclopentadienyl rings in nickelocene can be displaced. For example, the reaction with donor ligands can form $[NiL_4]$ complexes (Jolly and Wilke, 1974).

$$[(C_5H_5)_2Ni] + 4PPh_2Bu \longrightarrow [Ni(PPh_2Bu)_4] + 2C_5H_5 \qquad (4.3)$$

Treatment of nickelocene with lithium in the presence of cyclooctadiene leads to a displacement of the rings and a high yield formation of di(cyclooctadiene)nickel.

$$[(C_5H_5)_2Ni] + 2Li + 2C_8H_8 \longrightarrow [(C_8H_8)_2Ni] + 2LiC_5H_5 \qquad (4.4)$$

Nickelocene is also the parent compound for many mono-cyclopentadienyl nickel complexes, e.g. $[(\eta\text{-}C_5H_5)Ni(\eta^3\text{-allyl})]$ and $[(C_5H_5)NiR(L)]$. Ring protonation is observed on reacting nickelocene with secondary phosphines and thiols and also with HF or CF_3CO_2H, resulting in a cyclopentadienylnickel cation (**IV.9**) (a useful intermediate) following *exo* addition of the proton. A consequence of the 20-electron diradical structure of

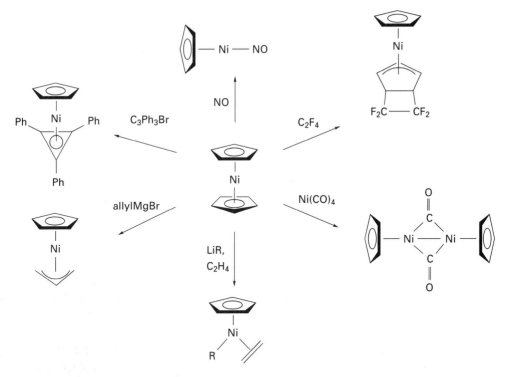

Scheme 4.19 The reactivity of nickelocene.

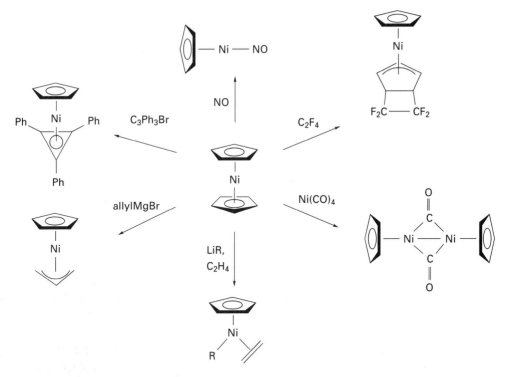

(IV.9)

nickelocene is that various electrophilic alkenes and alkynes undergo cyclo-addition to give stable products where the metal has an 18-electron configuration. Reactions with perhaloalkenes give 1,2-cycloadducts where the Cp ring has been converted into a η³-cyclopentenyl ligand with $F_2C=CF_2$ (**IV.10**). However, addition of electron-poor alkynes gives 1,3-cycloadducts,

(IV.10)

where a Cp ring has been transformed into a three-electron donor σ, π-nor-bornadiene ligand **IV.11** (where R = CO_2Me and the alkyne has added to the *endo* side of the Cp ligand).

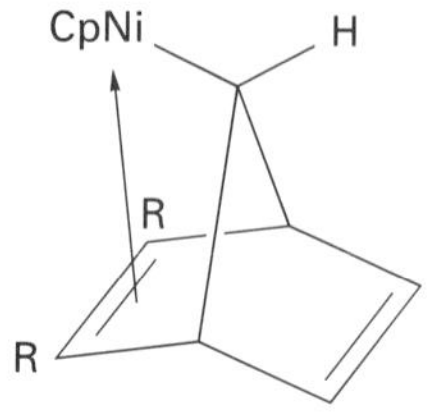

(**IV.11**)

Finally, on treating nickelocene with HBF_4 in propionic anhydride the first multi-decker sandwich complex was formed (**IV.12**). The triple-decker

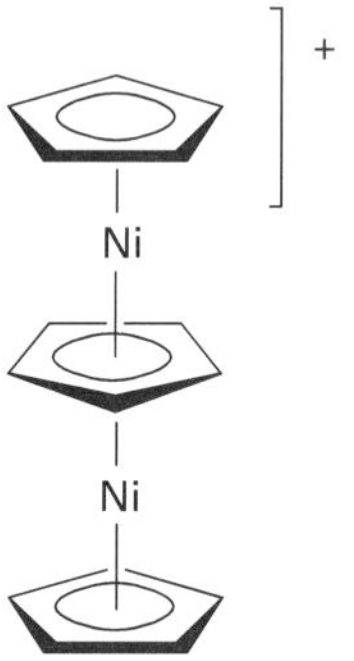

(**IV.12**)

sandwich cation $[(C_5H_5)_3Ni]^+$ contains three parallel C_5H_5 rings (conformed between eclipsed and staggered) with the nickel atoms sandwiched between them, the outer rings being closer to the nickel than the inner one. The compound can also be formed from the reaction of nickelocene with a variety of Lewis acids and generally many multi-decker sandwich compounds with larger or heterocyclic rings are known and are discussed in Chapter 5.

4.5 The chemistry of bent dicyclopentadienyl transition metal complexes

4.5.1 Dicyclopentadienyl metal hydrides

Dicyclopentadienyl metal hydrides are formed predominantly by the second (4*d*)- and third (5*d*)-row transition metals, i.e. Cp_2MH_3(M = Nb, Ta), Cp_2MH_2 (M = Mo, W), Cp_2MH (M = Tc, Re). MO considerations of bent metallocenes show that, for example, $[(C_5H_5)_2ReH)]$ (d^2) possesses two non-bonding electron pairs and acts as a base (comparable to the strength of an amine). Examples of the reactivity are shown in Scheme 4.20. The surprisingly strong basic properties were discovered on treating the hydride

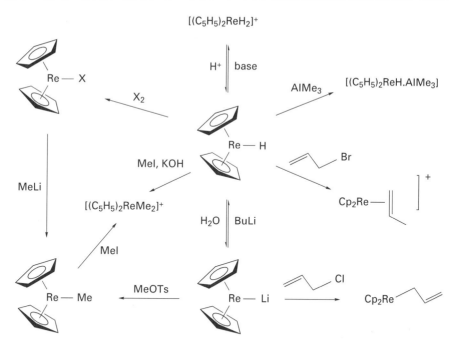

Scheme 4.20 The chemistry of dicyclopentadienylrhenium hydride.

with mineral acid when protonation at the metal centre was seen and indicated a similar basicity to that of the nitrogen atom in organic amines. This property manifests itself with Lewis acid adduct formation, i.e. with AlR_3, BF_3 and BCl_3. With halogens and organic halides the hydride gives cationic derivatives and with addition to alkynes, alkenyl derivatives are formed; it can also reduce ketones.

The reactivity of all the cyclopentadienyl metal hydrides is mainly determined by the Lewis basicity of the central metal and they can almost all be protonated:

$$[(C_5H_5)_2MH_2] \underset{OH^-}{\overset{H^+}{\rightleftharpoons}} [(C_5H_5)_2MH_3]^+ \qquad (M = Mo, W) \qquad (4.5)$$

$$[(C_5H_5)_2ReH] \underset{OH^-}{\overset{H^+}{\rightleftharpoons}} [(C_5H_5)_2ReH_2]^+ \qquad (4.6)$$

There is also a ready addition of Lewis acids, such as BF_3 (Scheme 4.21). Adduct formation of a different type occurs in the synthesis of titanocenyl boranate when Cp_2TiH and BH_3 can associate through three-centre, two-electron hydride bridges **(IV.13)**. The formation of Cp_2MH species which then dimerise is a characteristic feature of electron-poor metallocenes.

149

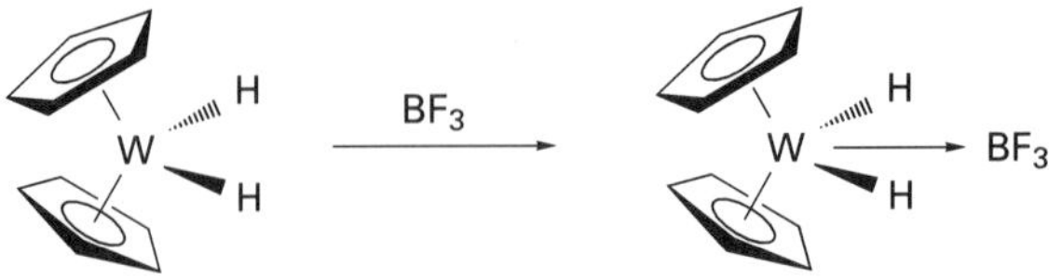

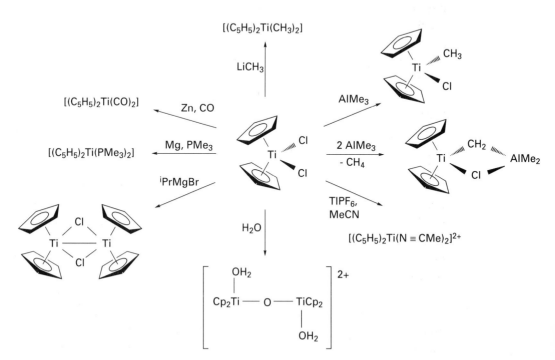

Scheme 4.21 The addition of a Lewis acid to a cyclopentadienyl metal hydride.

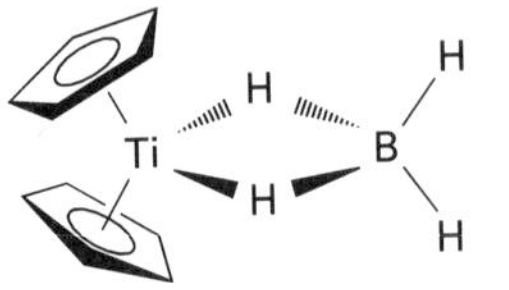

(IV.13)

4.5.2 Dicyclopentadienyl metal halides

Dicyclopentadienyl metal halides are ideal starting materials for ligand exchange and redox reactions. The variety of Cp–metal halides far outweighs that of the Cp–metal hydrides: e.g. $[Cp_2MCl_2]$ (M = Ti, Zr, Hf, V, Mo, W); $[Cp_2MBr_2]$ (M = Nb, Ta); $[Cp_2MCl]$ (M = Ti, V); $[Cp_3MCl]$ (M = Th, U). An example of a reaction scheme for a metallocene dihalide, e.g. $[Cp_2TiCl_2]$, is given in Scheme 4.22. It should be noted that only the non-Cp ligands

Scheme 4.22 The chemistry of titanocene dichloride.

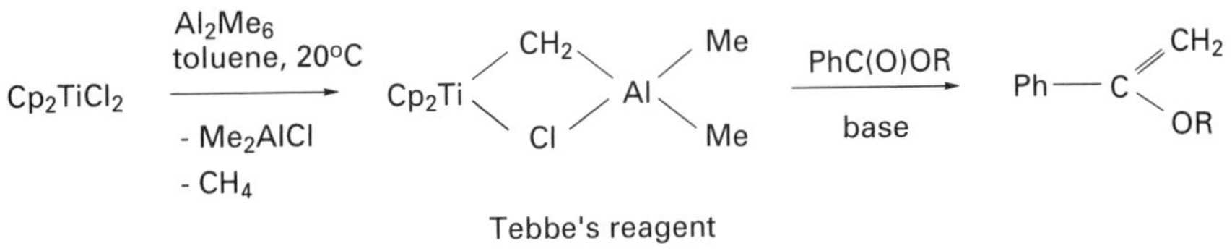

Scheme 4.23 Conversion of an ester to a vinyl ether using Tebb's reagent.

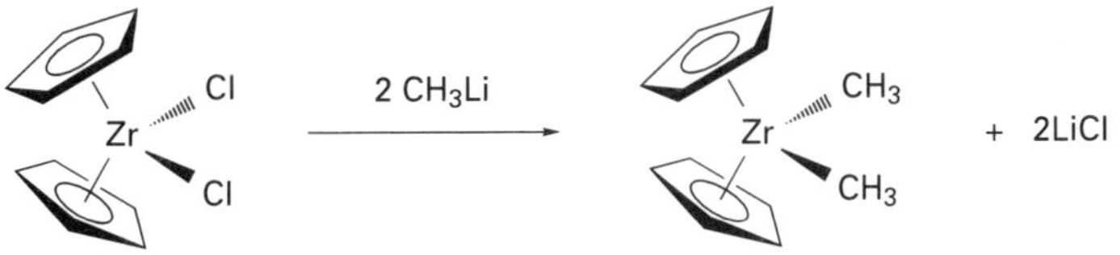

Scheme 4.24 The formation of zirconocene dimethyls.

participate in the reactions and the major reaction types observed are: (i) alkylations; (ii) reductions; and (iii) exchange reactions with donor ligands.

[Cp$_2$TiCl$_2$] is a convenient starting material for the preparation of many other organometallic compounds of titanium and has found application as a catalyst or catalyst component for a wide variety of hydrometallation and carbometallation reactions as well as numerous other transformations involving Grignard reagents. In organic synthesis a titanocene halide derivative, 'Tebbe's reagent' (Tebbe et al., 1978) has proved to be useful, particularly as an alternative to classical Wittig reagents (Scheme 4.23). The actual methylene transfer agent is the alkylidene complex [Cp$_2$Ti=CH$_2$] formed from Tebbe's reagent by loss of Me$_2$AlCl, and for the conversion shown in Scheme 4.23 conventional Wittig reagents are not suitable. Carboxylic esters and lactones can be converted in good yields to the corresponding enol ethers by treatment with Tebbe's reagent and it also gives good results with ketones. Of course in the field of Ziegler–Natta catalysis, these metallocenes have proved to be useful alkene polymerisation catalyst systems; this will be detailed in Chapter 6. In general, dicyclopentadienyl metal halides are excellent starting materials for the synthesis of Cp–metal alkyls and Cp–metal aryls (Scheme 4.24). The mixed Cp–metal chloride hydride [(C$_5$H$_5$)$_2$ZrCl(H)] has become useful in organic synthesis as 'Schwartz's Reagent'. Using this and via hydrozirconation and subsequent oxidative decomplexation, alkenes can be converted into substituted alkanes (see later) (Schwartz and Labinger, 1976).

The reactivity of the various Cp–metal halides is once again governed by the central metal and attainment of 18 VE structures. As seen in Chapter 3, after rehybridisation of the frontier orbitals a group 6 metallocene dihalide uses only two of the available three orbitals to bind the halides, leaving the third orbital as a lone pair pointing between the two substituents (and it can

(IV.14)

also take part in any back donation) (**IV.14**). On the other hand, group 4 metals only bind two ligands and having only four valence electrons their maximum oxidation state is M(IV). This leaves the 16-electron metallocene dihalide with an empty orbital (**IV.15**) rather than a filled one (IV.14). The

(IV.15)

differences in the chemistry of the group 4 and 6 metallocene compounds can therefore be accounted for. The former act as Lewis acids and tend to bind to π-basic ligands such as —OR but the latter act as Lewis bases and bind π-acceptor ligands such as ethylene. π-Bonding can be either stabilising if the ligand carries low-lying acceptor orbitals or destabilising if the ligand is composed of relatively high-lying donor orbitals. For example, in d^2 [(C$_5$H$_5$)$_2$MX$_2$] (X = halogen, OR, SR or NR$_2$) and as the a$_1$ orbital is occupied, a lengthening of the MX bond is seen. Trends have been noted in [Cp$_2$MoCl$_2$]$^+$ (d^1) (Mo—Cl = 0.239 nm) versus [Cp$_2$MoCl$_2$] (d^2) (Mo—Cl = 0.247 nm) and in [(C$_5$H$_4$CH$_3$)$_2$TiCl$_2$] (d^0) (Ti—Cl = 0.236 nm) versus [(C$_5$H$_4$CH$_3$)$_2$TiCl$_2$] (d^1) (Ti—Cl = 0.24 nm) and are identified as being due to π-anti-bonding.

4.5.3 Dicyclopentadienyl metal alkenes

These complexes are known for several of the early transition metals and are thought to be important intermediates in many reactions. [(C$_5$H$_5$)$_2$M(C$_2$H$_4$)] systems are known for Mo and W and the filled π-orbital of ethylene can act as a donor orbital and interacts strongly with the fragment 2a$_1$ orbital (Fig. 4.4) (Lauher and Hoffmann, 1976).

The empty alkene π* orbital is of b$_2$ symmetry and stabilises the metal b$_2$. This backbonding is very beneficial because there is a high energy metal donor b$_2$ orbital hybridised towards the ligand that also has an excellent overlap with the alkene π* orbital. These ethylene complexes possess one non-bonding orbital (1a$_1$) and are thus readily protonated to give [(C$_5$H$_5$)$_2$M(C$_2$H$_4$)H]$^+$ (M = Mo and W) species which are analogous to the neutral niobium and tantalum hydrides [(C$_5$H$_5$)$_2$M(C$_2$H$_4$)H]. Some complexes are known to insert

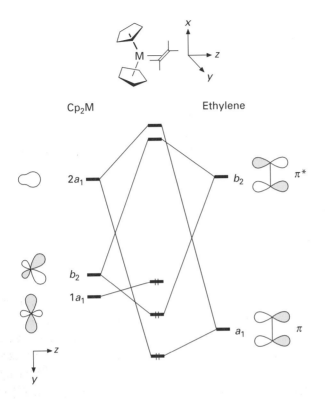

Fig. 4.4 The simple bonding picture of dicyclopentadienyl metal alkenes. (Lauher and Hoffmann (1976), with permission.)

an alkene into a metal hydride bond directly. The 'hydrozirconation' reaction has been developed which involves the reaction of [Cp$_2$ZrClH] with alkenes and, via anti-Markovnikov H-addition, immediate alkene insertion occurs (Fig. 4.5). The [Cp$_2$ZrClH] molecule has one open co-ordination site to co-ordinate an alkene. It is a d^0 complex which results in no stability being added to the metal–alkene bond due to donation into the empty π* orbital. When co-ordinated, the alkene and hydride ligands are in close proximity and the insertion can take place readily, driven by the energy gained from forming a new C—H bond.

Following on from hydrozirconation reactions, Zr(IV) alkyl complexes will readily react with CO to give carbonyl insertion into the metal alkyl bond

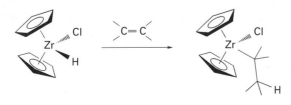

Fig. 4.5 An example of a hydrozirconation reaction.

153

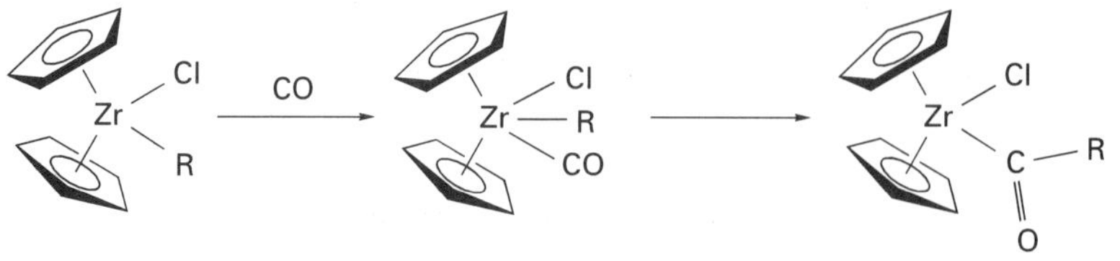

Scheme 4.25 Carbonyl insertion with zirconocene alkyls.

(Scheme 4.25). The formation of acyl derivatives from alkenes and CO is similar to the reactions which are part of the cobalt hydroformylation process. The insertion reactions can be understood as the dialkyl complexes are d^0 with one low-lying empty orbital, and the complex will therefore react with a good σ-donor with ligand attack being along the y axis of the molecule. The resulting species (again d^0) has three σ-bonding ligands and, importantly, the normal π-backbonding into the CO π^* orbitals is not a possibility here. The angle between the alkyl group and the CO ligand is less than 90° and, consequently, there is good overlap between the CO π^* and the alkyl σ-donor orbital and the insertion or migration reaction can readily take place.

4.6 The chemistry of cyclopentadienyl lanthanide and actinide complexes

There is considerable ionic character in these compounds (especially the lanthanides) which tends to dominate the chemistry. As such, they show an instantaneous and quantitative reaction with ferrous chloride in tetrahydrofuran solution to give ferrocene. The tricyclopentadienyl species are formally nine co-ordinate but readily form ten co-ordinate base adducts $[(C_5H_5)_3Ln \cdot L]$. These adducts have featured ammonia, THF, triphenylphosphine and pyrazine as donors. The $[Cp_3Ln \cdot L]$ species tend to be significantly more soluble in organic solvents than the base-free compounds. The actinide species act in a similar fashion, i.e. Cp_3U is a strong Lewis acid and forms adducts with a variety of Lewis bases such as THF, cyclohexyl isocyanide and nicotine.

The organoactinide compound uranocene is extremely reactive towards oxygen and often enflames in air. U^{4+} species have two electrons outside the noble gas core and these are readily lost to form the stable $+6$ oxidation state. Controlled oxidation results in precipitation of uranium oxides. It reacts only slowly with water or acetic acid and hydrolysis is very gradual. Reaction with a base can be considered to occur with the lowest vacant orbitals of the substrate. For uranocene, the lowest vacant orbitals are $5f$ orbitals on the metal and so these reactions liberate cyclooctadiene dianions which then protonate. An electrochemical study of uranocene has provided evidence for a uranocene cation (4.7). This species is unstable and quasi-reversible cyclic voltammograms could only be observed at subambient temperatures.

$$[(C_8H_8)_2U] \underset{}{\overset{-e^-}{\rightleftharpoons}} [(C_8H_8)_2U]^+ \longrightarrow products \qquad (4.7)$$

Unlike the situation with ferrocene, all reactions with strong electrophiles result in the complete decomposition of uranocene. Electrophilic reagents usually react with the highest occupied electrons in a substrate. In uranocene these are the two $5f$ electrons on the metal, therefore reaction with electrophilic reagents is at uranium with resultant decomposition of the complex. Bromine or iodine in CH_2Cl_2 give insoluble brown amorphous solids as 1 : 1 complexes in an irreversible reaction. Generally, the chemistry of uranocene itself is limited when compared to other metallocenes. This can be rationalised on the basis that both the HOMOs and LUMOs are predominantly metal orbitals. Therefore, reactions generally occur at the metal with consequent decomposition of the metal–ligand bonds.

4.7 The chemistry of cyclopentadienyl main group element complexes

The chemistry of these compounds is largely unknown but general principles of reactivity are being developed (Jutzi, 1990). The chemistry is determined by the acceptor characteristics of the particular central atom, by the role of the lone pair electrons (if present) and by the functionality of the cyclopentadienyl and other ligands.

4.7.1 The cyclopentadienyl ligand as a leaving group

Reactions with electrophiles

The most widespread reaction in main group cyclopentadienyl chemistry involves cyclopentadienyl transfer to other elements from the p, d and f blocks.

$$R_5C_5M + M'X \longrightarrow R_5C_5M' + MX \qquad (4.8)$$

Cyclopentadienyl compounds of the alkali metals, magnesium and thallium are often used for cyclopentadienyl transfer. Due to the varying nucleophilicity and tendency for redox processes, the choice of best transfer agent depends on the electronic requirements at the electrophilic centre M'. Indeed, in some situations, use of cyclopentadienyl transfer agents with low nucleophilicity is preferred, e.g. transfers at boron centres. In the metallocene chemistry of germanium, tin and lead, Cp—M bond cleavage by electrophiles EX has been utilised as a route to half-sandwich complexes. In the processes, oxidative addition and reductive elimination reactions may well be involved, e.g.

$$[(R_5C_5)_2M] \xrightarrow[-R_5C_5E]{+EX} [(R_5C_5)MX] \tag{4.9}$$

$$[(R_5C_5)M]^+X^-$$

(M = Ge, Sn, Pb)

Reactions with nucleophiles

The reaction of cyclopentadienyl π-complexes with strong anionic nucleophiles results in an overall nucleophilic substitution featuring the cyclopentadienide ligand as the leaving group.

$$[(R_5C_5)M] + Nuc^- \longrightarrow NucM-(R_5C_5)^- \tag{4.10}$$

For example, decamethylstannocene reacts with [bis(trimethylsilyl)methyl]-lithium to give bis[bis(trimethylsilyl)stannylene and lithium pentamethylcyclopentadienide (4.11). When using methyllithium the formation of the short-lived intermediate [bis(pentamethylcyclopentadienyl)methyltin]lithium species has been shown by trapping experiments (4.12).

$$[(Me_5C_5)_2Sn] + 2[(Me_3Si)_2CHLi] \longrightarrow [\{(Me_3Si)_2CH\}_2Sn] \tag{4.11}$$
$$+ 2[Me_5C_5Li]$$

$$[(Me_5C_5)_2Sn] + MeLi \longrightarrow [(Me_5C_5)_2Sn(Me)Li] \tag{4.12}$$
$$\text{then} \qquad \xrightarrow[-LiX]{+RX} [(Me_5C_5)_2Sn(Me)R]$$

In general, displacement of a cyclopentadienyl ligand leads to structures with open coordination positions, i.e. with high reactivity and limited stability. Nucleophilic substitution of cyclopentadienide ligands is therefore a potentially very important synthetic method in the chemistry of low-valent p-block elements.

4.7.2 Single-electron transfer reactions

Unlike the bonding in most of the cyclopentadienyl complexes of the d-block elements, the HOMOs and LUMOs in p-block element complexes are concentrated mainly on the π-ligands and not on the metal. Therefore in single-electron reduction or oxidation processes, the attack takes place at a cyclopentadienyl ligand and, due to the stability of the cyclopentadienyl radical or anion, cyclopentadienyl–element bond cleavage results.

$$[(R_5C_5)E] \quad \begin{cases} \xrightarrow{-e^-} R_5C_5^- + E \\ \xrightarrow{-e^-} R_5C_5^{\cdot} + E^- \end{cases} \tag{4.13}$$

As a result, stable radical cations or anions in cyclovoltammetric studies of sandwich or half-sandwich compounds have not been detected. Under mass spectrometric conditions, the relevant molecular ions readily fragment with loss of a cyclopentadienyl moiety. Treatment of cyclopentadienyl complexes with single-electron oxidising or reducing agents leads to decomposition. For example, group 14 metallocenes react with arene radical anions to give the cyclopentadienide anion and the corresponding group 14 element in its zero oxidation state. Thus, it is difficult to find synthetic strategies for main group cyclopentadienyl complexes that are based on reduction processes. Appropriate conditions have to be chosen to avoid further reduction of the desired compounds and indeed some syntheses start from higher-valent precursors.

Part B: Spectroscopic properties

4.8 Vibrational (infrared and Raman) spectroscopy of metallocenes

Infrared and Raman spectroscopies have proved to be important in the analysis of cyclic polyenyl metal sandwich species with particular use in elucidating covalent or ionic M–ring bonds and distinguishing between centrally and peripherally co-ordinated rings. The non-complex-bonded cyclic systems C_nH_n possess symmetry group D_{nh} provided they have a plannar, regular polygonal structure. The bands belonging to the irreducible representations and the infrared- and Raman-active vibrations are shown in Table 4.1 (Fritz, 1964).

It is clear that for C_nH_n systems with $n < 4$, regardless of the size of the ring, four infrared-active normal vibrations and seven Raman-active ones are to be expected. The frequencies of vibrations of the ligands themselves can be distinguished from those of the complex when the ligand rings are assumed to be rigid discs. Theoretical approaches can be used to predict the number of vibrations in a particular species. The total symmetry of a complex can be determined except for the presence of a centre of symmetry. For example, an analysis of the vibrational spectrum of ferrocene reveals that the molecule has either D_{5d} or D_{5h} symmetry but a decision between these two cannot be made. The number of vibrations for sandwich compounds can

Table 4.1 Irreducible representations and numbers of IR- and Raman-active normal vibrations of cyclic D_{nh} systems.

D_{3h}	D_{4h}	D_{5h}	D_{6h}	D_{7h}	D_{8h}
IR-active					
$\left.\begin{array}{l}A_2'' \\ E'\end{array}\right\}4$	$\left.\begin{array}{l}A_{2u} \\ E_u\end{array}\right\}4$	$\left.\begin{array}{l}A_2'' \\ E_1'\end{array}\right\}4$	$\left.\begin{array}{l}A_{2u} \\ E_{1u}\end{array}\right\}4$	$\left.\begin{array}{l}A_2'' \\ E_1'\end{array}\right\}4$	$\left.\begin{array}{l}A_{2u} \\ E_{1u}\end{array}\right\}4$
Raman-active					
$\left.\begin{array}{l}A_1' \\ E' \\ E''\end{array}\right\}6$	$\left.\begin{array}{l}A_{1g} \\ B_{1g} \\ B_{2g} \\ E_g\end{array}\right\}7$	$\left.\begin{array}{l}A_1' \\ E_1'' \\ E_2'\end{array}\right\}7$	$\left.\begin{array}{l}A_{1g} \\ E_{1g} \\ E_{2g}\end{array}\right\}7$	$\left.\begin{array}{l}A_1' \\ E_1'' \\ E_2'\end{array}\right\}7$	$\left.\begin{array}{l}A_{1g} \\ E_{1g} \\ E_{2g} \\ E_{3g}\end{array}\right\}7$

be predicted, i.e. for ferrocene there is a total of 21 atoms and so $3N - 6 = [3(21) - 6] = 57$ normal vibrations can be expected. Table 4.2 gives the spectral expectations for $[(C_5H_5)_2M]$ compounds with D_{5d} or D_{5h} symmetry. Many detailed investigations of the IR and Raman spectra of the iron group metallocenes have been made. There is still some disagreement over the assignment of some of the fundamental frequencies in the spectra of ferrocene and ruthenocene, but generally accepted assignments and ordering are detailed in Table 4.3 (Rosenblum, 1965). The variation in IR spectral bands for a series of mono-, di- and tricyclopentadienyl metal compounds has been discussed. The first four bands in Table 4.3 do not show any pronounced or regular variation but the C—H out-of-plane bending mode shows a progressive shift to lower frequency as the metal–ring bond becomes more ionic, i.e. at $\sim 800 \ \mathrm{cm}^{-1}$ for the covalent Fe, Ru and Os metallocenes but at $701 \ \mathrm{cm}^{-1}$ for potassium cyclopentadienide. Coupled with this, the intensity of the band near $1100 \ \mathrm{cm}^{-1}$ which is strong in covalent compounds was observed to decrease noticeably in the more ionic species.

Raman studies with polarisation measurements and patterns have shown that the force constant of the Cp—M bond shows a systematic increase in the

Table 4.2 Spectral expectations for $[(C_5H_5)_2M]$ with D_{5d} or D_{5h} symmetry.

D_{5d}	D_{5h}	$[(C_5H_5)_2M]$	C_5H_5	C_5	C—H	M—Ligand
A_{1g}	A_1'	4	3	1	2	Symmetrical stretch
A_{1u}	A_1''	2	1	0	1	Torsion
A_{2g}	A_2'	1	1	0	1	–
A_{2u}	A_2''	4	3	1	2	Assymmetrical stretch
E_{1g}	E_1''	5	4	1	3	Symmetrical ring tilt
E_{1u}	E_1'	6	4	1	3	Assymmetrical ring tilt, deformation
E_{2g}	E_2'	6	6	3	3	–
E_{2u}	E_2''	6	6	3	3	–

Table 4.3 Spectral bands and assignments of iron group metallocenes.

	Ferrocene (cm^{-1})	Ruthenocene (cm^{-1})	Osmocene (cm^{-1})
C—H stretch	3085	3100	3095
C—C stretch	1411	1413	1405
Ring deformation	1108	1103	1096
C—H deformation	1002	1002	995
C—H out-of-plane bend	811	806	819
Ring tilt	492	528	428
M—ring stretch	478	446	353
M—ring bend	170	185	

order $Cp_2Fe < Cp_2Ru < Cp_2Os$. Coupled with this, a slight shift of the C—C stretching bands and the C—C breathing bands to a longer wavelength seems to indicate a weakening of the bonding between the ring carbons proceeding from ferrocene to osmocene; this is because of the stronger bonding between the rings and the metal atom. Additionally, the position of the ring—Ru—ring symmetric stretching frequency at 332 cm^{-1} in ruthenocene compared to that in ferrocene at 303 cm^{-1} has been interpreted as being due to the tighter ring–metal bonding in the former complex. The two absorption peaks exhibited by ferrocene at around 1100 and 1000 cm^{-1} have proved invaluable in assigning structures to poly-substituted ferrocenes. A rule concerning these bands states 'ferrocene derivatives in which one ring is unsubstituted exhibit absorptions near 1100 and 1000 cm^{-1} whilst those having both rings substituted lack these absorptions'. This also seems to apply to ruthenocene and osmocene derivatives and very occasionally bands in this region have been observed arising from substituents of heteroannular derivatives. Therefore, the absence of any absorptions in this region is of more help in identifying a structure.

As mentioned earlier, ionicity in M–ring bonding can be detected from vibrational spectra. For instance, magnesocene exhibits very similar IR and Raman spectra to that of the $C_5H_5^-$ anion, i.e. bands at 1000 and 800–760 cm^{-1} are very strong in the IR spectra whilst the bands at 1100 cm^{-1} are very weak. Assignation of v_{Mg-Cp} at 219 cm^{-1} lends support to the thought that the M—Cp bond in magnesocene is mainly ionic in nature (although the question of covalent contribution to the bonding is still open to discussion on the basis of other data). The IR spectra of Cp_2M (M = Ba, Ca, Sr) have been interpreted as an indication of essentially ionic species and down the group the frequencies of the out-of-plane C—H ring vibrations decrease in line with an increase in the ionic character of the M—Cp bond. This is also the case for Cp_2V, where on the basis of the vibrational spectra D_{5d} symmetry is assigned together with partial ionic character of the V—Cp bond.

The IR and Raman spectra of nickelocene have been studied in detail. Both spectra resemble that of ferrocene with the main difference being that the nickel–ligand vibrations are found at lower frequency and that the order of Cp—ring tilt and the metal—Cp stretch vibrations is reversed, being at 208 and 255 cm^{-1} for nickelocene and at 390 and 306 cm^{-1} for ferrocene. A typical assignment of the Raman spectrum is: ν_{C-H} 3108(w), 3102(w), 3085(w); ν_{C-C} 1426(w), 1336(m); δ_{C-H} 1219(vw), 1002(w); ω_{pulse} 1109(s); ρ_{C-H} 1053(m), 780(m); ν_{M-Cp} 255(s); ν_{tilt} 208(m).

Vibrational spectroscopy can also distinguish between centrally and peripherally bonded Cp rings. The IR spectrum of a centrally bonded cyclopentadienyl ring features only two allowed C—H stretching bands because of the high symmetry (C_{5v}) of the ring. Peripherally bonded rings obviously have a much lower symmetry (C_s) with all of the C—H stretching frequencies being different and therefore allowed. The complexity of the IR spectra allow assignation of the rings as either centrally or peripherally bonded, i.e. the vibrational spectrum of Cp$_2$Hg is too complex to be consistent with a centrally bonded structure and so must feature a σ-bonded species. The localised Cp–ring differs from the delocalised Cp analogue via possession of two carbon–carbon double bonds and an aliphatic sp^3 C—H bond. The double bonds are easily identified from a stretching frequency of around 1600 cm^{-1}. The C—H stretching vibrations for the vinyl C—H groups are observed at c. 3100 cm^{-1} and the stretching frequency of the methylene C—H at c. 2900 cm^{-1}. The lowest energy C—H stretch is around 2900 cm^{-1} for M = H but as H is replaced by less electronegative 'metals', delocalisation of the rings increases as shown by the value of the angle between the ring plane and the M—C bond. The frequency of the lowest energy C—H stretching band changes to 2932 cm^{-1} for Ge, 2970 cm^{-1} for Sn and 3013 cm^{-1} for Hg in dicyclopentadienylberyllium and the frequency of the lowest energy band in ionic, centrally-bonded cyclopentadienyl compounds can be as low as 3027 cm^{-1}. The presence or absence of an aliphatic sp^3 C—H stretching band cannot unequivocally be used as a diagnostic tool to decide between central and peripheral structures. The frequency does though indicate the degree of delocalisation within the ring in compounds which are known to contain the peripherally bonded Cp—M group. Beryllocene lacks any bonds in the 1600–1500 cm^{-1} region which are normally assigned as carbon–carbon double bond stretching vibrations. The structure indicates that although one ring is peripherally bonded there are no full double bonds in the ring and the C—H bond stretches therefore resemble those of centrally bonded derivatives.

IR and Raman spectra of tricyclopentadienyl complexes of the lanthanides show vibrations typical of a η^5-co-ordinated ligand and that the intensity distribution reflects the considerable ionic character of the M—Cp bonds. The lines of vibrations allowed in the Raman spectra as for the free $C_5H_5^-$ ion are strong, whereas those forbidden such as β_{C-H} at 1000 cm^{-1} and ν_{C-C} at $c.$ 1440 cm^{-1} are weak. Similar intensity distributions in the Raman spectra have been reported in Cp_2Mg and Cp_2Mn, species also featuring considerable ionic $Cp-M$ bonding.

4.9 NMR (^1H and ^{13}C) spectroscopy of metallocenes

Nuclear magnetic resonance is the most widely applied tool in the study of metal sandwich compounds and organometallic species in general, giving information on nuclear structures in solution, as liquids and gases and, more recently, in the solid state. Proton magnetic resonance for diamagnetic organotransition metal compounds is observed in the chemical shift range $25 < \delta\ ^1H > -40$ ppm however for sandwich complexes the range is much more localised (Fig. 4.6). Viewed as aromatic systems, the chemical shifts in these species are anomalous since they lie at a much lower frequency than would be anticipated assuming a π-orbital ring current, typical of aromatic systems, to be associated with each ring. Proton resonance appears at a lower frequency of $\sim$ 3ppm compared to benzenoid compounds. Whilst a drift of electrons from the metal atom might provide some additional shielding, its magnitude would hardly be large enough to account for the observed chemical shifts. The effect of the metal atom itself obviously should not be neglected and its shielding influence must play some part. Protons which are fixed to metal-bonded carbon atoms display a small span of chemical shifts. The co-ordination shifts $\Delta\delta\ ^1H$ of η-C—H units are ruled by local effects (partial charge, rehybridisation) and by non-local effects (neighbouring group anisotropy of the metal, co-ordination-induced perturbation of diamagnetic ring currents) and usually lie in the range $1 < \Delta\delta\ ^1H < 4$ ppm and are always negative (i.e. shifts to low frequency). If symmetrically bonded, the sandwich structures exhibit very simple spectra due to the equivalence of the C and H atoms on the organic ligand via the free rotation of these rings, i.e. 'fluxionality' (see later). Clearly, the spectra will be more complicated with unsymmetrical bonding (e.g. η^1 of a C_5H_5 ring) or following substitution of the rings.

Despite having a low natural abundance (1.1%) and relatively low receptivity of the ^{13}C nucleus, ^{13}C NMR has become an important spectroscopic tool working separately or in tandem with ^1H NMR for several reasons: (i) $^{13}C-\{^1H\}$ spectra (proton-decoupled) in the absence of other magnetic nuclei display sharp singlets. As signals are spread widely in ^{13}C NMR (see Fig. 4.7) as opposed to ^1H NMR, ^{13}C spectra are particularly useful for analysis of elaborate compounds, mixtures of isomers and in the recognition of fine, structural details; (ii) coupling constants $J(^{13}C, ^1H)$ together with ^1H NMR

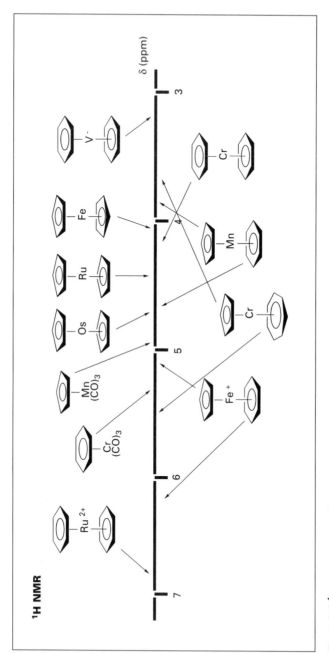

Fig. 4.6 δ ^{1}H values for some diamagnetic metal sandwich compounds.

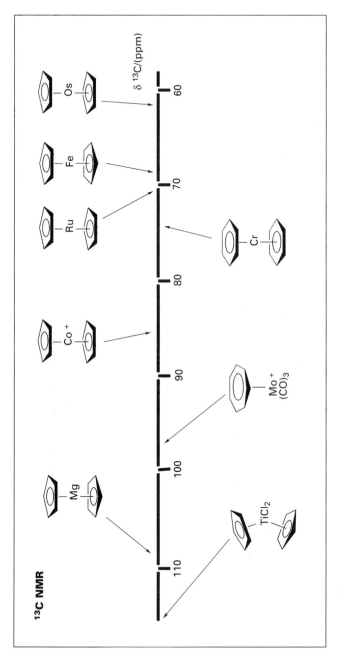

Fig. 4.7 δ ^{13}C values for some diamagnetic metal sandwich compounds.

values $J(^1H, {}^1H)$ help elucidate ligand structure and the hybridisation states of carbon atoms. The coupling constants $^1J_{M-C}$ contain information concerning the nature of the metal–carbon bond, i.e. for M—C σ-bonds considerably higher values of $^1J_{M-C}$ are observed as compared to M—C π-bonds due to the increased s-character in the σ-bonds; (iii) ^{13}C NMR is also particularly well-suited to the study of exchange processes. The range of chemical shifts in ^{13}C NMR spectra is about five times as large as in ^1H NMR and so considerably faster processes may be studied by ^{13}C NMR over the total dynamic range. In addition, the simple ^{13}C NMR spectra mean that bandshape analyses are easier and less cumbersome than for the respective ^1H NMR spectra.

Systematic surveys of transition metal cyclopentadienyl complexes have presented some general trends: (i) there appears to be a linear correlation between ^1H and ^{13}C chemical shifts for the η^5-C_5H_5 ligand suggesting that similar effects contribute to both chemical shifts; (ii) there are pronounced low frequency shifts and an increase in $^1J_{C-H}$ on complexation. The increased shielding is attributed to greater electron density at the ligand nuclei. The enhanced electron density is in turn associated with formation of a 'covalent delocalised multicentre bond'. The larger values of the one bond C—H couplings were rationalised on the basis of the 'attraction' of the carbon p orbitals to the metal d orbitals and consequent increased s electron density in the C—H bond; (iii) the δ ^{13}C of the Cp ligand is less shielded in the cyclopentadienyl derivatives of the titanium subgroup than in the $C_5H_5^-$ anion; (iv) for all other d-block metals, the η^5-C_5H_5 carbons are more effectively shielded than those in the $C_5H_5^-$ anion; (v) a maximum in ligand nuclei shielding occurs in the Fe subgroup.

In bent sandwich structures $[Cp_2MX_2]$ (M = Ti, Hf, Zr), the ^1H NMR chemical shifts of the Cp protons are indicative of freely rotating π-bound cyclopentadienyl groups, except when an inter-ring bridge prevents free rotation of the π-bound rings. Single resonances are observed in species such as Cp_2Zn and Cp_2Hg (δ 5.84, $J(^{199}Hg-{}^1H)$ 66 Hz in THF) with all the protons being equivalent on the NMR timescale down to temperatures of −100°C, with only a slight broadening. This is also the case for essentially ionic systems such as Cp_2Ca, with ^1H NMR spectra again showing the expected singlet resonances at temperatures down to −100°C.

Care should be taken when dealing with paramagnetic compounds as they can give rise to chemical shifts of several hundred ppm and the detection depends on certain restrictions imposed by the electron spin relaxation time. For instance, the ^1H and ^{13}C NMR spectra of nickelocene exhibit large chemical shifts {δ(^1H), +254.8, δ(^{13}C), −1300 ppm} with respect to ferrocene as a result of hyperfine coupling between the unpaired electron spin and the nuclear spin. This effect is due to the transfer of positive spin density from the metal to the ring carbon atoms and is accompanied by a change in the spin on transfer to the protons (Jolly and Wilke, 1974). There is still equivalence of the rings brought about by the free rotation although this is masked by a broadening of the lines due to the presence of unpaired electrons. This is also the case for cobaltocene which possesses a paramagnetic $\delta(^{13}C)$ shift of -541 ppm relative to ferrocene. By preparing the fully deuterated $[(\eta\text{-}C_5D_5)_2Co]$, an isotope shift $\Delta\delta(^{13}C)$ of $+7.4$ ppm has been found which is much larger than normally observed in diamagnetic molecules.

4.9.1 Fluxionality

Fluxional compounds differ from others in possessing more than a single configuration representing an energy minimum. Several such minima may be present and accessible with ordinary thermal energies. If the barrier between the various possible configurations is thermally accessible, the system is fluxional. The 'timescale' of a spectroscopic technique is important in observing a fluxional process and generally the higher the frequency of the technique the more precise are the molecular co-ordinates that can be obtained. Nuclear magnetic resonance techniques, with timescales of 10^{-1} to 10^{-9} s, have proved invaluable in the study of fluxional molecules. There are two limiting cases which depend on interconversion rates. One features a molecule where there are two distinct proton environments but no interconversion and gives two resonance lines separated by a difference (Δ) that shows the difference between the two sites. The other case shows that interconversion takes place with a large frequency compared to the separation Δ and the spin of the proton hardly loses phase from the time it is in a particular site until it is at that site again. Therefore, the proton response is that it is in an average environment so producing a single line at an average frequency. In an experiment, a molecule can be 'frozen' in a particular configuration thereby producing a distinct spectrum with individual lines for each site. Then by encouraging the fluxional behaviour of the molecule (normally by raising the temperature) the formerly split lines collapse to produce a single, average signal.

Fluxionality is characteristic of certain classes of organometallic compound and this phenomenon of stereochemical non-rigidity is seen in compounds containing conjugated cyclic polyenes such as cyclopentadienes, cycloheptadienes and cyclooctatetrenes. One obvious type of fluxional process is called '*internal rotation*' and involves free rotation of the polyene rings. For example, at room temperature the two cyclopentadienyl rings in ferrocene rotate rapidly relative to each other (Fig. 4.8). In general, the $\eta^5\text{-}C_5H_5$ ligands rotate very rapidly around the metal–Cp (centroid) axis

Fig. 4.8 Internal rotation in ferrocene.

giving rise to the typical sharp singlets in ^{1}H and ^{13}C NMR spectra (which is a feature of high diagnostic value in complex characterisation). In the solid state, the two Cp ligands in a metallocene complex may adopt staggered or eclipsed configurations but the rotational barrier which the Cp rings experience is very small ($\sim$ 8–9 kJ mol^{-1}) and it is practically impossible to 'freeze out' the C_5H_5 ring rotation in solution. To date, activation energies for rotation have only been determined in the solid state or from relaxation measurements and these energies (Table 4.4) appear to be dominated by intermolecular interactions rather than internal electronic constraints (Mann, 1982). In lightly substituted ferrocenes, e.g. [(η-C_5H_5)Fe(η-C_5H_4R)] (R = butyl, pentyl or CHO) and related sandwich complexes, e.g. [(η-C_5H_5)M(η^4-cod)] (M = Rh, Ir; cod = cyclooesta-1,5-diene) C_5H_5 ring rotation rates are again only measurable from NMR T_1 data. However, when the Cp rings carry numerous bulky substituents, the barriers to rotation increase as expected and rotation rates are reduced sufficiently to lie within the timescale of NMR chemical shift modulation. Examples of activation barriers ($\Delta G\ddagger$) to ring rotation for substituted metallocenes are given in Table 4.5. Compari-

Table 4.4 Activation parameters for the rotation of η-cyclopentadienyl groups.

Compound	E_a (kJ mol^{-1})
[(η-C_5H_5)$_2$TiCl$_2$]	2.1±0.8*
[(η-C_5H_5)$_2$Fe]	7.5±0.8*
	9.6*
	8.4*
[(η-C_5H_5)Fe(η-C_5H_4Et)]	14.2*
	34.3*†
[(η-C_5H_5)Fe(η-C_5H_4Bu^n)]	c. 7‡
[(η-C_5H_5)Fe(η-C_5H_4Pr^i)]	33.1*
	39.7*†
[(η-C_5H_5)Fe(η-C_5H_4Bu^t)]	35.1*
	39.7*†
[(η-C_5H_5)$_2$Ru]	9.6±0.8*
[(η-C_5H_5)$_2$Co]	7.5±0.8*
[(η-C_5H_5)$_2$Co]$^+$	7.5±0.8*
[(η-C_5H_5)$_2$Ni]	7.5±0.8*

* Solid state.
† Substituted ring.
‡ In solution.

Table 4.5 Measurements in solution of activation barriers ($\Delta G\ddagger$) to ring rotation for substituted ferrocenes* and ruthenocenes†.

Substituents	$\Delta G\ddagger$ (kJ mol^{-1})
n-Bu*	8.28§
—CMe$_2$Et*	8.57§
1,1′,3,3′-(SiMe$_3$)$_4$*	46.0¶
1,1′,3,3′-(CMe$_3$)$_4$*	54.81¶
	55.6**
1,1′,3,3′-(CMe$_3$)$_4$†	39.7**
1,1′,3,3′-(CMe$_2$Et)$_4$*	56.7**
1,1′,3,3′-(CMe$_2$Et)$_4$†	45.7**
1,1′,2,2′,4,4′-(SiMe$_3$)$_6$*	46.02¶
1,1′,2,2′-(SiMe$_3$)$_4$ 4,4′-(CMe$_3$)$_2$*	40.6¶
1,1′,2,2′,3,3′,4,4′-(i-Pr)$_8$*	56.8¶

§ T_1 measurements; C$_5$H$_5$ ring.

¶ Coalescence temperature measurements.

** Bandshape analysis measurement.

sons indicate a dependence on the metal (Fe $\gg$ Ru) and alkyl substituents (t-pentyl $> t$-butyl) with, as expected, a general increase in the energy with increase in number and steric size of the Cp ring substituents (Abel *et al.*, 1991). The fluxionality of sterically crowded cyclopentadienyl complexes has been reviewed (McGlinchey, 1992).

Although the pentahapto co-ordination of the cyclopentadienyl group is very common and stable, these ligands are far from rigid and able to adopt alternative co-ordination modes if required. Several types of Cp mobility have been observed and fluxionality is found when the ring is attached to the metal atom through some but not all of its carbon atoms and the metal–ligand bonding may 'hop' around the ring, sometimes known as 'ring-whizzing'. [(η-C$_5$H$_5$)(η^1-C$_5$H$_5$)Fe(CO)$_2$] was the first organometallic compound for which fluxional behaviour was detected. At room temperature, the ^{1}H NMR spectrum showed only two sharp lines. The single line arising from the pentahaptocyclopentadienyl group was expected but the single sharp line for the monohapto–C$_5$H$_5$ ligand illustrated that there must be a low energy pathway for rapid rearrangement that renders all of the protons of η^1-C$_5$H$_5$ magnetically equivalent. The two most reasonable pathways for proton averaging were postulated as being either a 1,2- or a 1,3-metal shift, i.e. a movement of a metal atom around the C_5 ring from one carbon to the carbons α or β to it (Fig. 4.9). Further experiments indicated that the 1,2-shift is the correct mechanism as, on cooling, the process slows and the expected multiplet for a fixed η^1-Cp species is seen (Fig. 4.10).

In contrast to transition metals, η^1-Cp bonding is common with main group elements, e.g. [(η^1-C$_5$H$_5$)Ge(CH$_3$)$_3$] and [(η^1-C$_5$H$_5$)$_2$Hg]. In fact, mer-

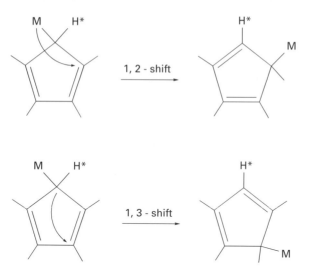

Fig. 4.9 Mechanism for proton-averaging around a cyclopentadienyl ring.

curocene could have two possible structures, one involving a divalent mercury σ-bonded to two C_5H_5 groups (monohapto complex; **IV.16**) and the second, an ionic structure analogous to the magnesium compound (pentahapto complex; **IV.17**). **IV.16** would be expected to exhibit three types of

(IV.16)

(IV.17)

hydrogen resonance in the ratio 1 : 2 : 2 whereas all of the hydrogen atoms are equivalent in **IV.17**. The NMR spectrum of the compound at $-70\,^{\circ}C$ consists of a single peak, showing equivalent protons. However, the IR spectrum is much more complex than is consistent with a highly symmetrical structure and it was thus postulated that the mercury–ring bonds undergo a constant precession and a sequence of haptotropic shifts (as shown in Fig. 4.10).

Another type of fluxional process occurs when a metal has two ligands which are chemically identical but differently bonded, e.g. in $[(\eta^5\text{-}C_5H_5)_2Ti(\eta^1\text{-}C_5H_5)_2]$ there are two types of metal-bound cyclopentadienyl groups which exchange (Fig. 4.11). The problem is that for steric reasons it is

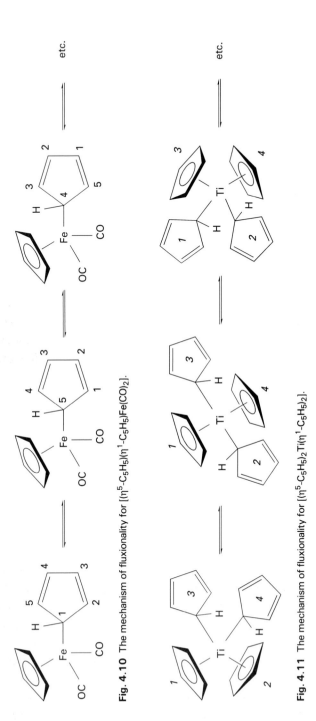

Fig. **4.10** The mechanism of fluxionality for $[(\eta^5\text{-}C_5H_5)(\eta^1\text{-}C_5H_5)Fe(CO)_2]$.

Fig. **4.11** The mechanism of fluxionality for $[(\eta^5\text{-}C_5H_5)_2Ti(\eta^1\text{-}C_5H_5)_2]$.

clear that the Cp ligands cannot all be η^5-bonded to titanium at the same time. ^{1}H NMR only distinguishes one type of proton at room temperature. At lower temperatures, *two* fluxional processes occur. Firstly, the mono- and penta-hapto (i.e. σ- and π-bonded) rings rapidly interchange their roles and, secondly, the point of attachment of each of the monohapto rings changes in a sequence of 'haptotropic shifts' or 'ring whizzes'. The result of the time-averaging processes is that all 20 protons become indistinguishable. The 1,2-shifts in the η^1-Cp ring occur at lower temperatures although in the case of $[(\eta^5\text{-}C_5H_5)_2(\eta^1\text{-}C_5H_5)Mo(NO)]$ there is only a small energy separation between the 1,2-shift process and the ring interchange. Two mechanisms have been proposed for the interchange of differently bonded rings (Scheme 4.26), one using X-ray data to involve η^4-C_5H_5 rings, and the processes are still a subject of debate.

By contrast, the much larger U^{4+} ion is able to bind four η^5-Cp ligands and therefore Cp_4U is not fluxional. Whilst molybdenum forms a stable 18 VE complex $[(\eta\text{-}C_5H_5)_2Mo(CO)]$, the tungsten analogue reacts with excess CO to give an apparently 20 VE species, $[(\eta\text{-}C_5H_5)_2W(CO)_2]$. Although the NMR spectrum shows that both Cp ligands are identical in solution, the solid state structure confirms that one Cp ring is only three-co-ordinate ($\eta^5 \rightarrow \eta^3$ ring slippage), and so the 18 VE count is maintained (Fig. 4.12). Fluxionality is also observed in the spectra of beryllocene. The ^{1}H NMR spectrum consists of a sharp singlet even down to $-135\,°C$. This indicates that the rings rotate relative to each other with the beryllium going back and forth from the two alternative positions at such a frequency that all the hydrogen atoms appear to be equivalent.

Fluxionality and ring whizzing is obviously not confined to cyclopentadienyl rings and there can be rearrangements of metals on π-systems containing C_nH_n rings. In all cases of η^m-C_nH_n ($m < n$) ring systems (excepting the η^6-C_8H_8 ring system), the mechanism of movement is a 1,2-shift. Activation energies for a number of systems have been studied and, in general, η^2-C_6H_6, η^3-C_7H_7 and η^4-C_8H_8 are highly fluxional whereas the neutral systems η^4-C_6H_6, η^5-C_7H_7, η^2-C_8H_8 and η^6-C_8H_8 are more or less static or possess high activation energies. This behaviour has been explained using the Woodward–Hoffmann rules and mechanisms detailing 'allowed or forbidden' 1,2-, 1,3- and 1,5-metal shifts (Mann, 1982). In many cases, C_8H_8 ligates as a mono-, di- or tri-olefin to a transition metal, thereby exhibiting varying degrees of non-planarity (Fig. 4.13). However, the co-ordination of conjugated double bonds in the C_8H_8 ligand to the metal often results in fluxionality. For example, for $[(\eta^4\text{-}C_8H_8)Fe(CO)_3]$ earlier interpretations suggested planar rings due to equivalence of the protons but X-ray studies showed it to be a 1,3-diene adduct. There were a series of suggestions of how the proton environments in the compound could be averaged: (i) a 1,2-shift

Mechanism (i)

$[(\eta^1-C_5H_5{}^*)_2Ti(\eta^5-C_5H_5)_2]$ ⇌ $[(\eta^1-C_5H_5{}^*)(\eta^4-C_5H_5{}^*)Ti(\eta^4-C_5H_5)(\eta^5-C_5H_5)]$

16 electrons   18 electrons

$[(\eta^1-C_5H_5{}^*)(\eta^1-C_5H_5)Ti(\eta^5-C_5H_5{}^*)(\eta^5-C_5H_5)]$

16 electrons

Mechanism (ii)

$[(\eta^1-C_5H_5{}^*)_2Ti(\eta^5-C_5H_5)_2]$ ⇌ $[(\eta^1-C_5H_5{}^*)(\eta^3-C_5H_5{}^*)Ti(\eta^5-C_5H_5)_2]$

16 electrons   18 electrons

$[(\eta^1-C_5H_5{}^*)(\eta^3-C_5H_5{}^*)Ti(\eta^3-C_5H_5)(\eta^5-C_5H_5)]$

16 electrons

$[(\eta^1-C_5H_5{}^*)(\eta^5-C_5H_5{}^*)Ti(\eta^3-C_5H_5)(\eta^5-C_5H_5)]$

18 electrons

$[(\eta^1-C_5H_5{}^*)(\eta^1-C_5H_5)Ti(\eta^5-C_5H_5{}^*)(\eta^5-C_5H_5)]$

16 electrons

Scheme 4.26 Two possible mechanisms for the interchange of differently bonded cyclopentadienyl rings.

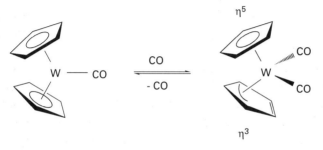

Fig. 4.12 Ring slippage in $[(\eta^5-C_5H_5)(\eta^3-C_5H_5)W(CO)_2]$.

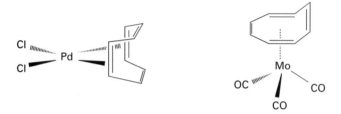

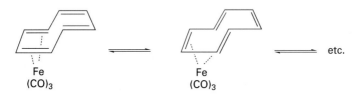

Fig. 4.13 Various metal-ligation modes of non-planar C_8H_8 rings.

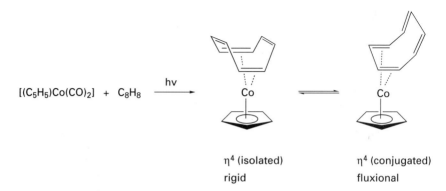

Fig. 4.14 The 1,2-shift mechanism for $[(\eta^4\text{-}C_8H_8)Fe(CO)_3]$.

circular ring migration; (ii) a structure bonded through transannular (1,5) double bonds interconverting through a flipping mechanism; and (iii) a transannular jump (1,5-shift) of the iron atom. NMR studies were inconclusive as the slow-exchange limit was not reached even at $-150\,^{\circ}C$ and it was only when the Ru analogue was formed that conclusive evidence for mechanism (i) could be put forward (Fig. 4.14). In cases where there is only partial co-ordination of the C_nH_n perimeter to the transition metal, isomerism can arise. For example, $[(C_5H_5)Co(C_8H_8)]$ is an inseparable mixture of $[(\eta\text{-}C_5H_5)Co(1,2\eta:5,6\eta\text{-}C_8H_8)]$ and $[(\eta\text{-}C_5H_5)Co(1\text{-}4\eta\text{-}C_8H_8)]$ which can slowly interconvert and the latter complex is fluxional in solution (Fig. 4.15).

$$[(C_5H_5)Co(CO)_2] + C_8H_8 \xrightarrow{\ h\nu\ }$$

η^4 (isolated)
rigid

η^4 (conjugated)
fluxional

Fig. 4.15 The isomerism of $[(C_5H_5)Co(C_8H_8)]$.

4.10 Electronic spectroscopy of metallocenes

One of the most powerful tools for locating the excited states and checking various features of the bonding theories of metallocenes is optical spectroscopy. Consequently, the electronic absorption spectra of various metallocenes have been widely studied. The ligand field model has proved to be useful in the interpretation of the electronic spectra (d–d and charge transfer (CT) transitions) of these systems, and a representative example is displayed in Fig. 4.16 (Sohn *et al.*, 1971). Three spin-allowed d–d transitions are proposed for ferrocene (ground state for d^6 ferrocene is $^1A_{1g}$-$\{(e_{2g})^4(a'_{1g})^2\}$) (two $^1A_{1g} \rightarrow {}^1E_{1g}$ and one $^1A_{1g} \rightarrow {}^1E_{2g}$ transitions). Low intensity bands in the visible range at 22 700 and 30 800 cm^{-1} can be seen for ferrocene, although the lowest energy visible peak is resolved into two bands giving an assignment of three spin-allowed d–d transitions. Three very weak spin-forbidden d–d absorption bands have been calculated and in the ferrocene absorption spectrum a strong charge transfer band (ε 51 000) at 50 000 cm^{-1} and two shoulders at 37 700 and 41 700 cm^{-1} can be seen. The transition energy decreases markedly with increasing positive charge of the metal atom and the energy of the main band in these metallocenes lies in the order $[Cp_2Co]^+ < Cp_2Fe < Cp_2Ru$. This is in order of decreasing metal electronegativity and indicates that the band arises from L $\rightarrow$ M charge transfer and is assigned to a $^1A_{1g} \rightarrow {}^1A_{2u}$ ($e_{1u} < e^*_{1g}$) transition.

A comparative study of the electronic spectra of some first-row transition series metallocenes Cp$_2$M (M = V, Cr, Mn, Fe, Co, Ni) and their 1,1′-dimethylmetallocene analogues has been made (Gordon, 1977). It was assumed that observed charge transfer bands were most likely to be due to

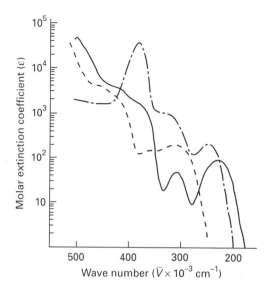

Fig. 4.16 Electronic spectra of d^6 metallocenes in solution at 300 K: Cp$_2$Fe (—) in isopentane, Cp$_2$Ru (– –) in isopentane and [Cp$_2$Co]ClO$_4$ (– ·) in aqueous solution. (Sohn *et al.* (1974), with permission.)

173

transitions from mainly ligand orbitals of odd symmetry, derived from the π-orbitals of the Cp rings, to predominantly metal $3d$ levels of even symmetry. The most likely ligand $\rightarrow$ metal (L $\rightarrow$ M) transitions are derived (using D_{5d} symmetry) from the one-electron excitations $e_{1u} \rightarrow e_{2g}$, $e_{1u} \rightarrow a'_{1g}$ and $e_{1u} \rightarrow e^*_{1g}$ depending upon the d-orbital occupations. There is also the possibility of metal $\rightarrow$ ligand (M $\rightarrow$ L) transitions, especially for d^7 and d^8 systems, where these would correspond to the excitation $e_{1g} \rightarrow e_{2u}$. For the neutral metallocenes the spectra of the parent and methyl-substituted systems are very similar, except for the case of manganese. Here, Cp_2Mn lies very close to the high–low-spin crossover point for d^5 systems, while for $(CpMe)_2Mn$ a thermal equilibrium between high- and low-spin species may be encountered. Generally, methyl substitution shifts the allowed bands to lower energies. For the V, Cr, Mn and Fe compounds the charge transfer bands appear to correspond only to ligand $\rightarrow$ metal excitations but for the Co and possibly Ni species, evidence is also found for metal $\rightarrow$ ligand ($e^*_{1g} \rightarrow e_{2u}$) transitions.

The electronic spectra of solutions of ruthenocene in halocarbon solvents (RX), but not in hydrocarbons or alcohols, show the formation of charge transfer complexes [$(C_5H_5)_2Ru \cdot RX$]. Photolysis of the CCl_4 CT complex gives [$(C_5H_5)_2Ru$]$^+$ in a primary photochemical process which proceeds via a triplet state. Lanthanide and actinide metallocenes show the influence of f orbitals in their electronic spectra. Uranocene exhibits weak bands attributed to f–f transitions in the region 850–1000 nm. Electronic absorption bands of thorocene at 2.75 eV and uranocene at 2.01 eV are assigned as ligand–metal charge transfer excitations, $e_{2g} (\pi) \rightarrow e_{3u} (f)$.

4.11 Mass spectrometry of metallocenes

Mass spectrometry of organic molecules has become well established and in recent years studies of the behaviour of organometallic species in the mass spectrometer has become more widespread. The effect of the metal on the fragmentation of the organic moiety has received considerable attention and the identification of metal-containing fragments is often facilitated by the isotope pattern of the metal. A comprehensive review of the mass spectra of transition metal metallocenes and related compounds has been reported by Cais and Lupin (1970). The major fragments observed are generally the parent molecular ion $C_{10}H_{10}M^+$ and those formed via dissociation of C_5H_5 fragments, i.e. $C_5H_5M^+$ and M^+. The comparatively weak, metal–ring bonding in magnesium and manganese derivatives is indicated by a significantly lower relative yield of the parent molecular ion for these compounds compared with the more covalently bonded species. For example, the total ion current carried by the ion [$(C_5H_5)_2Mn$]$^+$ is 18.5% while for [$(C_5H_5)_2M$]

(M = Fe, Co) the total ion current of the molecular ion is 60%. The increasing stabilities of the molecular ions $C_{10}H_{10}Ni^+$, $C_{10}H_{10}Fe^+$ and $C_{10}H_{10}Ru^+$ accurately parallels the increasing metal–ring bond strengths in these derivatives.

In the group 8 metallocenes, there is also elimination of neutral C_2H_2 fragments from the C_5 ring when attached to the metal. Characteristic of the group is the observation of the M + 1 and M + 29 ions formed by proton transfer from CH_5^+ and $C_2H_5^+$ (4.14) and electrophilic addition of $C_2H_5^+$ (4.15), respectively.

$$[(C_5H_5)_2M] + CH_5^+ \{C_2H_5^+\} \rightarrow [(C_5H_5)_2MH^+] + CH_4 \{C_2H_4\} \qquad (4.14)$$

$$[(C_5H_5)_2M] + C_2H_5^+ \rightarrow [(C_5H_5)_2MC_2H_5^+] \qquad (4.15)$$

As the group is descended, the relative abundance of the M^+ ions increases due to the relative rates of competing reactions, i.e. the rate of proton transfer from $C_2H_5^+$ decreases faster than the rate of electron transfer to $C_2H_5^+$ on moving from Fe to Os. It has also been noted that complexes of the first-row transition metals show a marked difference in their fragmentation compared to that of the second- and third-row elements whilst the 'isoelectronic' compounds Cp_2Os, Cp_2ReH and Cp_2WH_2 give rise to almost identical mass spectra. Mass spectrometry of Cp_2Ca shows high abundancies of ions Cp_2Ca^+ and $CpCa^+$ suggesting the presence of a simple sandwich species in the gas phase with a significant Cp—Ca interaction (cf. the solid state structure showing each Ca atom to be associated with four C_5H_5 rings). Smaller peaks due to Cp_2Ca^+ and $Cp_3Ca_2^+$ may arise from ion–molecule interactions between Cp_2Ca and $CpCa^+$.

Mass spectra for some tricyclopentadienyl lanthanides indicate that the strength of the lanthanide—cyclopentadienyl bond is greater for the lighter lanthanides than for the heavier derivatives. The mass spectrum of Cp_2V show a strong parent ion, doubly charged Cp_2V^{2+}, sequential loss of the Cp rings and evidence for dehydrogenation reactions.

4.12 Magnetic properties of metallocenes

For most metallocenes, electron configurations can be inferred from MO schemes (see Chapter 3) and these diagrams can correctly predict magnetic properties. However, there are a few examples where high-spin or low-spin forms cannot be predicted via these schemes. For instance, the magnetic moment of the Cp_2Fe^+ cation which exceeds the 'spin-only' value cannot be directly deduced from a rigid MO scheme as this suggests a ground state $^2A_{1g}$ whereas the ground state $^2E_{1g}$ is found experimentally. Any atom, ion or molecule that has one or more unpaired electrons is paramagnetic and so it, or any material in which it is found, will be attracted into a magnetic field. Substances that are diamagnetic contain no unpaired electrons and they are weakly repelled by a magnetic field. The magnetic susceptibility is a measure of the force exerted by the field on a unit mass of the specimen and is related to the number of unpaired electrons present per mole. The paramagnetism of a material containing unpaired electrons receives a contribution from the orbital motion of the unpaired electrons as well as from their spins. However, at times the spin contribution is so predominant (and the orbital angular momentum is quenched) that the measured susceptibility values can be interpreted in terms of how many electrons are present, i.e. the magnetic moment μ, due entirely to n unpaired electrons on the atom or ion, is given by

$$\mu = 2 \, [S(S+1)]^{1/2}$$

where S equals the sum of the spins of all the unpaired electrons, i.e. $n \times \frac{1}{2}$.

Numerous magnetic susceptibility measurements have been made on metallocene systems and have been summarised by Warren (1976). Clearly, paramagnetic compounds are of greatest interest with regard to magnetic measurements, e.g. values for tricyclopentadienyl lanthanides range from 10.3 BM for Cp_3Dy and 9.44 BM for Cp_3Er to 3.63 BM for Cp_3Nd. Detailed magnetic studies have shown chromocene to obey the Curie–Weiss law in the temperature range 83–293 K with a value of c. 3.27 BM. This is appreciably above the spin-only value (2.83 BM) for two unpaired electrons and so there must be a significant orbital contribution. Thus, a $^3\Delta$ ground state is suggested and such a configuration arises by removal of one electron from each of the completely filled e_{2g} and a_{1g} levels of the closed shell arrangement of ferrocene. Having one more electron than ferrocene, cobaltocene is paramagnetic ($\mu_{eff} = 1.70$ BM) and has attracted many theoretical and magnetic studies. The additional electron is predicted to be in the metal e_{1g} orbital rather than the e_2 ligand orbitals predicted for the $[Cp_2Fe]^-$ anion situation.

The magnetic moment of manganocene shows an unusual temperature dependence; as a solid at $T > 432$ K, in the melt and in solutions at room temperature, the complex has a magnetic moment of $\mu = 5.9$ BM as expected from an electron configuration $(e_{2g})^2(a'_{1g})^1(e^*_{1g})^2$ (ground state $^6A_{1g}$) – a 'high-spin' situation. Between 432 and 67 K with decreasing temperature, increasing antiferromagnetic behaviour is observed, corresponding to an increase of cooperative interactions. A dilute solution of $[(C_5H_5)_2Mn]$ (8%) in $[(C_5H_5)_2Mg]$ has a magnetic moment of $\mu = 5.94$ BM over the whole temperature range. Cp^*_2Mn gives a 'normal' sandwich structure and the magnetic moment ($\mu = 2.18$ BM) suggests a low-spin Mn (d^5) configuration (ground state $^2E_{2g}$). This is in contrast to 1,1'-dimethylmanganocene, which in the gas phase consists of an equilibrium mixture of high-spin and low-spin states. Clearly, manganocene and its derivatives are so close to the 'high-spin/low-spin crossover point' that the small intermolecular forces existing in frozen solutions or in molecular crystals are sufficient to produce the observed changes of the electronic ground state and the energy difference between the states is only 2.1 kJ mol^{-1}. Nickelocene is also a paramagnetic species ($\mu_{eff} = 2.89 \pm 0.15$ BM) with a magnetic moment in close agreement with the predicted spin-only value and it obeys the Curie–Weiss law. This is also the case for vanadocene which has three unpaired electrons (consistent with the $^4A_{2g}$ ground state) and a magnetic moment of 3.78 BM. The Curie–Weiss law is obeyed above c. 15 K but below this temperature there is evidence for long-range antiferromagnetic interactions in the solid state.

References

Abel, E.W., Long, N.J., Orrell, K.G., Osborne, A.G. and Sik, V. (1991) *J Organomet Chem*, **403**, 195.

Adams, R.D. and Selegue, J.P. (1982) in *Comprehensive Organometallic Chemistry* (eds E.W. Abel, F.G.A. Stone and G. Wilkinson) vol. 4, ch. 33, pp. 968–1057. Pergamon, Oxford.

Bennett, M.A., Bruce, M.I. and Mathesone, T.W. (1982) in *Comprehensive Organometallic Chemistry (eds E.W. Abel, F.G.A. Stone and G. Wilkinson) vol. 4, ch. 32.3, pp. 692–812. Pergamon Press, Oxford.*

Bublitz, D.E. and Rinehart, K.L. Jr (1969) *Org React*, **17**, 1.

Cais, M. and Lupin, M.S. (1970) *Adv Organomet Chem*, **8**, 211.

Davies, S.G., Green, M.L.H. and Mingos, D.M.P. (1978) *Tetrahedron*, **34**, 3047.

Davis, R. and Kane-Maguire L.A.P. (1982) in *Comprehensive Organometallic Chemistry* (eds E.W. Abel, F.G.A. Stone and G. Wilkinson) Pergamon Press, Oxford, vol. 3, ch. 26.2, pp. 954–1070.

Deeming, A.J. (1982) in *Comprehensive Organo-Merallic Chemistry* (eds E.W. Abel, F.G.A. Stone and G. Wilkinson), vol. 4, ch. 31.3, pp. 378–507.

Fritz, H.P. (1964) *Adv Organomet Chem* **1**, 254.

Gordon, K.R. and Warren, K.D. (1978) *Inorg Chem*, **17**, 987.

Jolly, P.W. and Wilke, G. (1974) *The Organic Chemistry of Nickel*, vol. 1, ch. 8. Academic Press, New York.

Jutzi, P. (1990) *J Organomet Chem* **400**, 1.

Lauher, J.W. and Hoffmann, R. (1976) *J Am Chem Soc*, **98**, 1729.

Mann, B.E. (1982) in *Comprehensive Organometallic Chemistry* (eds E.W. Abel, F.G.A. Stone and G. Wilkinson) vol. 3, ch. 20, pp. 91–171. Pergamon Press, Oxford.

McGlinchey, M.J. (1992) *Adv Organomet Chem*, **34**, 285.

O'Hare, D. (1996) in *Inorganic Materials* (eds D.W. Bruce and D. O'Hare) 2nd ed, ch. 4, pp. 171–244. Wiley, Chichester.

Rosenblum, M. (1965) *The Chemistry of the Iron Group Metallocenes–Part 1.* Wiley, New York.

Rosenblum, M. and Abbate, F.W. (1966) *J Am Chem Soc* **88**, 4178.

Schwartz, J. and Labinger, J.A. (1976) *Angew Chem*, **88**, 402; *Angew Chem Int Ed Engl*, **15**, 333.

Slocum, D.W. and Ernst, C.R. (1972) *Adv Organomet Chem*, **10**, 79.

Sohn, Y.S., Hendrickson, D.N. and Gray, H.B. (1971) *J Am Chem Soc*, **93**, 3603.

Tebbe, F.N., Parshall, G.W. and Reddy, G.S. (1978) *J Am Chem Soc*, **100**, 3611.

Watts, W.E. (1979) *J Organomet Chem Lib*, **7**, 399.

Watts, W.E. (1980) in *Organometallic Chemistry*, Specialist Periodical Reports, vol. 8 (eds E.W. Abel and F.G.A. Stone). Chemical Society, London.

Warren, K.D. (1976) *Struct Bonding*, **27**, 45.

5 Derivatives of Metallocenes

Following the discovery of ferrocene, interest was stimulated about the synthesis and characterisation of derivatives of this metallocene and other sandwich compounds. This chapter will feature a series of selected species ranging from bridged metallocenes (i.e. metallocenophanes) to bi- and poly-metallocenes and multi-decker sandwich systems.

5.1 Metallocenophanes

Metallocenophanes are a type of species that feature linking of the cyclopentadienyl (or polyarenyl) rings by the introduction of a heteroannular bridge (or bridges). Much attention has been given over the years to the study of the constituents and the effects of such a bridge. For example, it was suggested that bridging the rings would bring about a change in physical and chemical properties and that the metallocene could be structurally modified by pulling the rings together or by tilting them with respect to one another. Much of the work has focused on the iron group metallocenophanes (i.e. bridged ferrocenes and, to a lesser extent, ruthenocenes and osmocenes) where strained, ring-tilted [a] metallocenophanes have been found to undergo thermal ring-opening polymerisation (ROP). The resulting materials are rare examples of well-defined, high molecular mass, soluble polymers with transition metals in the main polymer chain and they can exhibit unusual physical and chemical properties. In recent years, bridged group 4 metallocenes have also come to the fore because of their use as catalysts in stereoselective olefin polymerisation (NB: in this context, metallocenes with a molecular bridge connecting the cyclopentadienyl rings are usually known as 'ansa-metallocenes'). Studies on these novel, metallocene-based catalysts have featured the development of new materials and the elucidation of a correlation between the catalyst structure and polymer properties. The versatility and applicability of these materials has led to an explosion of interest, and the applications of metallocenophanes are discussed in Chapter 6.

For further reading on metallocenophanes, consult articles by Smith (1964), Watts (1967), Deeming (1982), Herberhold (1995), Hisatome *et al.* (1995) and Zanello (1995).

5.1.1 Metallocenophane nomenclature

The metal forming the initial sandwich compound obviously denotes the type

of metallocenophane, i.e. iron in ferrocenophanes, osmium in osmoceno-
phanes, and in general metallocenophanes can be divided into two classes.
The first type involves those species where the rings of a single metallocene
nucleus are linked. These are trivially known as [a] metallocenophanes where
a denotes the number of exocyclic atoms in the bridge, e.g. the [2] ferrocenophane (**V.1**) and the [4] ferrocenophane (**V.2**). Those compounds with
more than one bridging group are known as [a] [b] . . . metallocenophanes
where a, b, etc. indicate the length of each bridge, in decreasing order of size,
e.g. the [4] [3] ferrocenophane (**V.3**) and the [4] [4] [3] ferrocenophane (**V.4**).
The second class of metallocenophanes includes derivatives in which rings of
two or more metallocene nuclei are linked so that each metallocene unit forms
a bridge with respect to one another. These are known as [a. b. c . . .] metallocenophanes with a, b, c etc. denoting the lengths of the bridges, e.g. the [4.4]
metallocenophane (**V.5**).

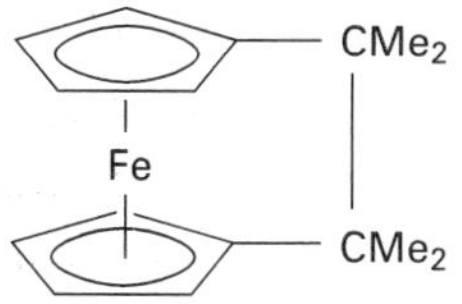

(V.1)

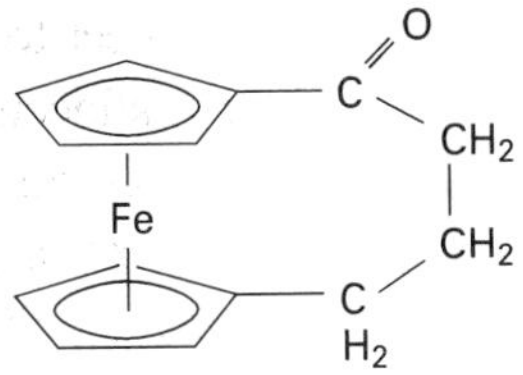

(V.2)

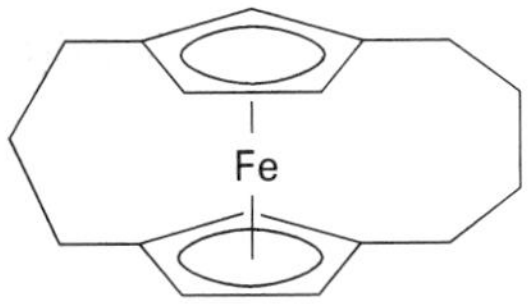

(V.3)

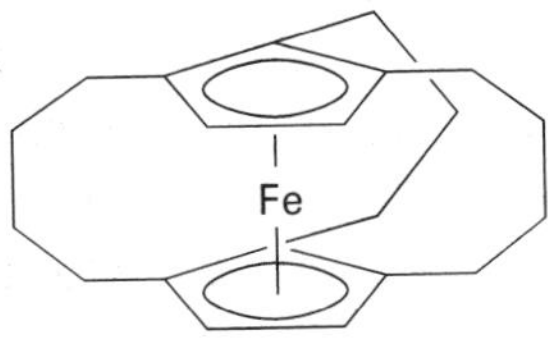

(V.4)

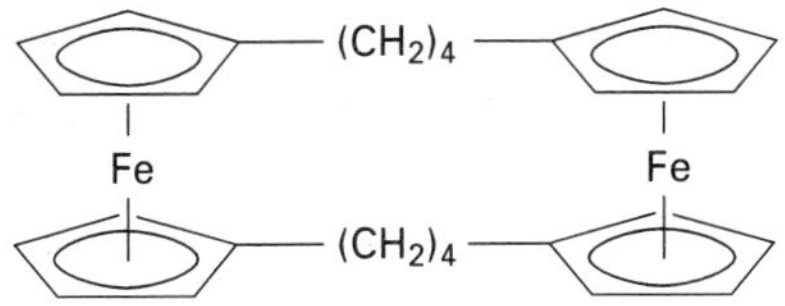

(V.5)

It should be noted that a uniform nomenclature for metallocenophanes
does not exist. The system described above, based on organic cyclophanes

(V.6)

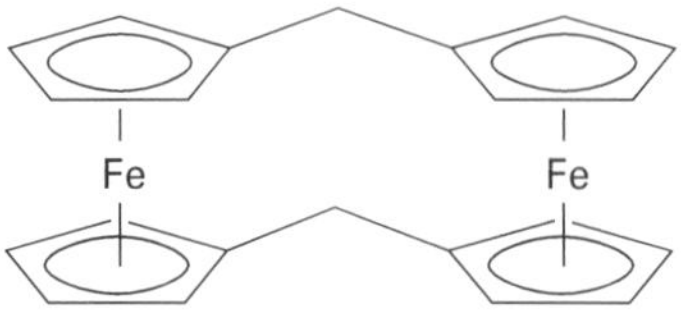

(V.7)

(Watts, 1967), seems to be the choice of the majority of researchers in the area. The exact location of the bridges cannot be specified using this system however and more complex terminology has to be invoked, i.e. in **V.6**, a triply-bridged ferrocene, its name according to metallocenophane nomenclature is [3](1,1′)[3](2,2′)[3](4,4′) ferrocenophane but by IUPAC rules it is unambiguously 1,1′; 2,2′; 4,4′-tris(trimethylene)ferrocene. Similarly, **V.7** is simply a [1.1] ferrocenophane but more completely could be described as [1.1](1,1′; 1′,1‴) ferrocenophane. Other nomenclature nuances are [1.1] metallocenophanes also being referred to as [1²] metallocenophanes and the terms [3₃] or [3] [3] [3] used when dealing with a number of (in this case, three) identical bridges in mononuclear metallocenophanes, **V.6**. The first class (mononuclear species) will be focused on with brief discussion of the second binuclear class of metallocenophanes later in the chapter.

5.1.2 Methods of synthesis

Various methods have been utilised to synthesise metallocenophanes and these have been well reviewed (see previous references). A brief outline of the more important methods follows.

Direct synthesis or C—C coupling

This method involves the use of an anhydrous metal(II) chloride and a dicyclopentadienide anion. An example is shown (Scheme 5.1) however the method is only preparatively useful for [2] metallocenophanes as yields fall sharply with increased bridge length.

Intramolecular acylation

[3] Metallocenophanes containing trimethylene bridges have been generally formed by this method. The bridging reaction is an electrophilic acylation at the C_5H_5 ligands under acidic conditions (Scheme 5.2) (Hillman *et al.*,

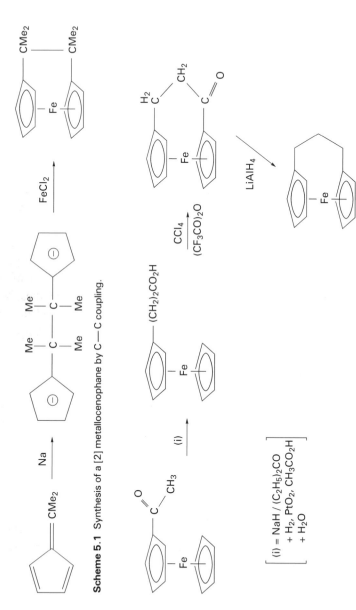

Scheme 5.1 Synthesis of a [2] metallocenophane by C—C coupling.

Scheme 5.2 Synthesis of a [3] metallocenophane by intramolecular acylation.

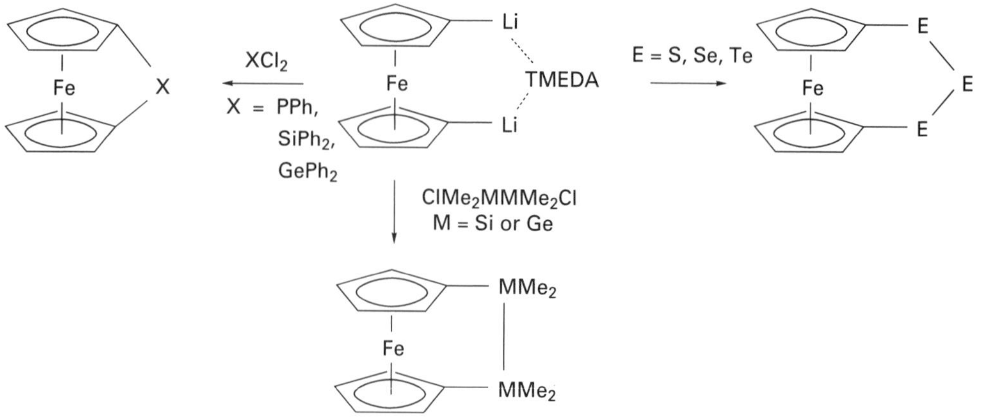

Scheme 5.3 Various bridge length metallocenophanes formed from the 1,1'-dilithiometallocene intermediate.

1978). Multiply-bridged species of this [3] metallocenophane have been formed, but synthesis becomes more difficult as the number of bridges is successfully increased. Multi-bridged [4] ferrocenophanes and their methylene bridge enlargement have been well investigated and [3] and [4] ruthenocenophanes have also been prepared by similar routes.

From 1,1'-dilithiometallocene·TMEDA

Coupling reactions involving suitable metal or non-metal dihalides or elemental chalcogenides and reactive dilithio species have proved useful for the preparation of [1], [2] or [3] metallocenophanes (Scheme 5.3) mainly where the metal has been iron or ruthenium. The key reagent is the 1,1'-dilithioferrocene·TMEDA adduct where a crystal structure determination showed a $[(C_5H_4Li)_2Fe]_3(TMEDA)_2$ system with the lithium-bridged ferrocene structure comprising the three ferrocenediyl units wrapped around a tetrahedron of inner lithium atoms, with two outer lithium ions co-ordinated to the TMEDA molecules (Butler et al., 1985).

Linkage of substituents on each cyclopentadienyl ring

This is the most generally applicable method with many examples known. If each ring carries a donor heteroatom then the species can act as a bidentate, chelating ligand forming such complexes as **V.8–V.10**. Certain diols will readily cyclise under acidic conditions to form ether-bridged ferrocenophanes (Scheme 5.4).

$$\text{(V.8)}$$

$$\text{(V.9)}$$

$$\text{(V.10)}$$

5.1.3 Chemical properties of metallocenophanes

In general, the properties of these species are similar to the simple metal-locenes but there are some anomalies that can be illustrated by comparing some chemical properties of [*a*] ferrocenophanes with non-bridged 1,1′-disubstituted ferrocenes. These arise because short bridges cause distortions which affect the reactivity of the molecule and, secondly, the bridge can be of steric hindrance to the approach of an attacking species.

Reductive cleavage

Lithium in propylamine will cleave ferrocene and its alkyl derivatives to the corresponding cyclopentadienides whilst trimethylene-[3] ferrocenophane is only partly decomposed and a hexamethylene-[3] [3] ferrocenophane is totally unaffected.

Acetylation

In contrast to [3] ferrocenophanes which behave similarly to 1,1′-dialkyl-

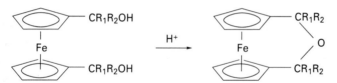

Scheme 5.4 Formation of a [3] ferrocenophane via cyclisation of a diol.

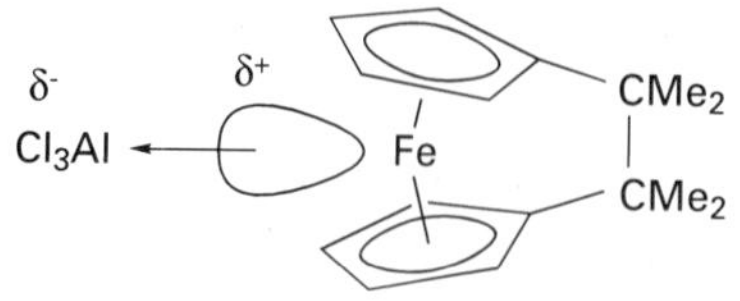

Fig. 5.1 A Lewis acid–ferrocenophane complex.

ferrocenes, [2] ferrocenophanes will not undergo acetylation and even under mild conditions extensive decomposition is prevalent. This behaviour of [2] ferrocenophanes has been suggested as being due to enhanced Lewis basicity of the iron atom. This causes the formation of a complex (Fig. 5.1) which is unstable and decomposes by ring–metal fission under the reaction conditions.

Lithiation

n-Butyllithium has no effect on dimethylene-[2] ferrocenophane but will readily lithiate alkyl ferrocenes under similar conditions.

Oxidation

Most polymethylene-[a] ferrocenophanes undergo straightforward oxidation with active MnO_2 to produce the corresponding [a] ferrocenophan-1-ones. However, dimethylene-[2] ferrocenophane reacts rapidly to give a complex mixture of products.

Protonation

Dimethylene-[2] ferrocenophane can be much more easily protonated than either ferrocene or simple non-bridged alkyl ferrocenes. This is thought to be due to an electronic redistribution in the ring-tilted compound increasing the Lewis basicity of the iron atom.

5.1.4 NMR spectra and fluxionality of metallocenophanes

Addition of a bridging group of atoms to a metallocene causes a loss of symmetry and affects the NMR spectra of these compounds. This is clearly shown in the ^{1}H NMR spectra of, for example, [2] metallocenophanes where the cyclopentadienyl ring protons give rise to two well-separated unsymmetrical triplets. The higher field signal has been assigned to the $H_{2,5}$ protons due to the fact that ring tilting results in an unsymmetrical distribution of electron density around the iron atom by causing hybridisation of the metal non-bonding orbitals. In this case, a differential shielding of the $H_{2,5}$ and $H_{3,4}$ protons is brought about.

Fluxionality is an interesting feature in these compounds and results from '*bridge reversal*', i.e. a rapid inversion of the bridgehead atoms. The possibility of fluxional processes occurring in [a] metallocenophanes was first recognised in the trimethylene-bridged complex $[Fe(C_5H_4)_2(CH_2)_3]$. The relatively simple room-temperature spectrum of the central methylene protons, comprising a single line, was postulated as being due to rapid reversal of the central methylene bridge group in direct analogy to the six-membered ring chair–chair conformation exchange in cyclohexane (Fig. 5.2). This bridge inversion process allows relative oscillation of the cyclopentadienyl rings within the molecule with the degree of this torsional freedom being controlled by the length of the bridge, i.e. less restriction with increased length. It was suggested that the process did not involve an energetically favourable planar transition state but was one involving rotation about the cyclopentadienyl ring–metal bonds and early investigations on methylene-bridged [a] ferrocenophanes indicated that there was a low energy barrier to bridge reversal in these complexes.

Chalcogen-bridged [3] metallocenophanes have been studied in depth and shown to possess larger energy barriers, the first example being the sulphur-bridged complex 1,2,3-trithia-[3] ferrocenophane. Trisulphide bridge reversal was postulated as being the explanation for the coalescence at high temperatures of an ABCD pattern for the Cp ring protons to an AA′BB′ pattern in the ^1H NMR spectrum. The energy of activation $\Delta G\ddagger$ for the bridge reversal was determined by NMR bandshape analysis as 80.4 kJ mol^{-1}, a striking increase from the value of 55.4 kJ mol^{-1} for the analogous species 1,2,3-trithia-cyclohexane. This seems to be predominantly due to the difference in the torsional energies of the relevant bridge bonds, i.e. markedly different torsional barriers for the Cp—iron bonds but also some lesser effects via bond rotations, variations in bridge length and the electronic influence of different ring substituents. From further studies of other chalcogen-bridged [3] ferro-

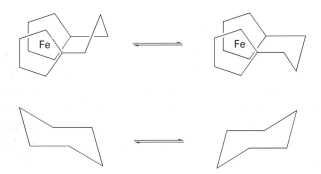

Fig. 5.2 The analogy between bridge reversal in [3] metallocenophanes and ring reversal in cyclohexane.

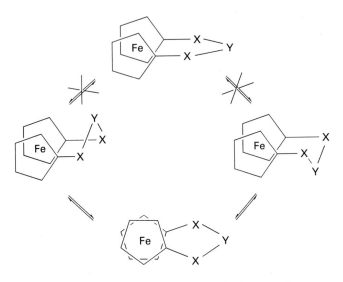

Fig. 5.3 Mechanisms for [3] metallocenophane bridge reversal.

cenophanes, the bridge reversal mechanism has been devised and is illus-
trated in Fig. 5.3. It shows a non-planar transition state involving staggered
Cp rings and results from rotation about the Cp—Fe and bridge C—X and
X—Y bonds (X and Y = bridging chalcogen and/or carbon atoms).

Accurate energy data for the bridge reversal of [3] ferrocenophanes with a
variety of chalcogen-bridging atoms have now been obtained using the
cyclopentadienyl protons as the probe nuclei (Table 5.1). The torsional
energies of the relevant bridge bonds are again the major contributors to the
total energy barrier. From Table 5.1, estimations can be made of the relative
magnitudes of torsional barriers about single bonds involving like and unlike
chalcogen atoms, i.e. the S—S torsion is 3.9 kJ mol^{-1} higher in energy than

Table 5.1 Bridge reversal energies in some [3] ferrocenophanes.

Bridge in the [3] ferrocenophane	Bridge length (nm)	Bridge reversal energy (kJ mol^{-1})
—CH$_2$—CH$_2$—CH$_2$—		40.4
—CH$_2$—O—CH$_2$—		39.7
—CH$_2$—S—CH$_2$—		34.6
—S—CH$_2$—S—		47.2
—S—CMe$_2$—S—		42.8
—S—S—S—	0.778	80.4
—S—Se—S—	0.804	72.6
—S—Te—S—	0.844	62.5
—Se—S—Se—	0.830	71.0
—Se—Se—Se—	0.856	67.2
—Se—Te—Se—	0.896	59.9

the S—Se torsion and 5.8 kJ mol^{-1} more than that of the Se—Se bond. The values are all noticeably higher than any bond torsion involving a carbon atom. The energies of activation also show a correlation with the total length of the bridge (C—X—Y—X—C) as determined by covalent radii, i.e. the longer the bridge length, the lower the energy barrier to bridge reversal. There is a discrepancy for the —Se$_2$Se— and —S$_2$Te— species where the S—Te bond length (0.241 nm) is not much more than that for Se—Se (0.234 nm). However, the crucial importance of the torsional barriers is demonstrated here, as although the bond lengths are similar the torsional barriers in the —S—Te—S— species are much less thereby reducing the value of $\Delta G\ddagger$ (Table 5.1)

These trends in relative torsion energies of the chalcogen—chalcogen single bonds are also followed in the novel multiply trichalcogen-bridged ferrocenophanes **V.11–V.13** which feature two separate bridges of three chalcogen atoms linking the cyclopentadienyl rings (Abel *et al.*, 1993). For instance, **V.11** exists in solution as a mixture of chair–chair and chair–boat

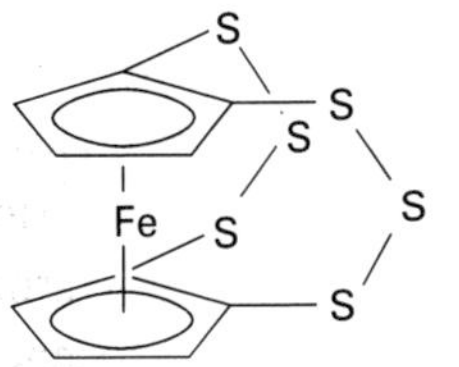

(V.11)

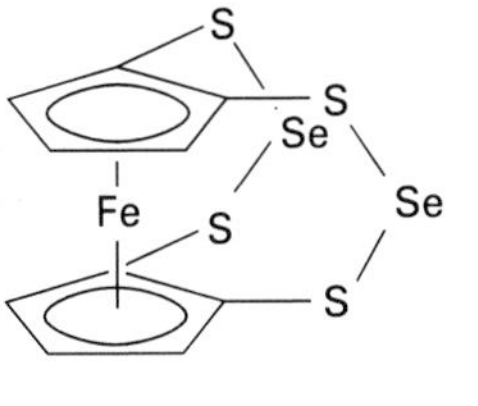

(V.12)

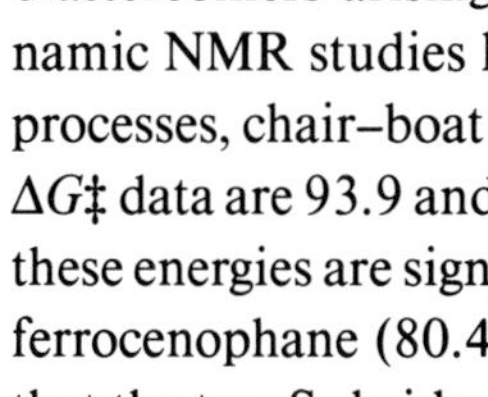

(V.13)

diastereomers arising from the orientations of the trisulphur linkages. Dynamic NMR studies have shown that the energies of the two bridge reversal processes, chair–boat to chair–chair and chair–boat to boat–chair, in terms of $\Delta G\ddagger$ data are 93.9 and 89.0 kJ mol^{-1}, respectively. It is interesting to note that these energies are significantly higher than for the singly trisulphur-bridged [3] ferrocenophane (80.4 kJ mol^{-1}). This is reasonable if one considers the fact that the two S$_3$ bridges are adjacent and this close proximity must hinder the reversal of either bridge. In addition, the bridge reversal mechanism (Fig. 5.3) is predicted to involve a conformation with staggered cyclopentadienyl rings, which would mean that both S_3 bridges change in a co-operative manner, and is likely to be a higher energy and improbable process.

In recent years, main group and transition metal moieties have been incorporated into the ferrocenophane bridge and fluxionality characteristics reported. Low barriers to ring reversal were found for compounds $[Fe(C_5H_4S)_2XMe_2]$ (X = C, Si, Ge, Sn) and $[Fe(C_5H_4E)_2Y]$ (E = S, Se; Y = PC_6H_5, AsC_6H_5) and postulated as being due to an increase in bridge length and the small torsional barriers associated with group 4 or group 5 element—chalcogen bonds. Formation of the transition metal complexes of the chalcogen-containing ligands 1,1'-bis(methylthio)ferrocene and 1,1'-bis(methylseleno)ferrocene (Fig. 5.4) (Abel *et al.*, 1991) converts the ferrocenyl moiety into a [3] ferrocenophane. Hence, the occurrence of both ferrocenophane bridge reversal and pyramidal chalcogen inversion (another intramolecular motion that can be monitored by dynamic NMR spectroscopy) can be expected (Abel *et al.*, 1984). In these systems, the very low torsional barriers of the bridgehead atoms result in the bridge reversal being too rapid to observe on the NMR timescale, which also affects the barriers to pyramidal chalcogen inversion being lower than in analogous compounds. The species $[Fe(C_5H_4Se)_2M(\eta-C_5H_4R)_2]$ (M = Zr, Hf; R = H, ^{t}Bu) is nonfluxional by a bridge reversal process which is presumably due to the steric interactions imparted by the bulky group 4-bonded cyclopentadienyl groups.

The effect on the bridge reversal energies of [3] metallocenophanes by differences in the inter-ring distances of the metallocenyl moiety has been examined. The considerably greater inter-ring separation in osmocene (0.371 nm) and ruthenocene (0.368 nm) in comparison to ferrocene (0.332 nm) leads to significant effects on the bridge geometry, i.e. the angle of the bridge (X—Y—X) will be greater in the osmo- and rutheno-cenophanes than in the analogous ferrocenophanes thereby rendering bridge reversal more difficult (as was established earlier in connection with relative bridge lengths). Magnitudes of $\Delta G\ddagger$ values increase in the order Fe < Ru < Os reflecting the increasing strength of the metal–ring bonding, and in the order $Se_3 < S_3$ due to the different torsional energies of the chalcogen–chalcogen bridge bonds.

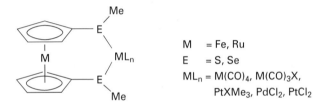

$$M = Fe, Ru$$
$$E = S, Se$$
$$ML_n = M(CO)_4, M(CO)_3X,$$
$$PtXMe_3, PdCl_2, PtCl_2$$

Fig. 5.4 Some chalcogen- and transition metal-bridged [3] metallocenophanes.

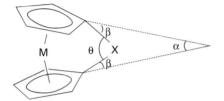

Fig. 5.5 Potential molecular deformations of metallocenophanes.

5.1.5 Structures of metallocenophanes

[a] *Metallocenophanes*

The most interesting structural feature of these species is the accommodation of different lengths of bridges by molecular deformation. Consequently, metallocenophanes can be strained or unstrained and as strained cyclic compounds they are potential monomers for ring-opening polymerisation (ROP). Strained metallocenophanes are currently of interest as precursors for polymers containing transition metals in sandwich units (Herberhold, 1995), and this will be discussed more fully in Chapter 6.

The molecular deformation occurs by one or more of the following ways (Fig. 5.5): (i) inclination, α, of the cyclopentadienyl rings (ring-tilt); (ii) bond angle distortion, θ, at the bridging atom(s); (iii) deviation, β, of the exocyclic bonds from the planes of the cyclopentadienyl rings. Clearly, some degree of distortion is required when the inter-ring distance of the ferrocene unit (0.332 nm) does not correspond with the length of the heteroannular bridge, i.e. greater distortion with greater differences in lengths. Some angles of tilt between the rings for a range of [1], [2], and [3] ferrocenophanes are listed in Table 5.2. It is clear that the angles increase quite sharply from [3] ferrocenophanes to two-atom bridged species. However, in the latter, ring-tilt is reduced by a movement of the exocyclic atoms out of the best planes of the cyclopentadienyl rings towards the iron, e.g. in **V.1** the bonds between the bridging carbon atoms and the rings are at an angle of 11° to the plane of the C_5 rings.

Table 5.2 Comparison of the data indicating ring strain in homologous [1] and [2] ferrocenophanes.

Bridging group (E)	Angle α (°)	Angle β (°) (average values)
SiMe$_2$	20.8	37.0
GeMe$_2$	19.0	36.8
Sn^{t}Bu$_2$	14.1	36.2
(CMe$_2$)$_2$	23.2	11.0
(SiMe$_2$)$_2$	4.19	10.8
(SnMe$_2$)$_2$	0.7	10.5

The angles of tilt in [1] ferrocenophanes are not as great as would be expected if the trends seen for the [2] and [3] species were extrapolated, i.e. angles of tilt (α) are predicted to be 36° and 39° for **V.14** (X = GePh$_2$) and

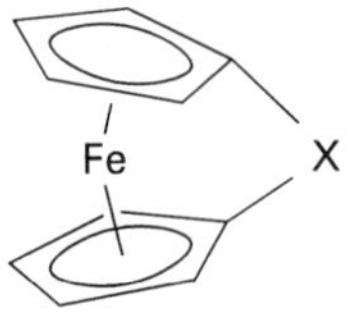

(V.14)

X = PPh, respectively). It appears that the exocyclic bond distortion is the dominant mode of deformation because large tilts of around 40° would bring the C1 and C1′ atoms much closer together than the normal separation distance found in ferrocene. Additionally, very small θ angles would be required thereby placing too much strain on the heteroatoms. The tilt angle (α) will increase as the covalent radius of the bridging element decreases and in fact [1] ferrocenophanes containing elements from the second row of the Periodic Table, e.g. B, C, N, O, have yet to be formed. To date, the smallest bridging element is sulphur and the novel thia [1] ferrocenophane **V.15** has

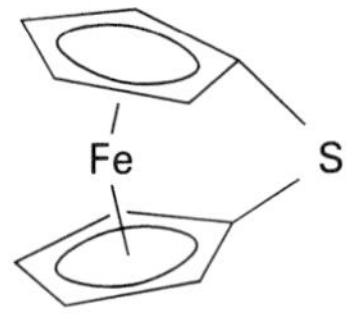

(V.15)

the largest recorded tilt angle of 31.1° and features an eclipsed conformation of the cyclopentadienyl rings.

Going further down the Periodic Table and therefore including larger elements, the ring strain is reduced. For example in **V.14** where X = GePh$_2$ or AsPh (this compound also features substitution of one CHMeNMe$_2$ group onto the Cp ring), α is reduced to 16.6° and 22.9°, respectively. To date, just one transition metal-bridged [1] ferrocenophane has been structurally characterised (**V.16**), where the tilt angle is 6.0° but a weak metal–metal bonding interaction is indicated by the relatively short Fe—Zr distance of 0.296 nm. Despite their larger covalent radii (and thus reduction in ring strain), incor-

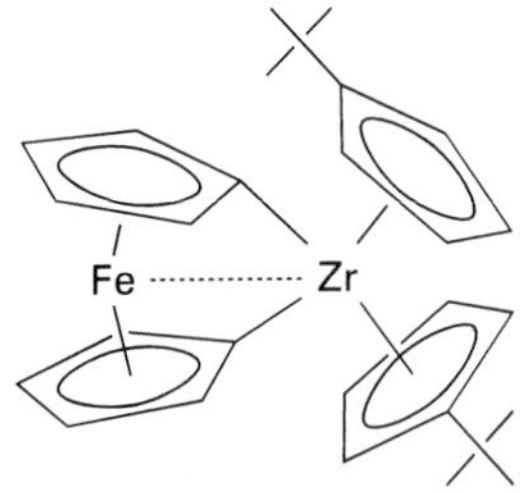

(V.16)

poration of heavier bridging elements has not been easy due to the weakness of the bonds between the *ipso*-carbon atoms and the bridging element. Although sila- and germa-[*a*] metallocenophanes are well known, [*a*] ferrocenophanes containing tin directly attached to the two cyclopentadienyl rings were unknown until very recently (Herberhold *et al.*, 1996; Rulkens *et al.*, 1996). Starting from 1,1′-bis(chlorodimethylstannyl)ferrocene or the analogous hydride 1,2-distanna [2]- (**V.17**) and 1,2,3-tristanna [3]-ferro-

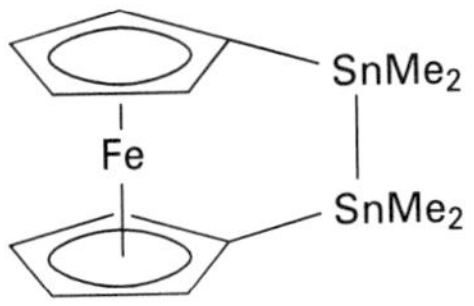

$$(V.17)$$

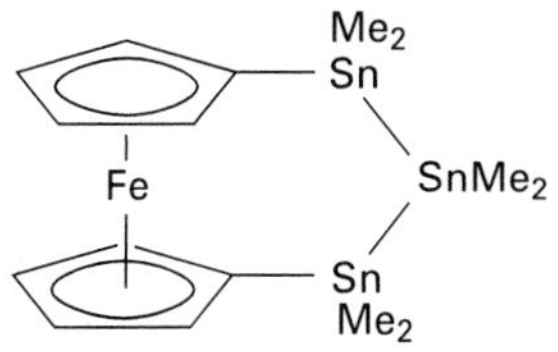

$$(V.18)$$

cenophanes (**V.18**) can be formed in reasonable yields. In **V.17** the cyclopentadienyl rings are virtually parallel (tilt angle α is 0.7°) and adopt a nearly eclipsed conformation and in this unstrained system the distance between the cyclopentadienyl rings (0.331 nm) is similar to that in ferrocene (0.332 nm). This clearly shows (Table 5.2) that the strain in [2] ferrocenophanes decreases as the E—E bond lengths increase in the order $E = CMe_2 < SiMe_2 < SnMe_2$ and this is also the case when examining the analogous [1] ferrocenophanes. From the reaction of dilithioferrocene·Two-thirds TMEDA with $^tBu_2SnCl_2$ in diethylether at low temperatures, stanna-[1] ferrocenophane (**V.19**) was isolated in 65% yield. X-ray diffraction studies

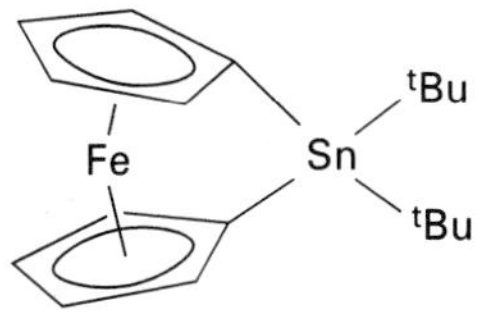

$$(V.19)$$

showed that the angle α was *c.* 14°, this being the smallest yet reported for a [1] ferrocenophane with a main group element in the bridge and significantly less than that for the silicon- and germanium-bridged analogues (Table 5.2). Interestingly, the Sn—Fe distance in **V.19** was 0.298 nm, only slightly less than in clearly linked Sn—Fe compounds, and indicates significant overlap of the orbitals of iron and tin to give a weak bond. Manners *et al.* have gone on to explore the ring-opening polymerisation (ROP) of **V.19** (see Chapter 6) and have demonstrated that it is the least ring-tilted [a] metallocenophane to undergo ROP to date. Indeed, previously prepared [2] metallocenophanes with tilt angles up to 13° were found to be resistant to ROP.

With multiple bridges, the angle of mutual inclination can be increased significantly though the substitutions must be made in adjacent positions to have any real effect. For example, in **V.20** the two non-adjacent trimethylene groups do not increase the ring tilt compared with other singly-bridged species, whereas in **V.21**, **V.22** and **V.23** the introduction of two or three

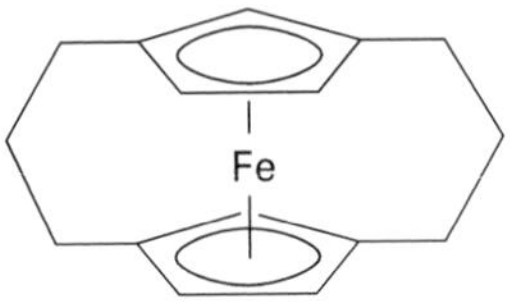

(V.20)

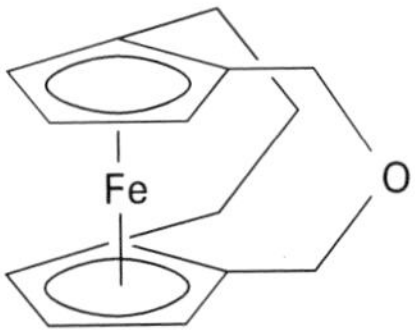

(V.21)

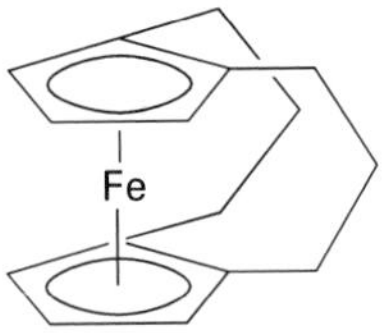

(V.22)

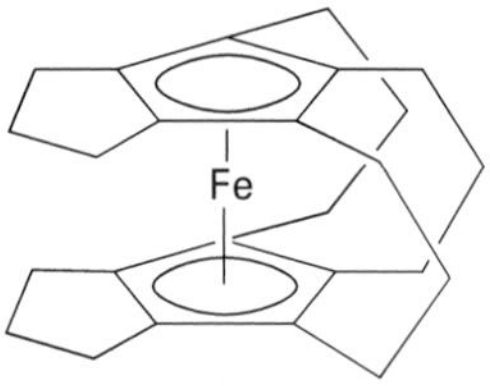

(V.23)

adjacent bridges increases the angle α to 15°, 13.1° and 12.5°, respectively. Some of the most visually (and chemically) attractive compounds in this area have been the intramolecularly multi-bridged ferrocenophanes. There has been much interest in this area but syntheses are often laborious, involving carbon-by-carbon chain extension and requiring careful avoidance of bridge rearrangements. Although the C_3 bridge is just too short to link the cyclopentadienyl rings in the metallocene without any strain, the formation of multi-bridged systems is still possible. [3] Ferrocenophanes with up to four bridges are known but attempts to form the fully-bridged [3] [3] [3] [3] [3] (or [3_5]) ferrocenophane have so far been to no avail. The additional strain from inclusion of the last bridge has resulted in cyclisation accompanied by a rearrangement of the existing adjacent bridge to give **V.23** instead. The simple [4] ferrocenophane was prepared in 1958 and, due to the decreased inter-ring strain, Hisatome *et al.* were able to form the first fully enclosed ferrocene system [3] [4] [4] [4] [4] ferrocenophane, containing one C_3 and four C_4 bridges. This was followed by a report from the same group on the synthesis of the much sought-after symmetrical [4] [4] [4] [4] [4] (or 4_5) ferrocenophane which, in analogy to superphane, has been dubbed 'superferrocenophane' **V.24** (Hisatome, 1992) and features a perfect shielding of the central iron atom from the external environment.

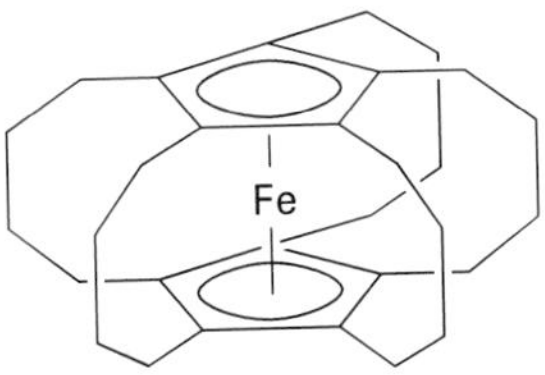

(V.24)

It should be noted that with longer bridges, movement of the exocyclic atoms can be made out of the planes of the rings away from the iron, e.g. outward displacement of the chalcogen atoms in **V.25** and **V.26** by around

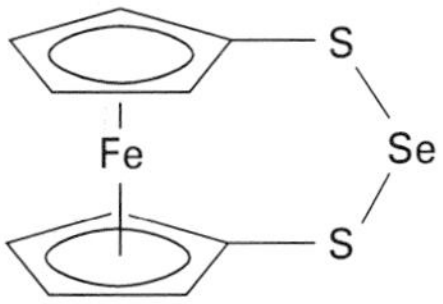

(V.25)

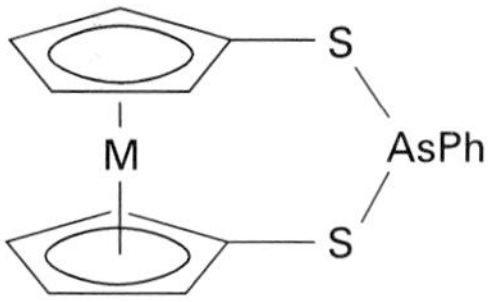

(V.26)

0.004 and 0.02 nm, respectively. Additionally, the slight strain in these types of compounds can be alleviated by a twisting of the rings. The inter-ring distance in the parent metallocenes expands on going from ferrocene to ruthenocene to osmocene and thus favours the formation of longer-chained derivatives, although for the synthesis of ruthenocenophanes far more rigorous conditions are required than for the ferrocene derivatives (Bruce, 1995). The S_3-[3] osmocenophane is the first and, to date, only osmocene derivative with linked rings. There are very few [1] or [2] ruthenocenophanes known and those that are feature far more distortion than the corresponding [1] and [2] ferrocenophanes showing an almost linear correlation (Herberhold, 1995). For example, in the series of structurally characterised carbon- and silicon-bridged [2] metallocenophanes (**V.27a–V.27c**) the ruthenium complexes are always more strained. Compounds **V.28** and **V.29** also show molecular distortions with the dihedral angles between the cyclopentadienyl

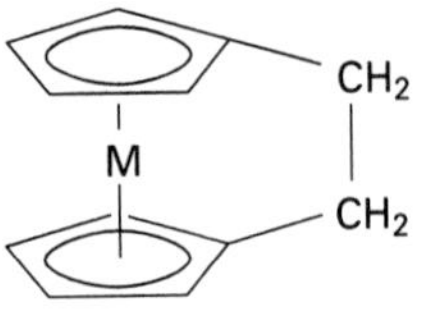

$$\begin{array}{c} \text{CH}_2 \\ | \\ \text{CH}_2 \end{array}$$

(**V.27a**)

M = Fe α = 21.6°

M = Ru α = 29.6°

$$\begin{array}{c} \text{CMe}_2 \\ | \\ \text{CMe}_2 \end{array}$$

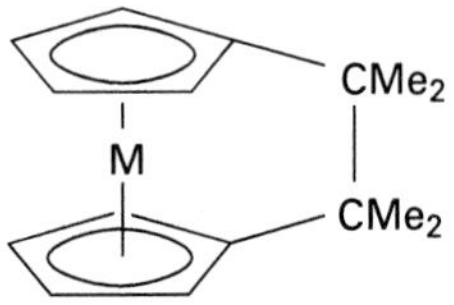

(**V.27b**)

M = Fe α = 23.0°

M = Ru α = 30.6°/31.5°
(two independent molecules
in the cell)

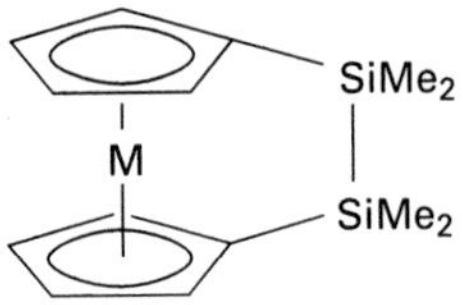

$$\begin{array}{c} \text{SiMe}_2 \\ | \\ \text{SiMe}_2 \end{array}$$

(**V.27c**)

M = Fe α = 4.2°

M = Ru α = 7.8°

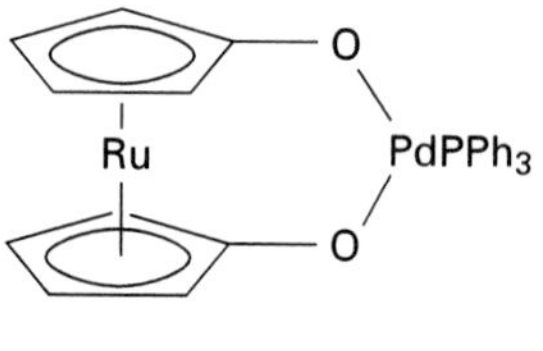

(**V.28**)

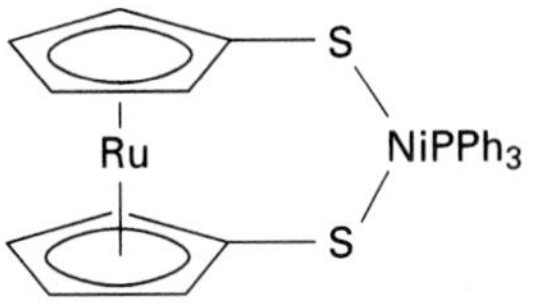

(**V.29**)

rings being 9.8(3)° and 13.7(2)°, respectively. Another explanation (apart from steric factors) for this increased ring-tilting lies in the proposed formation of a weak dative bond between the metallocenyl Ru atom and the metal atom in the bridge. The co-ordination of the non-bonding d electrons of the metal atom in the metallocene nucleus to the vacant orbitals of another transition metal has become a topic of interest in recent years. This type of interaction was first reported to occur in [2](1,1'-ferrocenophane) metal halide adducts and observed again in (1,1'-ferrocenedithiolato-S,S',Fe)-(triphenylphosphine)palladium(II) and platinum(II) (**V.30**). ^{1}H NMR spec-

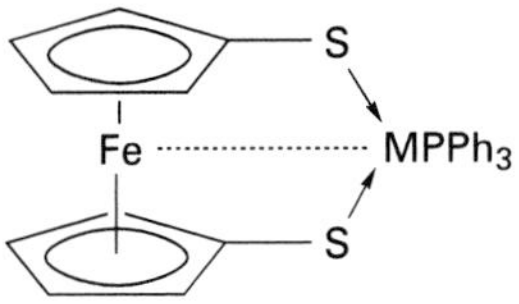

(**V.30**)

tra of the complexes showed a large separation ($c.$ 1.4 ppm) of the cyclopentadienyl ring proton signals, primarily because of the anisotropic effect of the dative bond between the two metal atoms. Further evidence was gained from X-ray studies which showed that the Fe—M distance in **V.30** was 0.294 nm (M = Pt) and 0.288 nm (M = Pd) and the Ru—M distance in **V.28** and **V.29** was 0.262 and 0.286 nm, respectively; these were in good agreement with other analogous known Fe—M (M = Pt, Pd) and Ru—M (M = Pd, Ni) bond distances. This led to the suggestion that the attainment of a stable 16-electron configuration around the bridgehead M(II) atom was arrived at by a weak dative M—M bond. The close proximity of the datively-bound metals can only be arrived at by some extraordinary structural distortions. For example, in **V.30**, the bond angle S—M—S is 165.4° (M = Pt) and 165.2° (M = Pd) which is much larger than that of other [3] metallocenophanes containing the metal atom at the 2-position in the bridge, e.g. in **V.31** and **V.10** the bond angles of S—Pd—S and As—Ni—As are 83.9° and 93.5°, respectively. A ring tilt of 21.0° in **V.30** (M = Pt) is also far greater than in comparable [3] metallocenophanes.

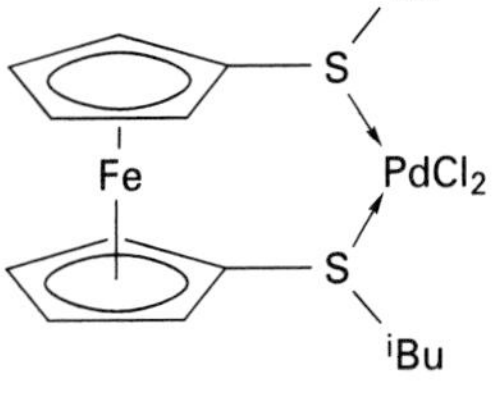

(**V.31**)

[a.b] *Metallocenophanes*

These types of compounds have been reviewed elsewhere (Mueller-

Westerhoff, 1986) but a brief discussion will be given here. In general, the binuclear complexes [a.b] metallocenophanes have been less well studied than their mononuclear counterparts, although with the availability of two or more co-ordination sites and the potentially novel redox properties, the former species may be useful via catalytic properties. With more than one metal centre, there is the possibility of different oxidation states being present leading to 'mixed-valence' compounds, currently much in vogue. Mixed-valence compounds can usually be identified by broad and relatively intense near-IR (NIR) absorptions resulting from the transfer of an electron from the metal atom with the lower oxidation state to the one with the higher oxidation state, i.e. $Fe^{II}/Fe^{III} \rightarrow Fe^{III}/Fe^{II}$. The electron-transfer process leaves both metals in a vibrationally excited state leading to NIR spectra and as their structural parameters can be easily and selectively varied [a.b] metallocenophanes are excellent study models.

Much of the work on [a.b] metallocenophanes has concentrated on the dimeric iron species but due to their unique properties [0.0] metallocenophanes are now known for all the first-row transition elements (**V.32,**

(V.32)

Scheme 5.5). Interestingly, no [0.0] metallocenophanes with two different metal atoms are known as novel synthetic strategies would have to be used but mixed-metal systems can be found in [1.1] metallocenophanes (**V.33**).

(V.33)

$$M^1 = M^2 = Fe$$
$$M^1 = Fe, M^2 = Ru$$
$$M^1 = Fe, M^2 = Co$$
$$M^1 = M^2 = Ru$$
$$M^1 = M^2 = Co$$

[1.1] Ferrocenophane was first reported as the product of reduction of the diketone $[(C_5H_4)_2C{=}O]_2Fe_2$ but a more general synthetic route uses difulvenyl metallocenes (Scheme 5.6) (Cassens et al., 1981). The difulvenyl metallocenes are available by two efficient routes and their conversion to

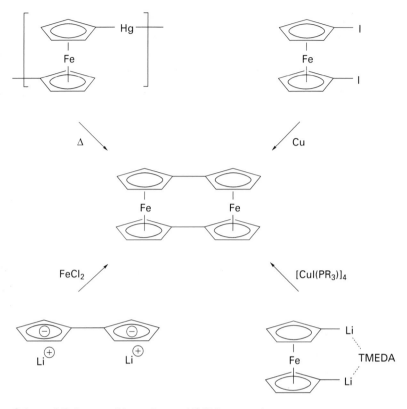

Scheme 5.5 Four possible syntheses of [0.0] ferrocenophane.

metallocenophanes involves reduction to the ligand dianion and reaction with metal(II) chlorides.

Carbon-bridged [1.1] metallocenophanes have attracted interest due to the isomerism and conformational flexibility possible. For example, [1.1] ferrocenophane can exist in *syn* or *anti* forms (Fig. 5.6). In both configurations, the inner protons in the 2-position inflict severe crowding. The *anti* isomer is quite rigid and therefore not able to relieve the overcrowding whereas the *syn* isomer experiences no such problems due to its flexibility and possible twisting. This is because of the low activation barrier for the rotation of the two ligands in the ferrocene units and as the rotations are linked via the methylene units, the molecule undergoes a synchronised motion from one *syn* conformation to its mirror image. This conformational flexibility is indicated by the ^{1}H NMR spectra where signals are averaged due to fluxionality. The motion is observed to occur rapidly at temperatures as low as $-80°$C, establishing an upper limit of 10 kJ mol^{-1} for the barrier to this *syn–syn* exchange. Addition of substituents on the rings or the bridges will however suppress or prevent this flexibility quite dramatically.

Examples of [1.1] ferrocenophanes with heteroatom bridges

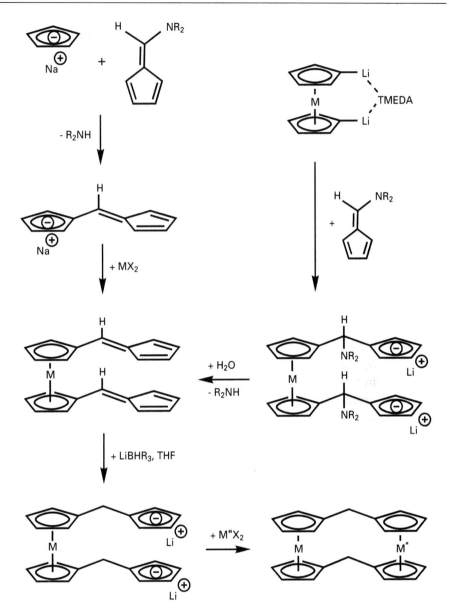

Scheme 5.6 The synthesis of [1.1] metallocenophanes via difulvenyl metallocenes.

$[(C_5H_4)_2X]_2Fe$, with $X = SiMe_2$, $PbPh_2$, $SnMe_2$, $Sn''Bu_2$ and $SnI''Bu$, have been reported. The tetrabutyldistanna derivative is interesting as the steric bulk of the *n*-butyl groups prevents the normal *syn* orientation. Owing to the greater length of the C—Sn bonds (0.215 nm) as opposed to a C—C bond (0.148 nm) there is less repulsion of the inner protons than in the *syn*-[1.1] metallo-cenophanes. For this reason, the tin species adopts the *anti* configuration and a nearly coplanar structure of the ferrocene units which is manifest by a simple

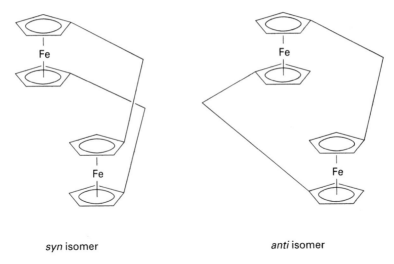

syn isomer   anti isomer

Fig. 5.6 The *syn* and *anti* configurations of [1.1] ferrocenophane.

^{1}H NMR spectrum showing a pairwise equivalence of the protons in the 2-and 5- as well as the 3- and 4-positions. There is a rich chemistry of the [1.1] metallocenophanes illustrated by: (i) the unusually stable carbanions formed by deprotonation of [1.1] ferrocenophanes; (ii) the stable carbocations formed by hydride abstraction from [1.1] metallocenophane bridges; (iii) the unusual reversible oxidation of [1.1] ruthenocenophane; (iv) the metal–metal interactions in mono- and di-oxidised [1.1] ferrocenophanes; and (v) in the utility of [1.1] ferrocenophanes as hydrogen generation catalysts (Barlow and O'Hare, 1996; Mueller-Westerhoff *et al.*, 1986).

Relatively few [1^n] metallocenophanes where $n > 2$, have been studied but a recent report featured the synthesis of some new [1^4] metallocenophanes (Barlow and O'Hare, 1996). The [1^4] ferrocenophane **V.34**, features alternat-

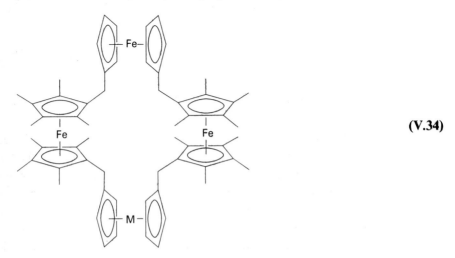

(V.34)

ing ferrocene and permethylated ferrocene units, but the methodology also permits synthesis of heteronuclear species comprising two permethylated ferrocenylene units, a ferrocenylene unit and a heterometallocenylene unit. Fewer still [2.2] and larger [a.a] ferrocenophanes are known and those that are have featured almost exclusively iron as the metallocene metal. An unusual [2.2] ferrocenophane (**V.35**) features the ferrocene units connected through two acetylene bridges and properties include oxidation to a mono- and di-cation. An important observation was that the unpaired electron in the mixed-valence monocation was fully delocalised, even though the iron atoms are more than 0.6 nm apart.

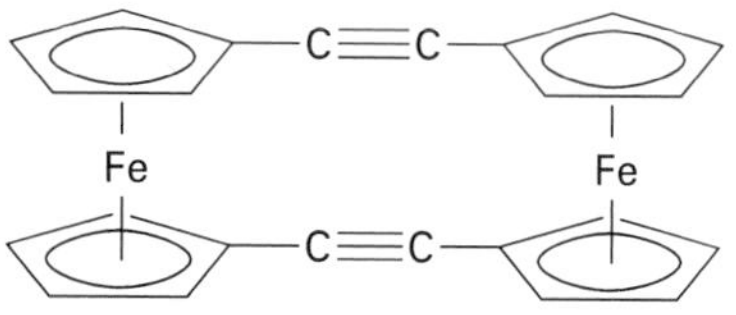

(V.35)

5.1.6 Metallocenophanes featuring transition metals other than iron and ruthenium

Recent studies into thermal ring-opening polymerisation and the formation of soluble polymers with transition metals in the main polymer chain have driven forward the synthesis of strained ring-tilted [a] metallocenophanes. Many of these polymers have been restricted to those containing iron and ruthenium but others featuring non-iron group metals are also of interest via the synthesis of strained [a] metallocenophanes with various electron counts. In this context, the air-sensitive 19-electron 1,1'-ethylenylcobaltocene **V.36**

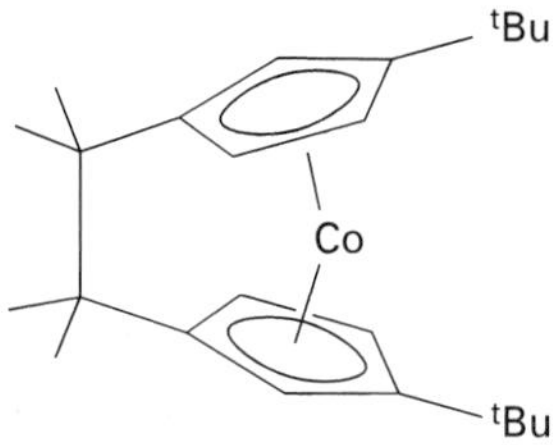

(V.36)

(and the 17-electron iron cation analogue) have been synthesised in low yields via the reductive coupling of 2-*tert*-butyl-6, 6-dimethylfulvene, using magnesium/CCl_4 as the reducing agent and reaction of the subsequent anion with $CoBr_2$ in THF. The former was the first isolated [2] cobaltocenophane, which could be oxidised by NH_4PF_6 to the corresponding air-stable, yellow salt, **V.36a**. This 18-electron species again features a substantial ring-tilt (24.8°) as was the case for other [2] ferrocenophanes. Some years previously, a series of 1,1'-disubstituted cobaltocenium salts were formed. From the reaction of di(cobaltocenium-1,1'-dithiolato)zinc(II) and aqueous sodium sulphide followed by ammonium hexafluorophospate, 1,2,3-trithia-[3] cobaltoceniumophane hexafluorophosphate (**V.37**) was isolated as a yellow powder in 60% yield. Its behaviour was exactly analogous to the iso-electronic 1,2,3-trithia-[3] ferrocenophane and, as expected, showed fluxional behaviour in NMR spectra.

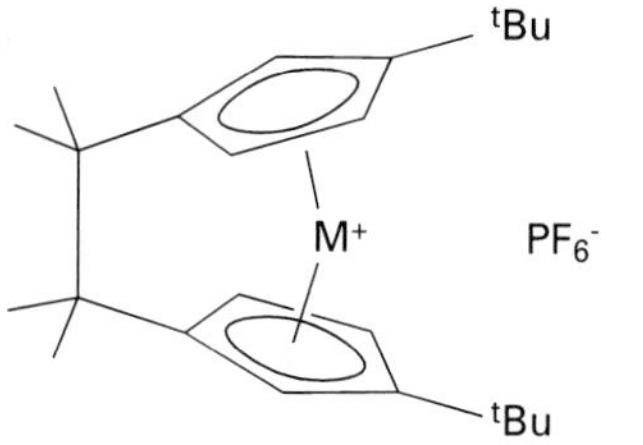

(V.36a)

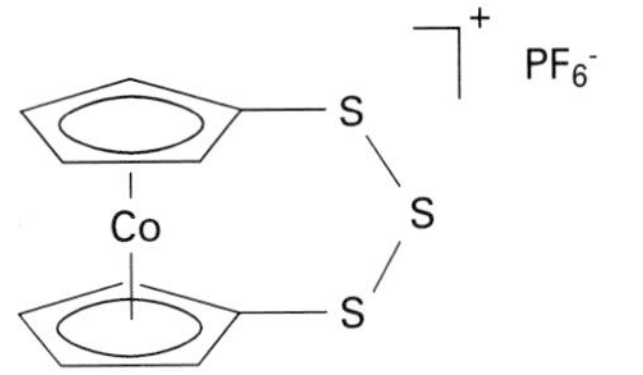

(V.37)

In efforts to characterise well-defined catalysts having multiple redox and reactivity states, recent reports have detailed the formation of the reversibly redox active hydrogenation catalyst precursor featuring a cobaltocene moiety **V.38**. The oxidised or reduced form of the ligand perturbs the electronic

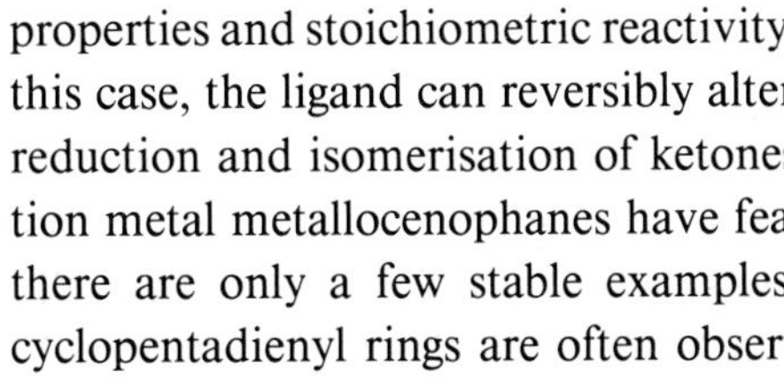

(V.38)

properties and stoichiometric reactivity of the bound transition metal and, in this case, the ligand can reversibly alter the rate of the rhodium(I)-catalysed reduction and isomerisation of ketones and alkenes. Other first-row transition metal metallocenophanes have featured manganese and chromium but there are only a few stable examples. In manganocene structures, tilted cyclopentadienyl rings are often observed and this seems to be because of steric effects rather than the more usual electronic effects of electrons residing in metal–ligand anti-bonding orbitals observed for other metallocene structures. On heating a mixture of $MnBr_2$ and solid $Li_2[C_5H_4(CH_2)_3C_5H_4]$ the [3] manganocenophane $[\{\eta^5,-\eta^5-C_5H_4(CH_2)_3C_5H_4\}Mn]$ is formed as a red–orange oil. Treatment of this oil with 3,5-dichloropyridine (Fig. 5.7) gives the

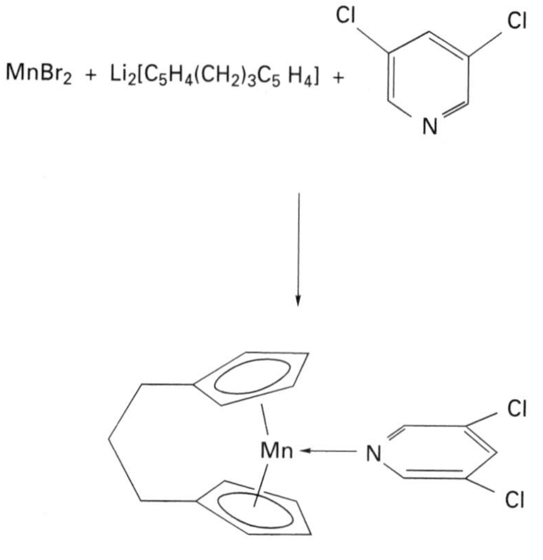

$MnBr_2 + Li_2[C_5H_4(CH_2)_3C_5 H_4] +$

Fig. **5.7** Synthesis of a [3] manganocene adduct.

1:1 adduct and the X-ray crystal structure of the species shows that both rings are bound in a symmetrical η^5 fashion to the manganese centre.

The only stable bridged polyenyl sandwich species for group 6 metals (particularly chromium) feature the di-(η^6-benzene) structure. The reactions of the di(arene) complexes are still rather unexplored because of difficulties in synthesis and handling, but lithiation of [(η-C_6H_6)$_2$Cr] with ^{n}BuLi and TMEDA gives the dimetallated species which can react with a variety of electrophiles (Scheme 5.7) (Morris, 1995). As with the ferrocene analogues, the substituted chromocenes can act as chelating ligands and bind moieties to construct a link between the benzene rings and form the metallocenophane structure.

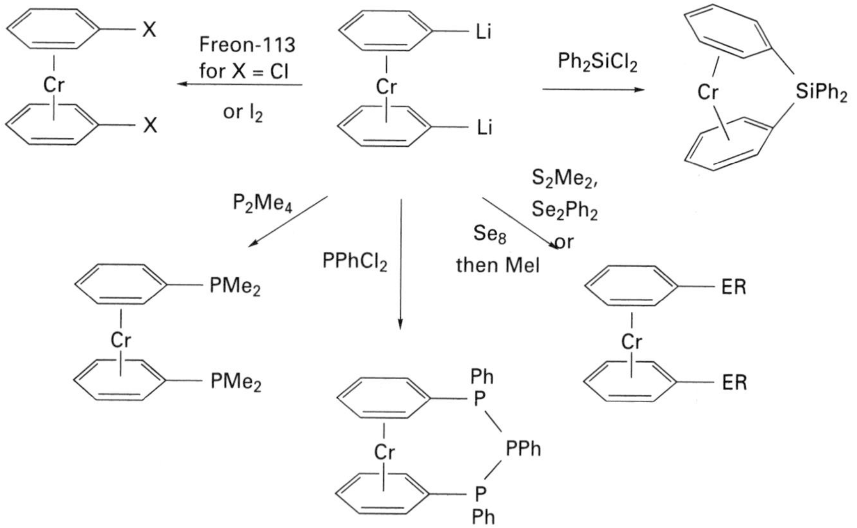

Scheme 5.7 Some reactions of the 1,1-dilithiodi(benzene)chromium intermediate.

5.1.7 Metallocenophanes featuring group 4 elements

The interest in metallocenophanes or *ansa*-metallocenes has been particularly significant in the development of the chemistry of early transition metal sandwich compounds. This has been due to the fact that complexes of this structural type have become important as highly stereoselective homogeneous alkene polymerisation catalysts (see Chapter 6). In recent years, much effort has been directed towards tailoring of the cyclopentadienyl ring system to improve catalytic activity and stereoselectivity (Binger and Podurbin, 1995; Bochmann, 1995). The first examples of zirconium and hafnium *ansa*-metallocenes were reported in the early 1970s and featured a trimethylene bridge, $[MCl_2\{(CH_2)_3(\eta\text{-}C_5H_4)_2\}]$, whilst $[TiCl_2\{(CH_2)_n(\eta\text{-}C_5H_4)_2\}]$ ($n = 1$ or 2; **V.39** (part of Fig. 5.8)) was formed in 1979. The standard synthetic method is reaction of the dilithium salt of the bridged dicyclopentadienyl anion with the metal tetrahalide, $TiCl_3$ or $[TiCl_3(THF)_3]$ followed by oxidation with aqueous HCl (Fig. 5.8). If a substituent R is introduced onto the cyclopentadienyl ring then *rac* and *meso* isomers can result (Fig. 5.9) and the isomeric ratio will vary according to the type of alkyl substituent, i.e. *rac:meso* ratio is 1.3 : 1 (R = Me), 1.5 : 1 (R = Et), 1.8 : 1 (R = ^{i}Pr) and 2.0 : 1 (R = ^{t}Bu). Disubstituted cyclopentadienyl derivatives similarly lead to *rac* and *meso* titanocenes, though the *rac* isomer is generally more favoured here. Substituents can be included on the bridge between the Cp rings to either influence the *rac:meso* ratio or to introduce chirality. For instance, the tetramethyl ligand $[C_2Me_4(C_5H_4)_2]^{2-}$ is obtained by reducing dimethylfulvene with sodium amalgam, and then reaction with $[TiCl_3(THF)_3]$ gives the red powder $[\{C_2Me_4(C_5H_4)_2\}TiCl_2]$. The isomeric ratios have been shown to be dependent on the nature of the cyclopentadienyl reagent (cyclopentadienyl Grignard/cyclopentadienyl lithium/cyclopentadienyl sodium), the metal moiety ($TiCl_4/[TiCl_3(THF)_3]$) and the temperature of the reaction.

Many other polyenyl ring systems have been utilised to date, such as indenes, fluorenes and substituted cyclopentadienes. These will not be discussed in great detail here, but an example with an indenyl ligand system features complexes with two chiral centres on the bridge (Scheme 5.8). The hydrogenated (*S*,-*S*)-bis(tetrahydroindenyl) ligand system is known as

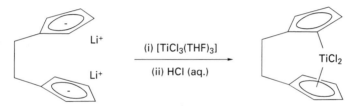

Fig. 5.8 Synthesis of the first *ansa*-titanocene.

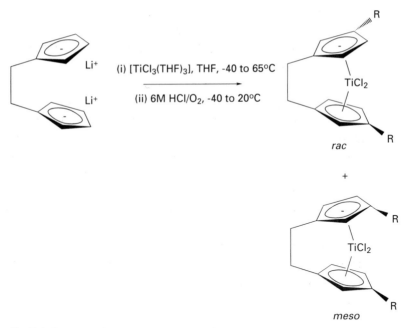

Fig. 5.9 Formation of *rac* and *meso* isomers of an *ansa*-titanocene.

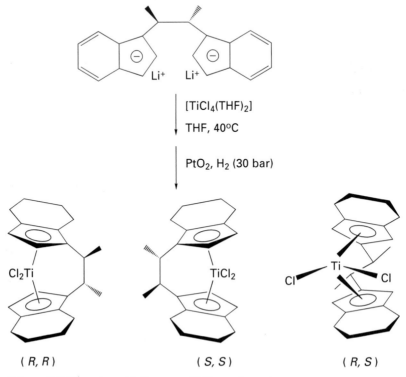

Scheme 5.8 The various chiral isomers of 'chiracene' complexes.

'chiracene' and lithium or potassium salts have been used for complex formation. The red 'chiracene' titanium dichloride complex was obtained as a
4.3 : 2.5 : 1 mixture of (R,S) (R,R) and (S,S) diastereomers. The isomers could
not be separated by fractional crystallisation but irradiation of the mixture
lead to extensive photoisomerisation to give up to 85% yield of the (R,R)
isomer. Bridged dicyclopentadienyl ligands are generally prepared from cyclopentadienes or fulvenes, i.e. moieties with preformed C_5 rings but an
unusual alternative route features a double Skattebøl rearrangement (Scheme
5.9). Three isomeric *ansa*-titanocenes were obtained and could be separated
by fractional crystallisation or conversion to 2-naphtholato complexes.

The introduction of substituents on, or bridges between, the cyclopentadienyl rings has an important effect on the '*co-ordination gap aperture*', i.e. the
largest possible angle (α) between two planes through the metal centre which
touch van der Waals surfaces of each of the cyclopentadienyl ligands. α decreases with increasing size of the cyclopentadienyl ligand and is generally
around 60–90° but with the introduction of small bridges such as CH_2 the
values can be as high as 100–110°, Also in simple metallocenes, the two
intersecting planes are perpendicular to the plane bisecting the Cl—Ti—Cl
angle. In the bridged systems, however, the conformation of the bridge and the
presence of substituents in α or β positions to the bridge results in a slanting of

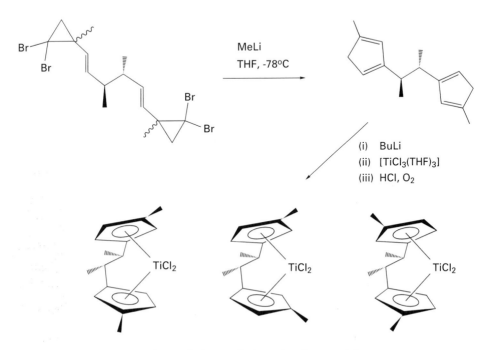

Scheme 5.9 An alternative synthetic route for formation of *ansa*-metallocenes.

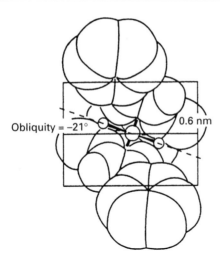

Obliquity = −21°   0.6 nm

Fig. 5.10 An example of the coordination gap aperture in *ansa*-metallocene dihalides, e.g. {ZrCl$_2$[Me$_2$Si(C$_5$H$_3$Ph-3)$_2$]}.

the tangential planes with respect to the MCl$_2$ plane. This angle between the intersection of the planes is termed the '*co-ordination gap obliquity*' and an example featuring [ZrCl$_2${Me$_2$Si(C$_5$H$_3$Ph-3)$_2$}] and an angle of 21° is shown in Fig. 5.10 (Burger *et al.*, 1993). This type of parameter has ramifications in 1-alkene polymerisations and asymmetric hydrogenations with chiral titanocenes as it can guide the enantiofacial preference of binding a prochiral substrate molecule to the metal centre.

The structures of *ansa*-metallocenes have also been closely studied due to the presence of configurational isomers and their effectiveness in the control of stereoselective reactions, notably 1-alkene polymerisations. One of the first stereoselective *ansa*-metallocenes to be studied was ethylene bis(4,5,6,7-tetrahydro-1-indenyl)titanium dichloride (**V.40**, see Figs 5.11 & 5.12 for the

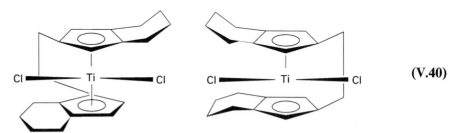

(V.40)

racemic and *meso* compound, respectively) and their X-ray structural determinations (Wild *et al.*, 1982). Figure 5.11 shows the *rac* isomer with the methylene-substituted edges of the rings bent away from the TiCl$_2$ moiety because of steric repulsion between the chloro ligands and the CH$_2$ groups. The axis normal to a C$_5$ ring plane forms an angle of 7.5 ± 1.5° with the plane bisecting the Cl—Ti—Cl angle. In Fig. 5.12, the *meso* isomer possesses

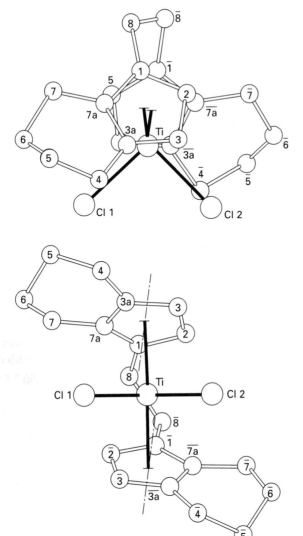

Fig. 5.11 Top and front views of *racemic*-**V.40**. (Wild *et al.* (1982), with permission.)

a twisted ethylene bridge and although the two tetrahydroindenyl units have slightly different positions in the solid state, interconversion can occur by minor conformational changes. Generally in *ansa*-metallocenes, the position of the bridge does not coincide with the plane bisecting the Cl—Ti—Cl angle and the whole ligand framework is shifted to one side of it. Repulsive steric interactions within the ligand framework are therefore minimised and so these metallocenophanes often exhibit deviations from C_2 symmetry in the solid state.

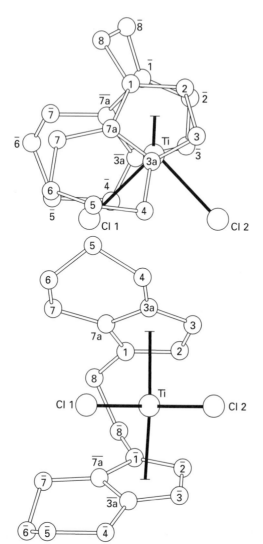

Fig. 5.12 Top and front views of *meso-V.40*. (Wild *et al.* (1982), with permission.)

5.1.8 Metallocenophanes featuring alkali and alkaline earth metals

X-ray structural studies of alkali metal cyclopentadienide complexes with Li, Na and K commonly reveal ionic polymeric chain structures, but in recent years the synthesis and characterisation of simple metallocene sandwich compounds have been reported (see earlier chapters) (Fig. 5.13). The position of equilibrium is difficult to predict due to solvation of the alkali metal. The possible existence of higher oligomeric CpMCpM structures or triple-decker anions CpMCpMCp⁻ complicates the solution equilibrium which is continuously changing due to Cp⁻ withdrawal. This Cp anion removal has been avoided by the synthesis of a novel anionic *ansa*-sodocene sandwich (or

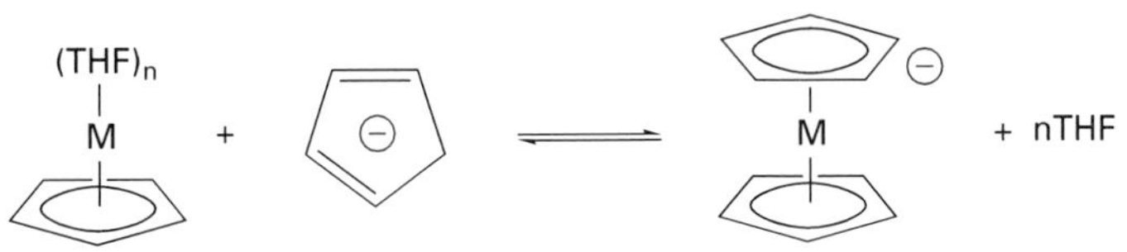

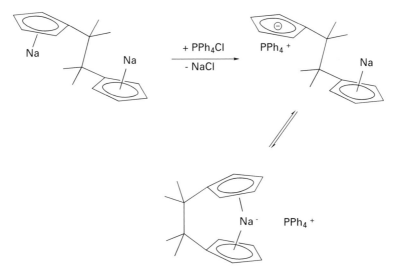

Fig. 5.13 Synthesis of alkali metal metallocenes.

Fig. 5.14 Synthesis of an anionic [2] sodocenophane.

[2] sodocenophane) (Fig. 5.14). From the addition of $Me_4C_2Cp_2Na_2$ in THF to a suspension of anhydrous Ph_4PCl there is immediate formation of a wine-red solution and from this red crystals of $[Me_4C_2Cp_2Na^-\cdot THF][PPh_4^+]$ were isolated. The X-ray analysis of the compound indicated a large degree of disorder of some of the atoms in the bridge and one of the cyclopentadienyl rings, because of a pivoting of the bridge and this ring around the other Cp ring. There is also the possibility of η^3-co-ordination of one of the Cp rings, via ring slippage. A similar *ansa*-metallocene or [2] metallocenophane has also been reported for calcium. Addition of 6,6-dimethylfulvene to a suspension of calcium granules in THF resulted in the formation in good yield of the colourless *ansa*-metallocene complex **V.41**. This non-solvated

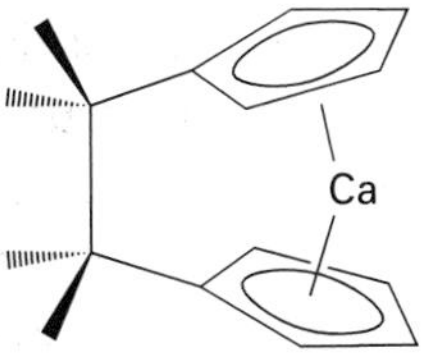

(V.41)

species is very air-sensitive and soluble in toluene and THF. In a similar fashion, strontium forms a colourless di(tetrahydrofuran) adduct but there is no reaction with barium. The *ansa*-calciocenes undergo interesting reactions with 1,4-diazadienes, e.g. reaction of **V.41** with *N,N'*-di(*tert*-butyl)glyox-aldialdimine in THF/toluene yields an air-sensitive, bright red 1:1 adduct **V.42**. Crystallisation of this adduct showed that the diazadiene was coordinated to the calcium atom in a *cis* arrangement, resulting in the calcium atom having a distorted pseudotetrahedral co-ordination geometry. The angle between the calcium atom and the centre of the rings was 119.0° but the N—C—N angle of 68.7° was substantially smaller than the tetrahedral angle.

(V.42)

5.1.9 Metallocenophanes featuring lanthanides and actinides

The bridging of two cyclopentadienyl rings by organic spacer groups results in chelating ligands which impose unusual properties especially on *f*-element organometallics. For instance, linking of the cyclopentadienyl rings will essentially prevent ligand redistribution processes and allow the otherwise impossible isolation of some lanthanide or actinide cyclopentadienyl species. A number of linkages between the cyclopentadienyl rings have been developed, ranging from simple methylene chains to those including donor atoms such as nitrogen and oxygen, and have allowed the isolation of unsolvated dicyclopentadienyl lanthanide halide derivatives. The dianion of 1,1'-trimethylene-dicyclopentadiene has been employed extensively in reaction with anhydrous lanthanide trichlorides to give disubstitution products of the type [{C$_5$H$_4$(CH$_2$)$_3$C$_5$H$_4$}LnCl(THF)]:

$$LnCl_3 + [NaC_5H_4(CH_2)_3C_5H_4Na] \xrightarrow{THF} [\{C_5H_4(CH_2)_3C_5H_4\}LnCl(THF)]$$
$$+ 2NaCl \qquad (5.1)$$

(Ln = La, Ce, Pr, Nd, Gd, Dy, Ho, Er, Yb, Lu)

Similar reactions have been utilised to include longer chains in the bridge such as pentamethylene [—(CH$_2$)$_5$—] and xylylene [—CH$_2$C$_6$H$_4$CH$_2$—] (Edelmann, 1995).

Lanthanide(II) *ansa*-metallocenes have been formed using the disodium salt of the trimethylene-bridged dicyclopentadienyl ligand and LnCl$_2$(Ln = Sm, Yb) in THF, or by another method involving the direct reaction of activated (by HgCl$_2$) lanthanide metal powders with 6,6-dimethylfulvene

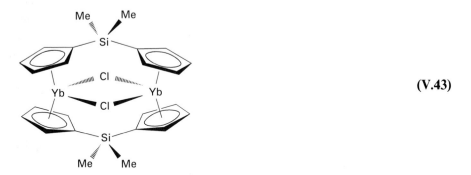

Ln = Yb, Sm

Scheme 5.10 Formation of *ansa*-metallocenes with lanthanide elements.

(Scheme 5.10). An unusual variation on the structure involving the ring-bridged cyclopentadienyl ligands has been reported. The compounds have the formula $\{[Me_2Si(C_5H_4)_2]Ln(\mu\text{-}X)\}_2$ (where Ln = Y, Yb; X = Cl, Br) and the ligands are not chelating but bridge the metal centres instead, as shown for **V.43** (Ln = Yb; X = Cl). Mass spectrometry can distinguish between the chelating or metal-bridging nature of the bidentate cyclopentadienyl ligands. Important catalytic behaviour has been noted for lanthanide hydrocarbyl and hydride compounds featuring tetramethylcyclopentadienyl units connected by Me_2Si or Me_2Ge links. These organolanthanide hydrocarbyls, $[\{Me_2Si(C_5Me_4)_2\}LnCH(TMS)_2]$ (Ln = Nd, Sm, Lu), are known as 'tied-back' and as shown via X-ray structural analysis exhibit C—H—Ln agostic interactions. Hydrogenolysis of these hydrocarbyls with H_2 gives the corresponding hydride dimers and these species can polymerise ethylene and oligomerise propylene and 1-hexene.

(V.43)

Bifunctional cyclopentadienyl ligands have been successfully utilised in organoactinide chemistry and allowed the isolation of $[Cp_2AnCl_2]$ derivatives otherwise inaccessible via the parent cyclopentadienyl species (Edelmann, 1995). For example, reaction of uranium tetrachloride with $\{Li_2[CH_2(C_5H_4)_2]\}$ gives $[U_2(\mu_3\text{-}Cl)_2(\mu\text{-}Cl)_3\{CH_2(C_5H_4)_2\}_2Li(THF)_2]$. There are problems of low solubility and proton abstraction in systems featuring unsubstituted Cp rings bridged by alkylene groups. Some of these problems can be overcome by using permethylated cyclopentadienyl rings and the bridged tetramethylcyclopentadienyl anion $[Me_2Si(C_5Me_4)_2]^{2-}$ has been used in the formation of soluble, thermally stable but reactive organoactinide hydrocarbyls, e.g.

$$ThCl_4 + Li_2[Me_2Si(C_5Me_4)_2] \xrightarrow{DME} \{[Me_2Si(C_5Me_4)_2]ThCl_2\} + 2LiCl \quad (5.2)$$

then

$$\{[Me_2Si(C_5Me_4)_2]ThCl_2\} + 2RLi \xrightarrow{DME} \{[Me_2Si(C_5Me_4)_2]ThR_2\}$$
$$+ 2LiCl \quad (5.3)$$

$(R = {}^nBu, {}^tBuCH_2, CH_2TMS, Bz, Ph)$

As with the lanthanide species, there is a certain 'opening up' of the coordination sphere around thorium through the linking of the cyclopentadienyl rings as witnessed by a decrease of $c.$ 20° in the ring centroid–Th–ring centroid angle compared to the corresponding non-bridged $[(Cp^*)_2ThR_2]$ complexes. This distortion of the bent metallocene geometry has a strong effect on the catalytic activity of the hydrido-derivative. Finally, the new metalloligands $[(C_5H_4PPh_2)_2UX_2]$ $(X = Cl, NEt_2, BH_4)$ have been used to synthesise bimetallic metallocenophanes. Treatment of the diphenylphosphinocyclopentadienyl derivatives with $[Mo(CO)_4(nbd)]$ (nbd = 2,5-Norbornadiene) gives a novel [3] uranocenophane, **V.44** (Edelmann, 1995).

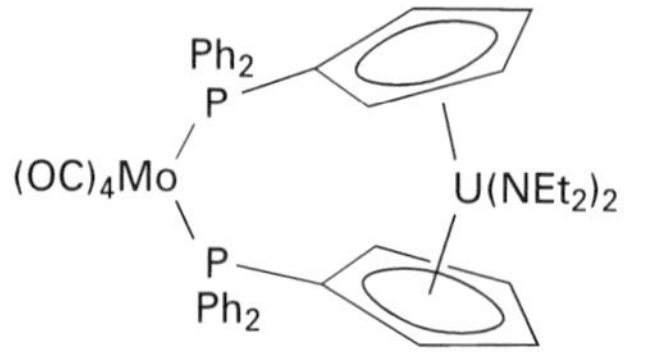

(V.44)

5.2 Polynuclear and heterobimetallic metallocenes

5.2.1 Ferrocene derivatives

These species show interesting mixed-valence properties. Upon one-electron oxidation of a compound with two or more equivalent ferrocene moieties, the electron vacancy could be localised on one ferrocene unit or completely delocalised. A series of flexible biferrocenyl cations have been studied and given interesting results, in particular the biferrocenyl triiodide complex **V.45**. Figure 5.15 shows ^{57}Fe Mössbauer spectra at varying temperatures and illustrates how the intramolecular transfer rate is dependent on the temperature (Cohn *et al.*, 1985). The observation of two doublets at room temperature indicates the presence of both Fe^{2+} and Fe^{3+} and demonstrates that intramolecular electron transfer is slow on the timescale (10^{-8} s) of the

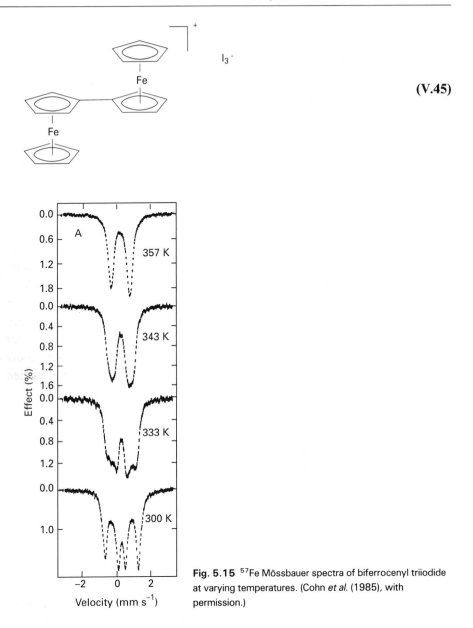

I_3^-

(V.45)

Fig. 5.15 ^{57}Fe Mössbauer spectra of biferrocenyl triiodide at varying temperatures. (Cohn *et al.* (1985), with permission.)

Mössbauer experiment and so this is a *mixed-valence* situation. However, as the temperature rises > 350 K, the merging of the signals into one doublet indicates that there is now only one type of iron centre present, it having an oxidation state of +2.5. So at this temperature the electron transfer is fast, illustrating an *intermediate valence* case. Such localisations are affected by the cation–anion crystal packing as in the localised dichloride **V.46** (X=Cl), where the anion is near one of the ferrocenyl groups which shows dimensions

213

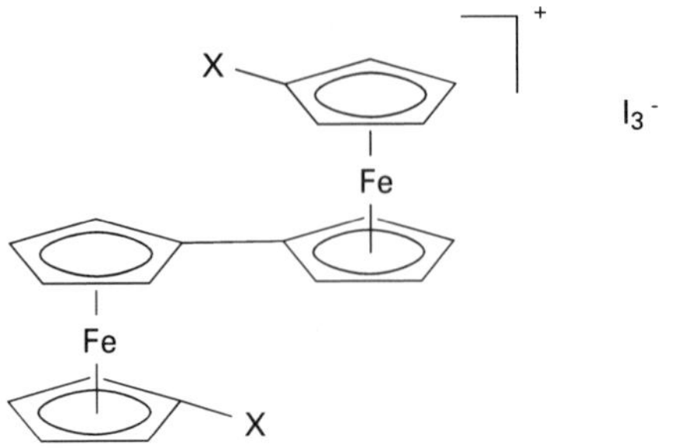

(V.46)

of a ferrocenium ion whereas the other resembles uncharged ferrocene. Consequently, the charge remains localised on the Mössbauer timescale even at 300 K.

As mentioned earlier, biferrocenophanes have been closely studied for their mixed-valence properties and interestingly in analogous compounds a trimethylene bridge was found to increase the intramolecular electron transfer rate by at least 10^3 compared to the pentamethylene-bridged species. Fully delocalised behaviour is rare to find and, to date, the only biferrocene cation other than [0.0] ferrocenophane$^+$ to show the delocalised characteristics on both Mössbauer and ESR timescales is the average valence species [2.2] ferrocenophane-1,13-diyne **V.35**. In this case, the unpaired electron density which is symmetric with respect to the bridge appears to be associated largely with the organic ligands. Conversely, the monocation of [1.1] ferrocenophane which has two ferrocenylene units joined by two methylene bridges shows distinctly localised behaviour. The crystal structure of 1,12-dimethyl[1.1] ferrocenophane$^+$ I_3^- exhibits distinct ferrocene and ferrocenium units (the cationic unit having characteristically longer Fe—C distances, nearer to the triiodide ion). This is also the case for the monocation of **V.47** which exhibits localised valences and the structural determination shows two structurally distinct iron environments with Fe—C distances in the ranges

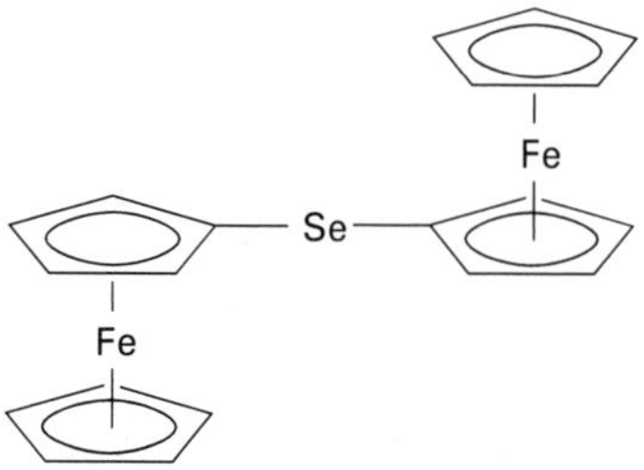

(V.47)

0.2016–0.2062 nm and 0.2053–0.2131 nm for the ferrocene and ferrocenium units respectively. Various cyclopentadienyl-substituted ferrocenes can be deprotonated and treated with $FeCl_2$ to form terferrocenyls and with partial oxidation exhibit interesting mixed-valence behaviour. Redox-active polymers containing ferrocenes as part of the backbone or as sidechains will be discussed in Chapter 6.

5.2.2 Ruthenocene derivatives

Biruthenocene has been obtained from the coupling of 1,1'-diiodoruthenocene and 1,1'-dilithioruthenocene but the best route to Rc_2 is from RcH in hot concentrated sulphuric acid where an intermediate radical species, possibly $[CpRu^{II}(C_5H_4{}^{\cdot})]$, is reduced by $TiCl_3$. In the solid state biruthenocene is disordered and adopts the transoid conformation with the mutual orientation of the Cp rings depending on intermolecular interactions. The mixed species ferrocenylruthenocene FeRu is formed via Ullmann coupling of a large excess of iodoferrocene with iodoruthenocene and more closely resembles biferrocene and biruthenocene. There is basically no interaction between the metal centres in this neutral compound but oxidation gives a diamagnetic species, probably $[FcRc''Rc''Fc][BF_4]$ featuring an Ru—Ru bond. Two different iron centres result from a temperature-dependent intramolecular electron-transfer equilibrium (Fe^{II}—Ru^{III} ↔ Fe^{III}—Ru^{II}). Oligoruthenocenylenes (**V.48**; $n = 2$–4) have been obtained as white crystals in low yields from the coupling of 1,1'-diiodoruthenocene with 1,1'-dilithioruthenocene·TMEDA. The low yields seem to be a result of the low anion nucleophilicity of the lithio compound and, with the low solubilities, no polymeric products were formed.

(V.48)

5.2.3 Vanadocene and rhodocene derivatives

Paramagnetic decamethylbivanadocene has been synthesised according to Scheme 5.11 and the product isolated in 25% yield as brown microcrystals. Vanadocenes and their co-ordination complexes have been used as starting materials for the synthesis of heterobimetallic complexes, e.g. the reaction of decamethylvanadocene and $V(CO)_6$ gives $[Cp_2^*V(\mu\text{-OC})V(CO)_5]$. The 18 VE ions $[Cp_2Rh]^+$ and $[Cp_2Ir]^+$ are very stable, unlike the neutral monomers Cp_2Rh and Cp_2Ir which to date have only been observed in matrix isolation. At room temperature they dimerise immediately (Fig. 5.16) and EPR measurements and MO calculations indicate that the unpaired electron in these complexes is located mainly on the C_5H_5 rings.

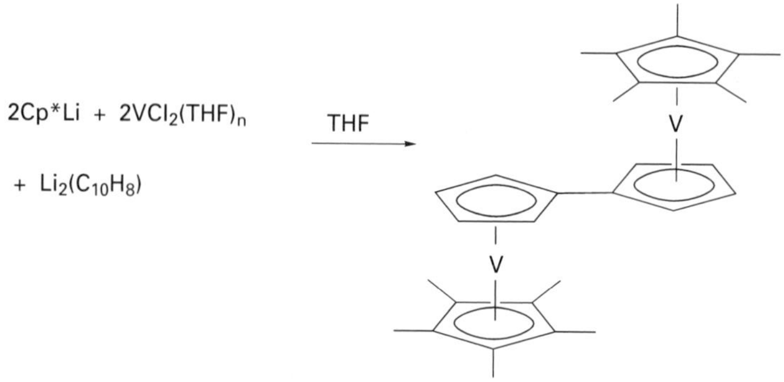

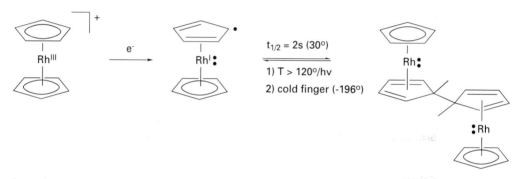

$2Cp^*Li + 2VCl_2(THF)_n$

$+ Li_2(C_{10}H_8)$

THF

Scheme 5.11 Formation of decamethylvanadocene.

Fig. 5.16 The dimerisation of rhodocene.

5.3 Multi-decker sandwich compounds

These types of materials are important for use as building blocks for creating materials with useful electronic properties. The possibility that transition metals might form triple-decker sandwich compounds was indicated by mass spectrometric results on $[(C_5H_5)_2Ni]$ and $[(C_5H_5)_2Fe]$ but it was some time before stable species were isolated and characterised. The first triple-decker sandwich compound featuring cyclopentadienyl ligands was formed in 1972. This unexpected product was obtained upon heating nickelocene with HBF_4 in propionic anhydride and analysed as the 34 VE cation $[(\eta^5\text{-}C_5H_5)_3Ni_2]^+$ (Scheme 5.12). The triple-decker arrangement was confirmed crystallographically with the rings arranged parallel to each other in a conformation between eclipsed and staggered, the outer ring being closer to the nickel atom than the inner one. The three rings were also within 3.97° of co-planarity.

The compound can also be formed from the reaction of nickelocene with a variety of Lewis acids, e.g. $[Ph_3C]BF_4$, $[Pd(C_3H_5)]BF_4$ and $[(C_5H_5)_2Ni]BF_4$, and the intermediacy of the 14-electron species $[(\eta\text{-}C_5H_5)Ni]^+$ was supported by its isolation from the reaction of $[Ni(\eta^4\text{-}C_5H_6)(\eta\text{-}C_5H_5)]^+F^-$ with Lewis acids and its further reaction with nickelocene to give **V.49** (product of

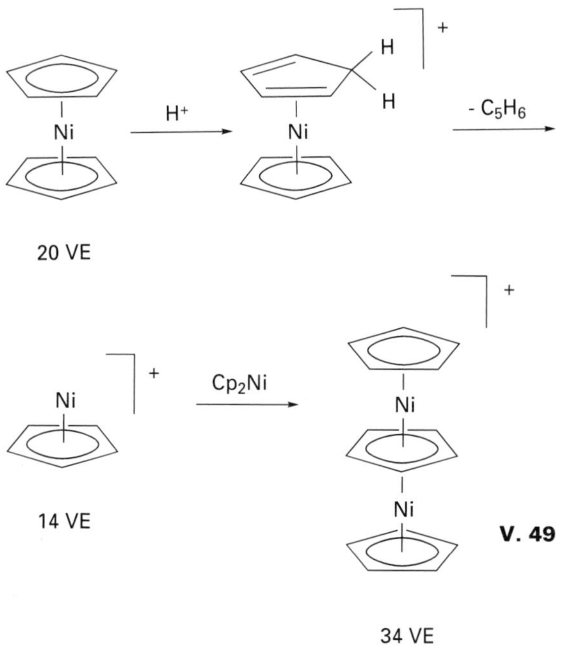

20 VE

14 VE

34 VE

Scheme 5.12 Synthesis of the first triple-decker sandwich compound.

Scheme 5.12). The electronic structure of **V.49** has been studied and it is thought that the orbitals are arranged in six low-lying levels and two e-levels of higher energy and, of these, the low-lying and slightly anti-bonding e' orbitals are occupied (Lauher *et al.*, 1976). This is illustrated in Fig. 5.17 upon examination of the interactions between the frontier orbitals of the $[(C_5H_5)_2Ni]^+$ fragments at the correct internuclear distance for a triple-decker sandwich compound and the central $C_5H_5^-$ ligand. In the $[(C_5H_5)Ni]^+$ fragment there are three filled low-lying molecular orbitals which are essentially non-bonding and have predominantly z^2 (a_1) and x^2-y^2, xy (e_2) character. Looking at how the two frontier orbitals of the two $[(C_5H_5)_2Ni]^+$ fragments interact in the absence of the central ring, the e_1 orbitals are only slightly split because they are of mainly d character and there is little net overlap whilst the overlap of the a_1 orbitals is more significant due to the higher proportion of metal s and p character and in- and out-of-phase combinations give rise to bonding a_1' and anti-bonding a_2'' orbitals. Considering the π-orbitals of the central cyclopentadienyl ligand, the a_2'' and e_1'' orbitals interact strongly with the corresponding orbitals of the $[(C_5H_5)Ni]^+$ fragments. These MO calculations indicate that triple-decker systems with 34 and 30 VE should display exceptional stability. In the 30 valence electron species, the bonding e_1'' molecular orbital which is delocalised over all three rings and the metal atoms is occupied. There is also a

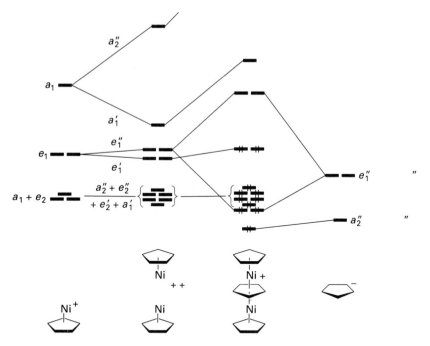

Fig. 5.17 A fragment molecular orbital analysis of the bonding in the triple-decker sandwich compound $[(\eta\text{-}C_5H_5)_3Ni_2]^+$. (Lauher *et al.* (1976), with permission.)

low-lying level of e_1' symmetry which is anti-bonding with respect to the outer rings and essentially non-bonding in relation to the central ring. In the 34 VE $[(C_5H_5)_3Ni_2]^+$ ion, this orbital is occupied by four electrons. Most triple-decker systems have 30 or 34 VE but intermediate numbers are known and in fact the first triple-decker species with a bridging benzene ligand $[(C_5H_5V)_2C_6H_6]$ **V.50** has only 26 valence electrons. There have been very

(V.50)

few examples of multi-decker species featuring purely unsubstituted cyclopentadienyl rings but one recent example was a thallium derivative, the first homometallic main group metal triple-decker sandwich anion (Fig. 5.18) (Armstrong *et al.*, 1995). This was formed from the reaction of $(C_5H_5)Li$ with

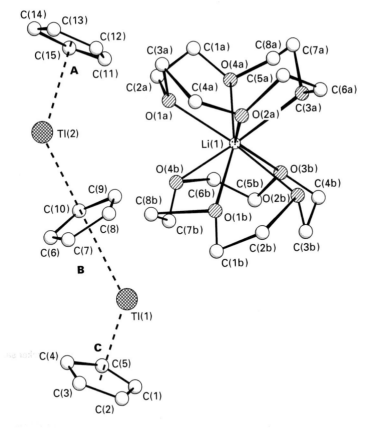

Fig. 5.18 The molecular structure of $[(\eta\text{-}C_5H_5)Tl(\mu\text{-}C_5H_5)Tl(\eta\text{-}C_5H_5)]^{-}[Li(12\text{-crown-4})_2]^{+}\cdot THF$. (Armstrong *et al.* (1985), with permission.)

$[(\eta\text{-}C_5H_5)Tl]$ in the presence of 12-crown-4 and a low temperature X-ray crystallographic study showed the structure to be the ion-separated complex $[(\eta\text{-}C_5H_5)Tl(\mu\text{-}C_5H_5)Tl(\eta\text{-}C_5H_5)]^{-}[Li(12\text{-crown-4})_2]^{+}\cdot THF$. The anion contained two Tl^{I} centres bridged almost symmetrically by a bent $\mu\text{-}C_5H_5$ bridge with the terminal C_5H_5 ligands being bonded more closely to the Tl centres. The anion can be viewed as the first homologue of the thallocene anion $[(\eta\text{-}C_5H_5)_2Tl]^{-}$ and a molecular segment of the polymeric structure of $[(\eta\text{-}C_5H_5)Tl]_\infty$.

A general approach to the 30-electron triple-decker sandwich molecules is via attack on electron-rich metallocenes by an electrophilic $[CpM]^{+}$ species. For instance, reactions between $Cp_2^{*}M$ (M = Ru, Os) and $[CpRu(NCMe)_3]^{+}$ gave 30-electron triple-decker sandwiches $[Cp^{*}MCp^{*}RuCp]^{+}$ whilst with $[CpFe(\eta^{6}\text{-arene})]^{+}$, $[Cp^{*}MCp^{*}FeCp]^{+}$ was formed (Scheme 5.13). The ruthenium and osmium compounds are stable in air and soluble in polar solvents. In the Ru_2 complex, the metal atom is 0.187 nm away from the inner Cp* group but only 0.175 nm from the outer Cp* group. Reorientation barriers for triple-decker sandwiches are affected by hydrogen bonding of the central ring to the $[PF_6]^-$ ion.

The heavy group 8 metals also give rise to triple-decker complexes featuring only the $C_5Me_5^-$ (Cp*) ligand which is often used due to its unusual electronic properties and the frequently excellent crystallisation behaviour of the resulting complexes, e.g.

$$[(C_5Me_5)_2Ru] + [(C_5Me_5)Ru(AN)_3]^+PF_6^- \rightarrow [(C_5Me_5)_3Ru_2]^+PF_6^- \qquad (5.4)$$

In addition, a paramagnetic cobalt triple-decker species **V.51** with 33 valence

$$(V.51)$$

electrons has been prepared via metal–vapour techniques. The distance from the Co atom to the central Cp* ring is once again longer than that of the metal atom to the outer ring. Due to the paramagnetism and different spin transfer mechanisms, interesting ^1H NMR shifts are seen, e.g. the signals for the two terminal Cp* rings are shifted to high frequency whereas that for the bridging Cp* ring is shifted to lower frequency relative to the solvent signal. There are some interesting examples of triple-decker sandwich species featuring two cyclopentadienyl ligands and one cyclooctatetraene bridging unit. The most thoroughly studied complexes here are $[(C_5H_5M)_2C_8H_8]$ (M = V, Cr, Fe, Ru, Co, Rh) and they can feature *syn-* and *anti-*variants. There are various methods of preparation:

$$2CrCl_2 + 2NaC_5H_5 + K_2C_8H_8 \xrightarrow[\text{(ii) 150°C}]{\text{(i) 25°C}} [(C_5H_5Cr)_2C_8H_8] \qquad (5.5)$$

$$2[C_5H_5Ru(CH_3CN)_3]^+ + K_2C_8H_8 \xrightarrow[-40°C]{\text{THF}} [(C_5H_5Ru)_2C_8H_8] \qquad (5.6)$$

$$[ClRhC_8H_8RhCl]_x + 2C_5H_5Tl \longrightarrow [(C_5H_5Rh)_2C_8H_8] \qquad (5.7)$$

In the complexes $[(C_5H_5Co)_2C_8H_8]$ and $[(C_5H_5Ru)_2C_8H_8]$ (36 VE), both metal atoms are *anti*-co-ordinated and each metal is bonded to a different pair of non-conjugated double bonds (η^4:η^4) of the boat-shaped C_8H_8 ring. If oxidised, the rhodium complex forms a stable dication (34 VE) where two

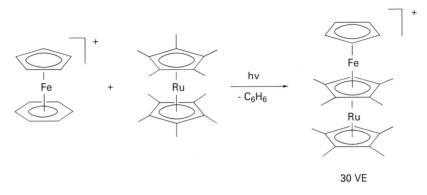

30 VE

Scheme 5.13 Formation of a mixed Fe—Ru triple-decker sandwich compound.

carbon atoms of the C_8H_8 ring are co-ordinated to both metal atoms (Scheme 5.14). The neutral isoelectronic $[(C_5H_5Ru)_2C_8H_8]$ displays the same 'slipped triple-decker' structure which on oxidation converts into a 'fly-over' structure in which the ring has opened and a metal–metal bond is formed (Scheme 5.14). Conversely, the 'fly-over' chromium complex $[(C_5H_5Cr)_2C_8H_8]$ (32 VE) is thermally converted into its isomer (30 VE) with *syn*-bridging cyclooctatetraene ($\eta^5:\eta^5$) (Figure 5.19).

In summary, for these types of complexes, decreasing the number of valence electrons leads to more closed structures. As was seen for the oxidation of $[(C_5H_5Rh)_2C_8H_8]$ (36 → 34 VE) dramatic structural changes are also observed for $[(C_5H_5Ru)_2C_8H_8]$, the species showing an unprecedented ring-opening and concurrent formation of a metal–metal bond. A further increase in electron deficiency, as observed in the chromium com-

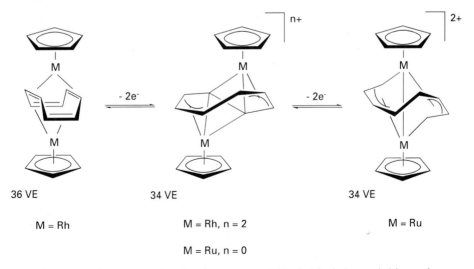

36 VE	34 VE	34 VE
M = Rh	M = Rh, n = 2	M = Ru
	M = Ru, n = 0	

Scheme 5.14 Interconversion of cyclooctatetraene-bridged triple-decker sandwich complexes.

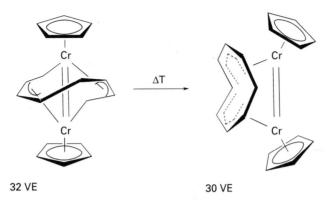

32 VE   30 VE

Fig. 5.19 The conversion of a 'fly-over' chromium complex to a *syn*-bridging cyclooctatetraene species.

plexes $[(C_5H_5Cr)_2C_8H_8]$ $(32 \rightarrow 30$ VE) completes the change from *anti*- to *syn*-co-ordination of the two C_5H_5M units and effects a higher metal–metal bond order.

There have now been large numbers of multi-decker sandwich compounds formed containing larger rings (with or without substituents) and/or heterocycles and whilst this is not really within the scope of this book, some are worth briefly mentioning here. Unsaturated ring systems in which individual (or all) carbon atoms have been exchanged for heteroatoms such as N, P, O, S or B can form transition metal π-complexes under certain conditions. This can lead to stabilisation of the heterocycles, previously unknown in their free state. The ligands $C_5H_5^-$ (cyclopentadienyl), $C_4H_4N^-$ (pyrrolyl), $C_4H_4P^-$ (phospholyl) and $C_4H_4As^-$ (arsolyl) are iso π-electronic and so synthesis of π-complexes containing these heterocycles is often analogous to that of corresponding complexes, e.g. the formation of 1-phosphaferrocene and 1,1'-diphosphaferrocene (Svara and Mathey, 1986) (Scheme 5.15). Considering triple-decker systems, a fascinating example features the hexaphosphabenzene P_6 ligand in a bridging role **(V.52)** (Scherer *et al.*, 1985). However, by far the most studied heterocyclic systems in multidecker sand-

(V.52)

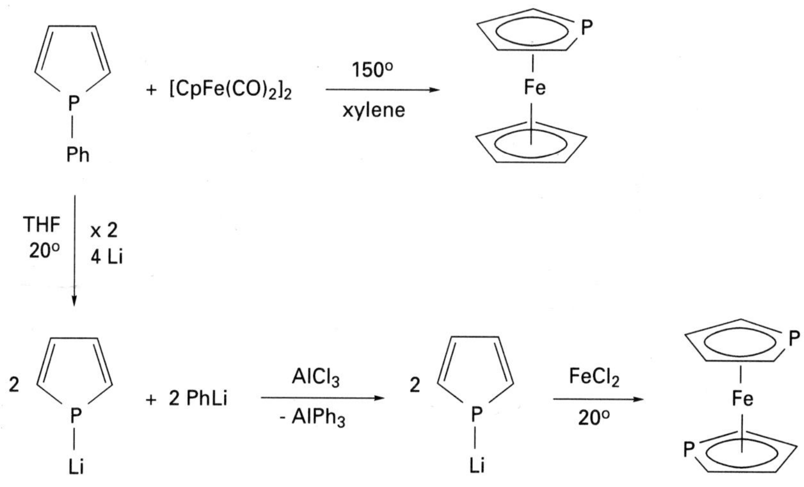

Scheme 5.15 The synthesis of 1-phosphaferrocene and 1,1'-diphosphaferrocene.

wich compounds have been the boron heterocyclic ligands (Herberich, 1995). The boron ligands have been very successful in multi-decker compound synthesis as it is an advantage if either the metal fragment (CpM) or the bridging ligand is electron deficient and now tetra-, penta-, hexa- and poly-decker complexes with boron ligands are known, e.g. **V.53** (Siebert, 1985). The presence of boron has been shown to greatly stabilise the structures due to its inherent electron deficiency. The fact that boron also readily shares

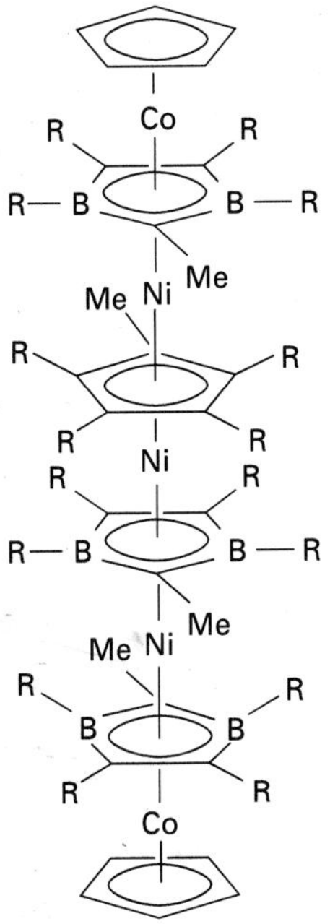

(V.53)

223

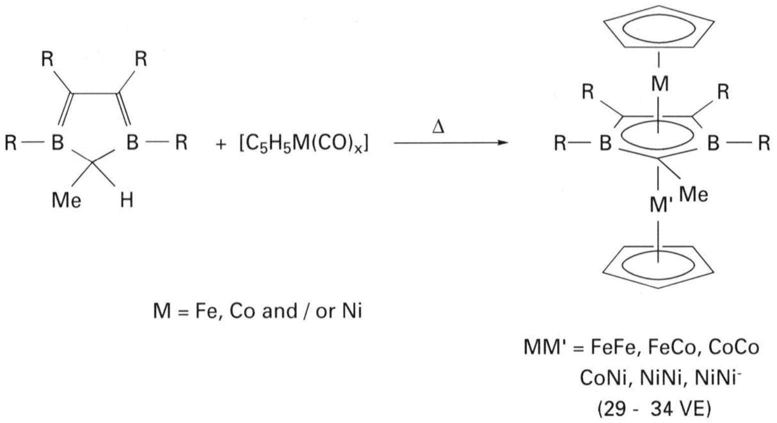
R
R
R— B
B — R
Me
H
+ [$C_5H_5M(CO)_x$]
Δ
M
R
R
R— B
B — R
M'
Me
M = Fe, Co and / or Ni
MM' = FeFe, FeCo, CoCo
CoNi, NiNi, $NiNi^-$
(29 - 34 VE)

Scheme 5.16 The formation of various metal triple-decker sandwich compounds involving a boron heterocyclic bridging ligand.

its valence electrons with neighbouring atoms also results in extensive electron delocalisation. This field has advanced in recent years, largely through the efforts of the research groups of Siebert (for boron ligands with three carbon atoms and two borons), Grimes and Hawthorne (for carborane multi-deckers) and Herberich (for multi-deckers incorporating C_4B rings).

In principle, all metals that possess the ability of polyhapto bonding to π-electron systems can also form complexes with boron ring systems but, in practice, the most commonly used metals are those of the triad from manganese to nickel (Scheme 5.16). One of the earliest observations was that these five- and six-membered rings with more than one boron atom featured a pronounced tendency to form multi-decker complexes, normally using (cyclopentadienyl)metal carbonyls as the 'stacking reagents'. In addition to the triple-decker species, the 1,3-diborolyl ligand lends itself as a bridging moiety in higher decker systems. The ultimate extension of this series has been reached by the preparation of poly-decker sandwich complexes **(V.54)**, the nickel complex being a semiconductor and the rhodium complex an insu-

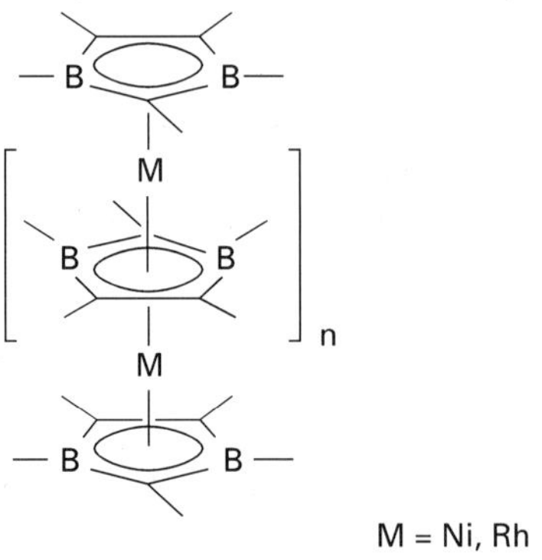
B
B
M
B
B
n
M
B
B
M = Ni, Rh

(V.54)

lator. Research continues into looking at these large, conducting materials for a wide variety of applications (e.g. transistors, electrochromic materials and conducting polymers).

References

Abel, E.W., Bhargava, S.K. and Orrell, K.G. (1984) *Prog Inorg Chem*, **32**, 1.

Abel, E.W., Long, N.J., Orrell, K.G., Osborne, A.G. and Sik, V. (1991) *J Organomet Chem*, **405**, 375, and references therein.

Abel, E.W., Long, N.J., Orrell, K.G., Sik, V. and Ward, G.N. (1993) *J Organomet Chem*, **462**, 287.

Armstrong, D.R., Edwards, A.J., Moncrieff, D. *et al.* (1995) *J Chem Soc Chem Commun* 927.

Barlow, S. and O'Hare, D. (1996) *Organometallics*, **15**, 3885, and references therein.

Binger, P. and Podubrin, S. (1995) in *Comprehensive Organometallic Chemistry* (eds E.W. Abel, F.G.A. Stone and G. Wilkinson) vol. 4, ch. 7, pp. 439–464. Pergamon Press, Oxford.

Bochmann, M. (1995) in *Comprehensive Organometallic Chemistry* (eds E.W. Abel, F.G.A. Stone and G. Wilkinson) vol. 4, ch. 4, pp. 221–271. Pergamon Press, Oxford.

Bruce, M.I. (1995) in *Comprehensive Organometallic Chemistry* (eds E.W. Abel, F.G.A. Stone and G. Wilkinson) vol. 7, ch. 10, pp. 615–617. Pergamon Press, Oxford.

Burger, P., Hortmann, K. and Brintzinger, H.H. (1993) *Makromol Chem, Macromol Symp*, **66**, 127.

Butler, I.R., Cullen, W.R., Ni, J. and Rettig, S.J. (1985) *Organometallics*, **4**, 2196.

Cassens, A., Eilbracht, P., Nazzal, A., Prössdorf, W. and Mueller-Westerhoff, U.T. (1981) *J Am Chem Soc*, **103**, 6367.

Cohn, M.J., Dong, T.-Y., Hendrickson, D.N., Geib, S.J. and Rheingold, A.L. (1985) *J Chem Soc Chem Commun*, 1095.

Deeming, A.J. (1982) in *Comprehensive Organometallic Chemistry* (eds E.W. Abel, F.G.A. Stone and G. Wilkinson) vol. 4, ch. 31.3, pp. 377–512. Pergamon Press, Oxford.

Edelmann, F.T. (1995) in *Comprehensive Organometallic Chemistry* (eds E.W. Abel, F.G.A. Stone and G. Wilkinson) vol. 4, ch. 2, pp. 11–195. Pergamon Press, Oxford.

Herberhold, M. (1985) *Angew Chem Int Ed Engl*, **34**, 1837.

Herberhold, M. (1995) in *Ferrocenes: Homogeneous Catalysis, Organic Synthesis, Materials Science* (eds A. Togni and T. Hayashi) pp. 219–278. VCH, Weinheim.

Herberhold, M., Steffi, U., Milins, W. and Wrackmeyer, B. (1996) *Angew Chem*, **108**, 1927; *Angew Chem Int Ed Engl*, **35**, 1803.

Herberich, G.E. (1995) in *Comprehensive Organometallic Chemistry* (eds E.W. Abel, F.G.A. Stone and G. Wilkinson) vol. 1, ch. 5, pp. 197–216. Pergamon Press, Oxford.

Hillman, M., Gordon, B., Weiss, A.J. and Guzikowski, A.P. (1978) *J Organomet Chem*, **155**, 77–86.

Hillman, M. and Fujita, E. (1978) *J Organomet Chem*, **155**, 87–98.

Hillman, M. and Fujita, E. (1978) *J Organomet Chem*, **155**, 99–108.

Spaulding, L.D., Hillman, M. and Williams, G.J.B. (1978) *J Organomet Chem*, **155**, 109–116.

Hisatome, M., Watanabe, J., Kawajiri, T., Yamakawa, K. and Iituka, Y. (1990) *Organometallics*, **9**, 497.

Hisatome, M. (1992) *Rev Heteroatom Chem*, **6**, 142.

Hisatome, M., Watanabe, J., Yamakawa, K., Kozawa, K. and Uchida, T. (1995) *Bull Chem Soc Jpn*, **68**, 635.

Lauher, J.W., Elian, M., Summerville, R.H. and Hoffmann, R. (1976) *J Am Chem Soc*, **98**, 3219.

Manners, I., Rulkens, R. and Lough, A.J. (1996) *Angew Chem*, **108**, 1929; *Angew Chem Int Ed Engl*, **35**, 1805.

Morris, M.J. (1995) in *Comprehensive Organometallic Chemistry* (eds E.W. Abel, F.G.A. Stone and G. Wilkinson) vol. 5, ch. 8, pp. 472–549. Pergamon Press, Oxford.

Mueller-Westerhoff, U.T. (1986) *Angew Chem*, **98**, 700; *Angew Chem Int Ed Engl*, **25**, 702.

Rulkens, R., Lough, A.J. and Manners, I. (1996) *Angew Chem*, **108**, 1929; *Angew Chem Int Ed Engl*, **35**, 1805.

Scherer, O.J., Sitzmann, H. and Wolmershauser, G. (1985) *Angew Chem*, **97**, 358; *Angew Chem Int Ed Engl*, **24**, 351.

Siebert, W. (1985) *Angew Chem*, **97**, 924; *Angew Chem Int Ed Engl*, **24**, 943.

Svara, J. and Mathey, F. (1986) *Organometallics*, **5**, 1159.

Smith, B.H. (1964) *Bridged Aromatic Compounds*. Academic Press, New York.

Watts, W.E. (1967) *Organometallic Chem Rev A*, **2**, 231.

Wild, F.R.W.P., Zsolnai, L., Huttner, G. and Brintzinger, H.H. (1982) *J Organomet Chem*, **232**, 233.

Zanello, P. (1995) in *Ferrocenes: Homogeneous Catalysis, Organic Synthesis, Materials Science.* (eds A. Togni and T. Hayashi) pp. 317–432. VCH, Weinheim.

6 The Uses and Importance of Metallocenes

This chapter illustrates the utilisation and versatility of metallocenes and their derivatives. Industrial and academic research has resulted in metallocenes being able to fulfil a range of roles and an overview of the important and topical areas will now be given. Clearly, the application of ferrocene and its derivatives has been the most well developed over the last few decades (recently reviewed by Togni and Hayashi, 1995) but it is the role of bent metallocenes in olefin polymerisation which is currently the most topical and important subject.

6.1 Olefin polymerisation

6.1.1 General introduction

Metallocene catalysts have been used to produce polyolefins commercially since the early 1990s, with particular interest being given to the production of polyethylene and polypropylene. These two polymers alone account for virtually 35% of all thermoplastics and elastomers; but metallocenes can also polymerise bulky monomers such as styrene and norbornadiene to form polymers whose physical properties compete with high performance engineering plastics such as nylon, polycarbonates and polyesters. Studies on novel, metallocene-based catalysts have focused on the development of new materials as well as gaining an understanding of the basic reaction mechanisms responsible for the growth of a polymer chain at a catalyst centre and the control of its stereoregularity. In contrast to heterogeneous catalysts, polymerisation by a homogeneous metallocene-based catalyst occurs at a single type of metal centre with a defined co-ordination environment. Thus, a correlation between the metallocene structure and polymer properties such as molecular weight, stereochemical microstructure, crystallisation behaviour and mechanical properties can be made. These catalysts can give efficient control of regio- and stereoregularities, molecular weights and distributions and co-monomer incorporation thereby greatly expanding the range and versatility of technically feasible types of polyolefin materials. This broad applicability is clearly of interest to the plastics industry. The field is currently one of the foremost areas of chemical interest and a wealth of material has been published in recent years. This alone could fill this book but for further detailed reading consult Brintzinger *et al.* (1995), Sinclair and Wilson

(1994), Kaminsky (1994), Möhring and Colville (1994) and Coates and Way-mouth (1995).

6.1.2 The history of and introduction to Ziegler–Natta polymerisation

The discovery of transition metal catalysts for alkene polymerisation by Ziegler and Natta in 1955 forms the foundation of today's polyalkene industry. During experiments to syntheses long-chain aluminium alkyls by treating triethylaluminium with ethene under pressure, Ziegler found that transition metal halides have a dramatic effect on the course of the reaction. For instance, nickel salts led to the dimerisation of ethene to butene but titanium tetrachloride catalysed ethene polymerisation to give a relatively high melting linear polymer. Ziegler initially used mixtures of $TiCl_4$ and $AlEt_3$ to give finely divided but polymeric $TiCl_3$ which acts as a heterogeneous catalyst. Catalysis occurs at crystal defect sites where the metal is co-ordinatively unsaturated and the whole process can be carried out at low pressures.

$$CH_2{=}CH_2 \xrightarrow[\text{25°C, 1 bar}]{TiCl_4/AlEt_3} \text{polyethylene (molecular weight } 10^4\text{--}10^5) \qquad (6.1)$$

Natta then applied the catalyst to the stereospecific polymerisation of propene. He established a link between the stereochemical structure and the bulk properties of the polymer.

$$CH_2{=}CH{-}CH_3 \xrightarrow[\text{25°C, 1 bar}]{TiCl_4/AlEt_3} \text{polypropylene}$$

$$\text{(molecular weight } 10^5\text{--}10^6) \qquad (6.2)$$

The general polymerisation of alkenes with metal halides activated by aluminium alkyls is now known as **Ziegler–Natta polymerisation.** Various mechanisms for the reaction have been put forward, but it is generally accepted that the reaction involves a heterogeneous catalyst with the active species being a metal alkyl possessing a vacant co-ordination site *cis* to the alkyl ligand (Scheme 6.1). The surface structure is thought to be responsible for the stereoregular propene polymerisation by imposing steric constraints on the mode of monomer co-ordination to the metal prior to insertion into the metal–alkyl bond. The observation by Natta in 1957 that species formed from $[Cp_2TiCl_2]/AlCl_3$ can act as homogeneous catalysts for ethene (but not propene) polymerisation has recently been re-examined and since then the area has undergone a renaissance.

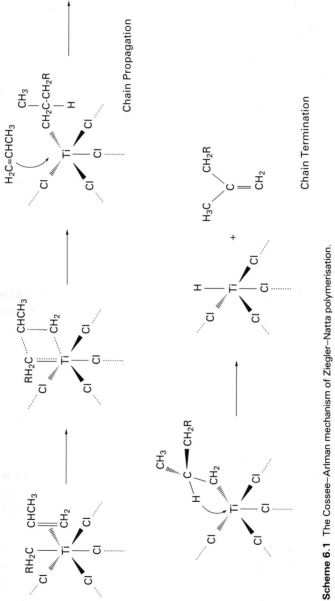

Scheme 6.1 The Cossee–Arlman mechanism of Ziegler–Natta polymerisation.

6.1.3 The discovery and mechanism of metallocene catalysis and types of catalyst

The studies on titanocene dihalide catalysts initiated by Natta showed the potential of these systems but it was not until the late 1970s and early 1980s that the situation changed dramatically when Sinn and Kaminsky (1980) discovered that partially hydrolysed trimethylaluminium (TMA) activated group 4 metallocenes for the polymerisation of both ethene and α-alkenes. Subsequent studies have shown that this reactivity resulted from the formation of methylaluminoxane (MAO) $\{[MeAlO]_n\}$ by the hydrolysis of TMA (Fig. 6.1). These units can condense together to form cyclic structures (with 5–12 Al atoms) which are soluble in hydrocarbons. This works best for Al : H_2O ratios of between 2 : 1 and 5 : 1, whilst the activity drops if the ratio falls below 1 : 3 because all the alkyl groups are then hydrolysed. Highly active alkene polymerisation catalysts are obtained when MAO is mixed with group 4 metallocenes in the molar ratio of about 1000 : 1. The reason for such a high concentration of MAO is thought to be in order to shift an equilibrium involving a metal alkyl and MAO to the active species – an ion pair (i.e. a metallocene cation that is co-ordinatively unsaturated and is stabilised by a bulky, non-co-ordinating MAO anion). The high catalytic activity reported suggests that the MAO-derived anion is labile and is easily displaced by the monomer prior to insertion. Basically, the aluminium alkyl functions in three ways: (i) it alkylates the transition metal; (ii) it acts as a Lewis acid and creates a vacant coordination site by abstracting a halide or alkyl ligand from the transition metal; and (iii) it acts in a cleaning role, by mopping up impurities from the monomer and the reaction medium.

The most common metal centres used in metallocene catalysts for olefin polymerisation are, in descending order, Zr, Ti and Hf while other significant research has featured the neutral group 3 complexes (Sc, Y) and organolanthanides and actinides. Considering the group 4 derivatives, the active species has been identified as the co-ordinatively unsaturated cationic $[\eta\text{-}C_5H_5)_2MR]^+$ (14 VE) system which mimics the catalytic sites of surface-alkylated $TiCl_3$ of the classical Ziegler–Natta process. The same cationic species are obtained in the absence of aluminium alkyls from metallocene

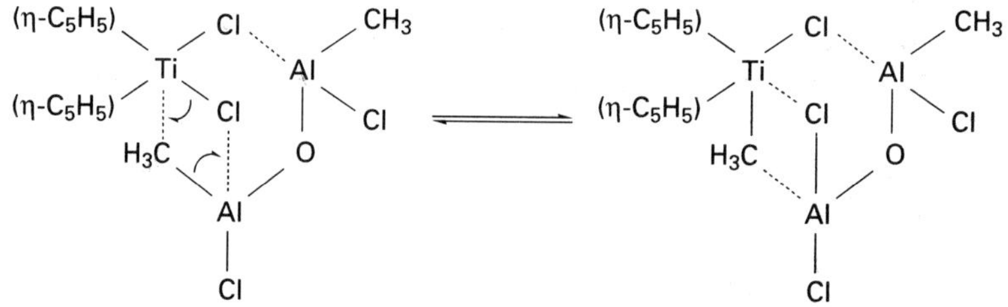

Fig. 6.1 The interaction of methylaluminoxane with the titanocene dichloride catalyst.

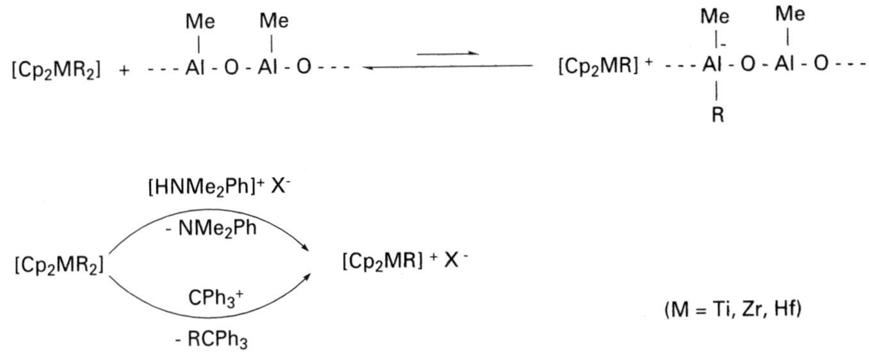

$$[Cp_2MR_2] + \text{---Al-O-Al-O---} \quad \xrightarrow{\qquad} \quad [Cp_2MR]^+ \text{---Al-O-Al-O---}$$

(with Me groups on the Al atoms on the left, and Me, Me, R groups on the Al atoms on the right)

$$[Cp_2MR_2] \xrightarrow[\substack{CPh_3^+ \\ -RCPh_3}]{\substack{[HNMe_2Ph]^+\ X^- \\ -NMe_2Ph}} [Cp_2MR]^+\ X^-$$

(M = Ti, Zr, Hf)

Scheme 6.2 Different methods of formation of the active $[Cp_2MR]^+$ species.

dialkyls by protolysis or R^- abstraction with CPh_3^+ (Scheme 6.2). With non-co-ordinating counterions such as $B(C_6H_5)_4^-$ and $B(C_6F_5)_4^-$ these cationic catalysts have displayed very high activity; however, these counterions can sometimes weakly co-ordinate (via a η^2-benzyl interaction) to cationic d^0 metal centres, thus preventing the olefin substrate from filling the empty sites required to undertake a catalytic growth reaction. In various ligand-stabilised complexes $[Cp_2MR(L)]^+$ (where L = THF, RCN, PR$_3$), the dissociation of the ligand L has to be facile in order to catalyse alkene polymerisation. The use of these weakly basic and poorly co-ordinating anions to generate active catalysts has been developed because, unlike MAO, which is needed in vast excess, they can be used in stoichiometric quantities with the metallocene.

The mechanism of polymer chain propagation is similar to that proposed for heterogeneous catalysts. In both cases, two co-ordination sites are required, i.e. one for the alkyl group and one for binding of the unsaturated substrate (Scheme 6.3). In general, the activity increases in the series M = Ti $\ll$ Hf < Zr but catalysis only occurs if the metal is d^0 (oxidation state IV); but why is this? In this state the metal cannot stabilise the intermediate olefin adduct by back-bonding. The d-orbitals of early transition metals are high in energy and, if occupied, they would lead to substantial stabilisation from back-bonding which would greatly increase the activation barrier of the alkyl migration step. A d^2 complex (**VI.1**) therefore resists all efforts to force alkyl migration to the C=C double bond even though it is structurally similar to the active intermediate ethene adduct. d-orbital energies decrease from left to right across the periodic table, and so back-bonding becomes less important in this direction. Therefore, co-ordinatively unsaturated alkyls of cobalt and

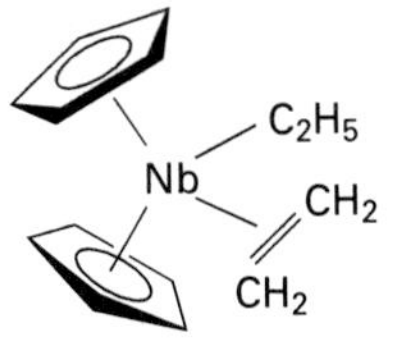

(**VI.1**)

(structure showing Nb with two Cp ligands, bonded to C_2H_5, CH_2, and CH_2)

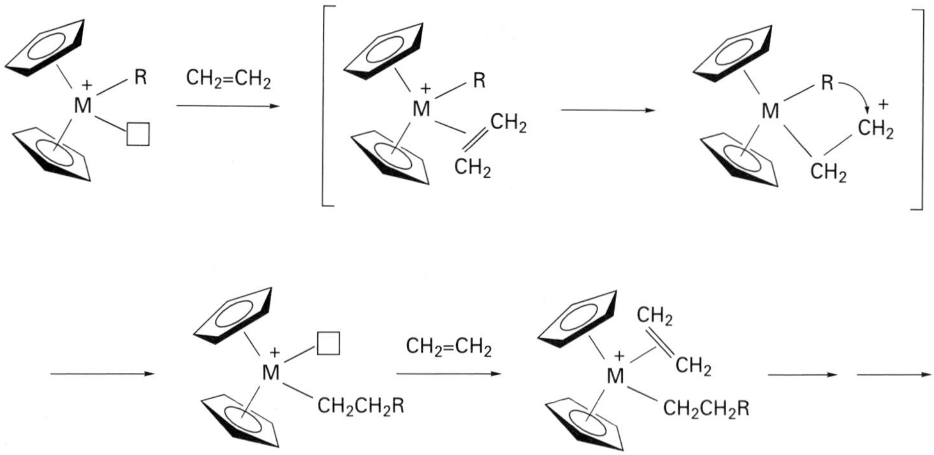

Scheme 6.3 The mechanism of olefin polymerisation using metallocene catalysts.

nickel are able to polymerise ethene though the activity is much lower than with d^0 systems.

Within the mechanism of polymerisation, the likelihood of chain termination is clearly crucial. The use of molecular hydrogen leads to formation of a metal hydride and thus effects chain termination. In the absence of hydrogen, alkyl chain growth is usually terminated by β-H elimination but if aluminium alkyls are present, transfer of the polymer chain from the transition metal to aluminium can also take place. The kinetic parameters k_p (rate of chain propagation) and k_t (rate of termination) clearly influence the rates and extent of polymerisation (Scheme 6.4). For example, if $k_t \simeq k_p$ then ethene is dimerised to butene as occurs when using $NiX_2/AlEt_3$ mixtures. These types of nickel catalysts can be made selective for the dimerisation, oligomerisation or polymerisation of ethene by a suitable choice of ligands. With d^0 complexes $[Cp_2MR]^+$ (M = Ti, Zr, Hf) β-H elimination is not easy and $k_p \gg k_t$. Some isoelectronic neutral scandium, yttrium and lanthanide complexes Cp_2MR and Cp_2^*MR also show high polymerisation activity for the same reason.

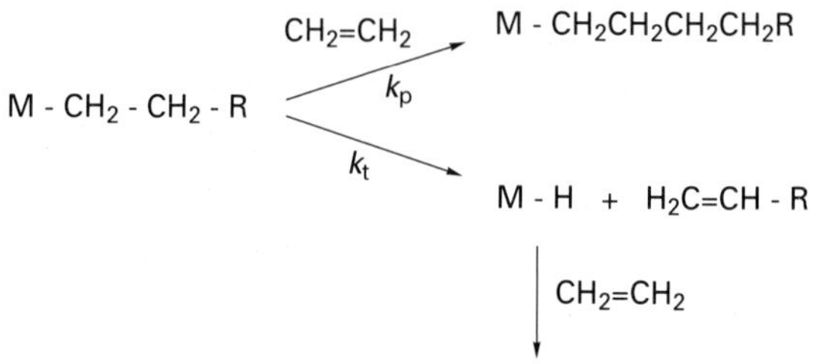

Scheme 6.4 The mechanisms of chain propagation and termination.

6.1.4 General features of metallocene catalysts

There are four main features that distinguish metallocene catalysts from all the other systems used in the polyolefin industry:

1 *They can polymerise almost any monomer irrespective of its molecular weight or steric bulk.* The patent literature of 1994 describes α-olefin polymerisation with metallocenes using more than 60 different monomers. Many of these were very difficult to polymerise with conventional Ziegler–Natta catalysts. Pre-metallocene Ziegler–Natta catalysts do not easily incorporate α-olefin co-monomers (e.g. the reactivity ratios for ethylene and 1-butene are greater than 1000) and, for example, the alternative Phillips system (CrO_3/ SiO_2) only works with ethene. By linking the metallocene's two cyclopentadienyl rings together via a bridge (i.e. forming a metallocenophane or *ansa*-metallocene), the catalytic site can be opened up and the reactivity ratios greatly reduced. This allows a greater variety of α-olefins to be polymerised and can aid co-polymerisation. Alternatively, the metallocene catalysts can be made more specific by increasing the bulk of the ligands on non-bridged metallocenes. This results in very high substrate specificity, and the overall ability to control these reactivity ratios enables the metallocene catalysts to accept such a wide range of monomers.

2 *They produce extremely uniform polymers and co-polymers of relatively narrow molecular weight distribution and narrow compositional distribution.* This is due to metallocene catalysts possessing a single catalytic site and as every catalytic site is identical this allows the metallocene to produce extremely uniform homo- and co-polymers with molecular weight distributions of 2 or less. If a broader molecular weight distribution is required it can be effected with exceptional precision by using two or more different metallocene catalysts in the same reaction in a controlled ratio, thereby giving more than one catalytic site.

3 *They can control vinyl unsaturation in the polymer produced.* If no chain transfer agents are present then the chain can be terminated by β-hydride (or β-methyl) elimination. This leaves a vinyl double bond at the end of every polymer chain which can be used later for further functionalisation or polymerisation. The control over the degree of vinyl unsaturation enables specific functionalisation and opens up further uses for the materials.

4 *They polymerise α-olefins with very high stereoregularity to give isotactic or syndiotactic polymers.* Polymer tacticity is controlled by the catalyst's stereochemistry and metallocene catalysts (i.e. *ansa*-metallocenes) can produce either isotactic or syndiotactic polymers. In general, the more sterically crowded the catalyst complex the more stereospecific the polymerisation (Fig. 6.2).

Metallocenes were the first systems to produce 98% syndiotactic polypro-

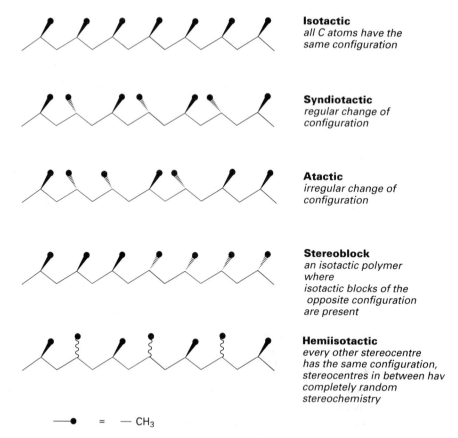

Isotactic
*all C atoms have the
same configuration*

Syndiotactic
*regular change of
configuration*

Atactic
*irregular change of
configuration*

Stereoblock
*an isotactic polymer
where
isotactic blocks of the
opposite configuration
are present*

Hemiisotactic
*every other stereocentre
has the same configuration,
stereocentres in between hav
completely random
stereochemistry*

$\bullet$ = $-CH_3$

Fig. 6.2 Different types of polymer tacticity.

pylene. Isotactic polypropylenes made using metallocenes also show significant promise for improved properties, i.e. better stiffness and stiffness/impact balance. Thus the physical properties of a desired polymer can be predetermined and the reaction catalyst tailored to meet these needs accordingly. In particular, syndiotactic polystyrene and other aromatic polymers comprise a new industrial research area that has been made possible by using metallocene catalysts.

6.1.5 'Tailoring' of metallocene catalysts

Extensive research has been devoted towards '*catalytic tailoring*' with modification of the catalyst system leading to specific changes in catalytic activity and product characteristics. This requires an understanding of the physical properties involved in ligand modification and there are several important factors affecting catalytic performance:

1 *Transition metal–olefin interaction*. The olefin has a basic character with

respect to the metal and therefore acts as an electron donor. The σ- and π-bonds formed between the metal and olefin destabilise and activate the olefin for insertion. Olefin co-ordination also destabilises the M—R (R = alkyl) bond. The stability of olefin co-ordination to M decreases with increasing olefin size due to steric factors and the energies of the olefin orbitals involved in bonding to the metal.

2 *Metal–alkyl bond stability.* Fine adjustment of the M—R bond is possible by variation of ligand electronic effects. The M—R bond should be unstable to permit facile opening and olefin insertion to form a new metal–alkyl bond but also strong enough to favour catalytic lifetimes. The strength of this bond depends on R itself and the stability decreases in the order Me > Et > $(CH_2)_n CH_3$. Olefin co-ordination is also another method of weakening the M—R bond in preparation for migratory insertion.

3 *Influence of other ligands.* Considering the Cp rings and substituents attached to them, then if the ligand system used is a good electron donor it will reduce the positive charge on the metal. This weakens the bonding between the metal and all other ligands, particularly the already unstable M—R bond, making insertion more facile. Conversely, this will also stabilise high oxidation state complexes and make co-ordination of the incoming olefin more unfavourable and so a balance must be found between these effects.

4 *Steric effects of the other ligands.* Bulky ligands will aid stereospecific olefin co-ordination and polymerisation. Steric effects influence the co-ordination of bulky monomers and selectivity towards different monomeric species in a polymerisation reaction can be maintained. As mentioned earlier, the co-ordination site can be opened or closed by controlling the angle that the cyclopentadienyl rings tilt away from each other. Shortening or lengthening the bridge in *ansa*-metallocenes can lead to much improved monomeric stereoselectivity.

By far the most exciting development in recent years has been the ability to induce stereoselectivity in the growing polymer chain and this has led to the breakthrough in industrial applications for metallocene catalysts. Examples of this will be examined in the next two sections.

6.1.6 Isotactic polymers using metallocene catalysts

The modification of group 4 metallocenes to produce catalysts capable of isospecific polymerisation developed slowly but dramatic successes were reported in the mid-1980s. Prior to this time, catalysts using achiral $[Cp_2MCl_2]$ systems could only produce achiral polypropylene; but then Ewen reported the use of metallocene-based catalysts for the isospecific polymerisation of propene whilst Brintzinger used chiral *ansa*-metallocenes (**VI.2**) to polymerise ethene with a very high degree of stereoselectivity (i.e. almost all C atoms were

(VI.2)

M = Ti, Zr

isotactic). Brintzinger found that as well as its stereoselective properties, the catalyst also had a remarkable activity and gave favourable molar weight distributions. It was shown that using a combination of *meso* and *racemic ansa*-metallocenes, a mixture of atactic and isotactic polymer chains were produced and soon it was recognised that the isotacticity of the polymer was directly related to the chirality of the metallocene that produced it. This was confirmed when isotactic polypropylene was formed using an isomerically pure sample of a chiral zirconocene analogue.

Zambelli has studied the origin of enantiofacial selectivity of alkene insertion. The insertion of propene into an M—Et bond proceeds with a high degree of stereoselectivity, whereas its insertion into M—Me proceeds with no observable stereoselectivity and this was attributed to the higher steric constraints of the ethyl group. Corradini has carried out studies in conformational modelling and found that the polymer chain is forced into an open region of the metallocene thus relaying the chirality of the metallocene to the incoming monomer through the orientation of the β-carbon of the alkyl chain (Scheme 6.5). These chiral metallocenes have C_2 symmetry with both possible reaction sites being homotropic and therefore selective for the same alkene enantioface. The result is a polymerisation reaction that yields an isotactic polyalkene.

Variation of the metallocene structure so as to affect the degree of stereoselectivity of the polymerisation reaction has been thoroughly investigated. Improvements in stereoselectivity occur on introducing a dimethylsilyl bridge (**VI.3**) or methyl substituents onto the aromatic rings. Brintzinger and co-workers have tailored a series of *ansa*-zirconocenes (**VI.4**) and found that all the catalysts yield highly isotactic polymers. Compound **2** was found to be better than **1** in terms of stereoselectivity because of its higher stereorigidity.

(VI.3)

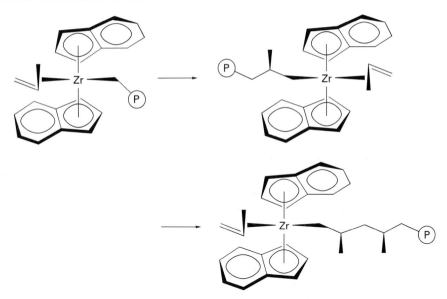

Scheme 6.5 The production of isotactic polymers using chiral metallocene catalysts.

1 R^1 = ^{t}Bu, R^2 = H, X = Me$_2$C-CMe$_2$

2 R^1 = ^{t}Bu, R^2 = H, X = Me$_2$Si   **(VI.4)**

3 R^1 = ^{t}Bu, R^2 = Me, X = Me$_2$Si

4 R^1 = ^{i}Pr, R^2 = Me, X = Me$_2$Si

Compound **3** managed to achieve an even higher isotacticity, because of its increased steric bulk whilst compound **4** showed less stereoselectivity due to decreased size. Overall, compound **3** produced the longest and most isotactic polymers and compound **1** the least. Bercaw recently formed the first single-component, iso-specific Ziegler–Natta catalyst which comprised a substituted *ansa*-metallocene on an yttrium metal centre **(VI.5)**. The system forms highly isotactic polyalkenes and has led to further research into yttrium

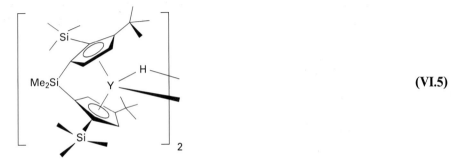

(VI.5)

derivatives. There have also been reports on unbridged metallocene catalysts in which the ligands are held in chiral arrangements by appropriately placed substituents.

The final barrier to the commercialisation of metallocenes for the production of isotactic polyolefins is expense. The synthesis of metallocenes involves multi-step procedures with the separation of pure racemic isomers, as is required for isotactic polymerisation, being a costly processing stage. However, the cost and performance of the metallocene catalyst systems for polypropylene are now in the competitive domain of industry something which is coupled with the better physical properties of the resulting polymers. In particular, stiffness is at least 30% higher than conventional isotactic polypropylenes and impact co-polymers have a much better impact/stiffness balance.

6.1.7 Syndiotactic polymers using metallocene catalysts

The first case of production of a syndiotactic polymer came in 1962 when Natta and Zambelli reported a heterogeneous vanadium-based catalyst mixture which produced partially syndiotactic polypropylene. Although this process suffered from low stereoselectivity and low activity, interest was generated in the concept of producing syndiotactic polymers. In 1988, Ewen reported the use of zirconium (or, less importantly, hafnium) metallocene catalysts (**VI.6**) to bring about the syndiospecific polymerisation of propene

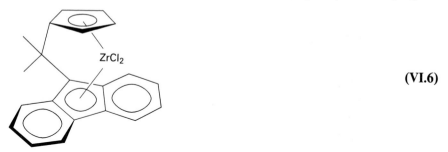

$$ZrCl_2$$

(**VI.6**)

and higher aliphatic α-alkenes. The regularly alternating insertion of alkenes at the heterotropic sites of the C_s-symmetric (achiral) Zr complex results in the formation of a syndiotactic polymer (Scheme 6.6). There is a chance of defect incorporation, as with isospecific C_2-symmetric catalysts through occasional mis-insertion of the incorrect enantioface.

The properties of syndiotactic polypropylene differ substantially from those of the isotactic analogue in several ways. Most importantly, syndiotactic polypropylene homopolymers possess a much higher room temperature impact strength and often higher optical clarity. These attractive properties must be weighed against the difficult processing characteristics which as yet cannot be controlled adequately. Highly syndiotactic polystyrene has also

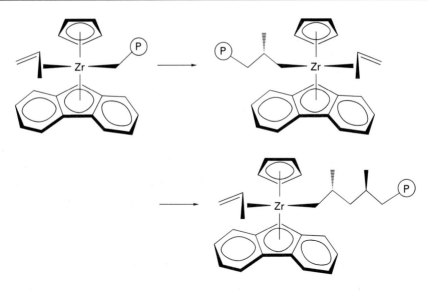

Scheme 6.6 The production of a syndiotactic polymer using zirconocene catalysts.

been made, it being crystalline and having a melting point of 270°C. The catalyst used is primarily pentamethylcyclopentadienyl titanium trimethoxide and the polymer is a potential competitor for engineering plastics such as nylon, polycarbonate and polyesters.

6.1.8 The commercial importance and future of metallocene catalysts

Many patents published in this field are based on catalysts for ethylene and propylene-based polymers and new patents are issued at around 120–200 per year. This is an amazing statistic as ten years ago metallocene catalyst R & D was negligible. This change is largely due to research efforts by industrial giants such as Dow, Exxon, BP, Fina, Hoechst, Du Pont and Mitsui. Indeed, multimillion pound plants featuring metallocene catalysts are already on stream with many more to come. In 1996 the global capacity for metallocene-based polymers was predicted to rise by almost 900 Mt and by the year 2000, metallocene-based polymers could account for more than 10% of the global thermoplastics and elastomers market or more than 20 Mt of polymer. This is a clear indication of the importance placed on these catalysts and though metallocene catalysts cost from a few thousand to several thousand dollars per kg, many times the cost of traditional Ziegler–Natta catalysts, they are significantly more productive. Some situations produce one or two orders of magnitude more polymer per kg of catalyst and, as these catalysts can often be substituted in existing processes, then the catalyst cost per kg of polymer produced starts to approach that of traditional catalysts. This ratio is continually being improved together with the vastly superior properties and/or processibilities.

The future looks bright for these systems as produced from metallocene-based initiators polyolefins are not restricted by the current immutable property relationships that are universally understood and are used to define polyolefins. For example, there is a fixed relationship between molecular weight distribution and melt flow characteristics in linear polyethylenes, but with polymers produced from metallocenes there is not and does not need to be a relationship. In many cases, the behaviour and physical properties of metallocene-derived polyolefins have yet to be fully explained. This leads to new possibilities and almost any combination of physical properties may be possible. Further scope is predicted in the enantioselective synthesis of optically active macromolecules, the polymerisation of cyclic monomers and the formation of functionalised polymers and block co-polymers. In addition there is still much to be learnt about the nature of the catalyst itself, with group 3 and lanthanide species coming more to the fore, but all in all the research in this area is set to remain intense.

6.2 Metallocene-containing magnetic materials (electron-transfer donor–acceptor salts)

Magnets are a well-known feature of the world today, being found in a myriad of everyday applications, encompassing frictionless bearings, medical implants, acoustic devices, copy machines, magneto-optical disks, motors, generators and communication equipment. In future years, magnets are expected to play a key role in the development of 'smart materials' based upon magnetic switches, sensors and transducers. The discovery of molecular materials displaying properties usually connected with the metallic phase, such as superconductivity and ferromagnetism, has stimulated activity in the field. Molecular materials are, in general, lighter, soluble, transparent and may have particular optical properties which could open up new, unique and more economic opportunities for the processing of the materials (Miller and Epstein, 1994, 1996; Togni, 1995).

Charge-transfer (CT) complexes containing a donor (D) and an acceptor component (A) capable of forming stable radical cations and anions respectively and displaying a solid state stacked structure can be electrical conductors with potential metallic character. This field of study was transformed in 1973 by the discovery of the high electrical conductivity and metallic behaviour of the one-dimensional CT complex formed by the donor tetrathiafulvalene (TTF) (**VI.7**) and the acceptor tetracyano-p-quinodimethane (TCNQ) (**VI.8**). Metallocenes and their alkylated derivatives have been intensively studied in connection with the magnetic properties of their CT complexes and this has led to the discovery of the first organometallic compound displaying bulk ferromagnetic properties.

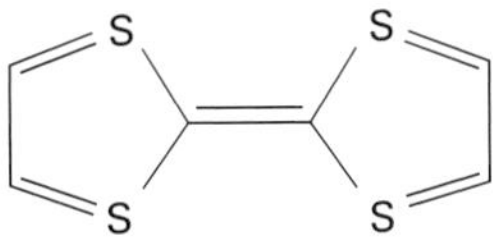

(VI.7)

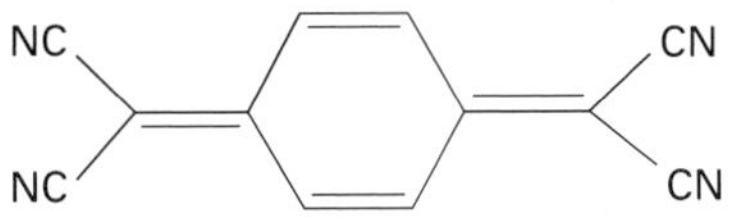

(VI.8)

Charge-transfer complexes need to fit several criteria in order to display stacking in the solid state, i.e. (i) the molecules should be planar or composed of planar fragments; (ii) they should form stable radical species in which the energy gap between HOMO and LUMO is fairly small; and (iii) the molecules should possess extended π-systems and should be able to approach one another at distances closer than the sum of the van der Waals radii, thereby increasing intermolecular overlap. Compared to many organic donors, ferrocene is a more compact cylindrical molecule featuring two planar rings and occupying parallel planes. Ferrocenes are also easily oxidised with the resulting ferrocenium salts being very stable and their electron-donating ability can be fine-tuned by the number and nature of the substituents.

6.2.1 Decamethylferrocene with organic acceptors

Ferrocene itself is relatively difficult to oxidise and thus forms only weak CT complexes but decamethylferrocene (VI.9) (being $c.$ 0.5 V easier to oxidise

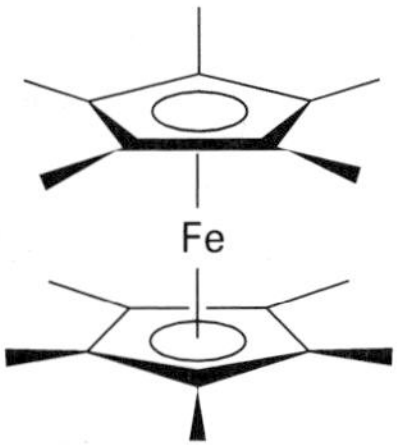

Fe

(VI.9)

and resistant to substitution reactions and ring exchange) has more potential. In 1979, $[Cp_2^*Fe^{III}]^{\cdot+}$ $[TCNQ]^{\cdot-}$ was shown to exhibit a high moment ferromagnetic state and to possess a linear chain structure. The moment μ significantly deviates to positive values at low temperatures with respect to the characteristic temperature-independent Curie behaviour of ferrocenium salts with spinless anions. The compound was the first example in which neither a one-, two- nor three-dimensional covalently bonded network structure was present and led to further studies to identify the steric/electronic features necessary to stabilise ferromagnetism and ultimately design a molecular species-based ferromagnet. Many other types of acceptors, both organic and inorganic, have been studied and with the view that a smaller radical anion would have a greater spin density which could lead to increased spin–spin interactions, $[TCNE]^{\cdot-}$ was selected. The first bulk organometallic (3D) ferromagnet $[Cp_2^*Fe^{III}]^{\cdot+}$ $[TCNE]^{\cdot-}$ was thus prepared and found to possess one-dimensional parallel ... $D^{\cdot+}A^{\cdot-}D^{\cdot+}A^{\cdot-}$ ... chains (Fig. 6.3) with spins residing on each isolated cation and anion. The physical properties of the salt have been thoroughly studied and it displays a Curie temperature T_c of 4.8 K. Below this temperature, the salt exhibits the onset of spontaneous magnetisation in zero applied field, thereby acting as a ferromagnet and is 36% more magnetic than iron metal on a mole basis. In contrast to most organic superconductors the application of pressure increases the critical temperature. (NB: the *critical temperature* is the temperature above which the material ceases to function as a magnet.) The critical temperature can be raised to 8.8 K by replacing iron with manganese which contains two unpaired spins. Removing a few spins from the compound by substituting iron with spinless cobalt significantly reduces the critical temperature. In comparison to the TCNE species, TCNQ salts possess lower critical temperatures but when the metal is three-spin chromium, both salts have an unexpected and not fully understood substantial drop in critical temperature (Fig. 6.4). This behaviour does not follow $T_c \propto S(S+1)$ and suggests the competition of more than one mechanism.

The analogous, but stronger acceptor $TCNQF_4$ forms a 1:1 CT complex with $[Cp_2^*Fe]$ that is isostructural with the above salts and displays a DAAD dimeric arrangement of donors and acceptors and also shows very similar magnetic properties. The salt undergoes a double reduction and leads to a 2 : 1 phase which is composed of parallel DADDAD chains. Incorporation of other organic acceptors such as $C_4(CN)_6$, $TCNQI_2$ or $DDQCl_2$ leads to ... $D^{\cdot+}A^{\cdot-}D^{\cdot+}A^{\cdot-}$ ... structured complexes with dominant ferromagnetic

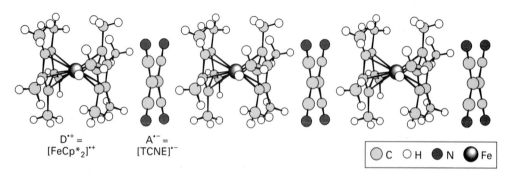

$D^{\cdot+} =$
$[FeCp*_2]^{\cdot+}$

$A^{\cdot-} =$
$[TCNE]^{\cdot-}$

○ C ○ H ● N ● Fe

Fig. 6.3 The chain structure of $[Cp_2^*Fe^{III}]^{\cdot+}$ $[TCNE]^{\cdot-}$.

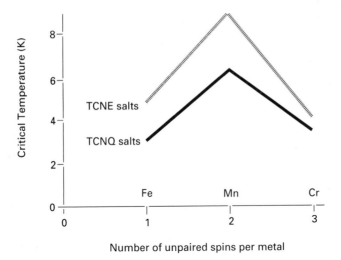

Fig. 6.4 The variation of T_c with change in metal in $[Cp_2^*M]^{\cdot+}[TCNE]^{\cdot-}$.

coupling. The $[Cp_2^*Co^{III}]^{\cdot+}[TCNE]^{\cdot-}$ complex with a diamagnetic donor has been prepared and exhibits essentially the Curie susceptibility anticipated for $[TCNE]^{\cdot-}$. This demonstrates the necessity of having both the D and A radicals for stabilising ferromagnetic coupling in the linear chain alternating $\ldots$ $D^{\cdot+}A^{\cdot-}D^{\cdot+}A^{\cdot-}\ldots$ motif. Attempts to prepare $[Cp_2^*M^{III}]^+$ (M = Ru, Os) salts of $[TCNE]^{\cdot-}$ have not yet led to compounds suitable for comparison with the highly magnetic Fe^{III} phase. Replacement of Fe^{III} in $[Cp_2^*Fe^{III}]^{\cdot+}$ $[A]^{\cdot-}$ (A = TCNE, TCNQ, $C_4(CN)_6$) with doublet Ni^{III}, triplet Mn^{III} and quartet Cr^{III} leads to compounds exhibiting antiferro-, ferro- and ferro-magnetic coupling, respectively.

6.2.2 Decamethylferrocene with inorganic acceptors

Decamethylferrocene has also been used as a donor for the formation of CT complexes with inorganic acceptors. These acceptors are usually composed of late transition metal complexes containing planar ligands (Fig. 6.5) and have at least two reversibly accessible oxidation states with the reduced form present in the CT complex usually being a radical anion. There is no magnetic ordering for bis(dithiolato)metallate salts of decamethylferrocenium but ferromagnetic coupling is observed with Curie–Weiss θ constants ranging from 0 to 27 K. Many examples have been studied but only $[Cp_2^*M][M'\{S_2C_2(CF_3)_2\}_2]$ (M' = Ni, Pt; M = Fe, Mn) and $[Cp_2^*Fe][Mo\{S_2C_2(CF_3)_2\}_3]$ have a one-dimensional chain structure. The M = Fe complexes have the largest θ values, the largest effective moments and the most pronounced magnetic field dependencies of the susceptibility.

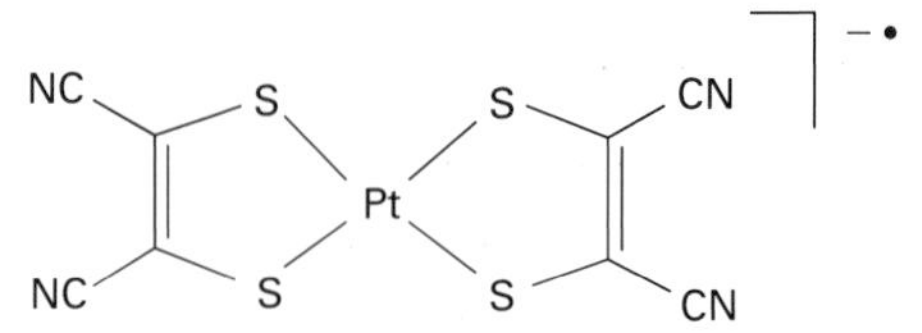

[Pt{C$_2$S$_2$(CN)$_2$}$_2$]

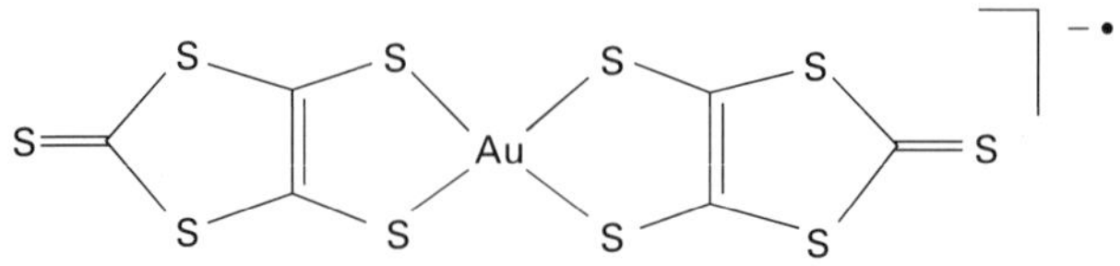

[Au(dmit)$_2$]

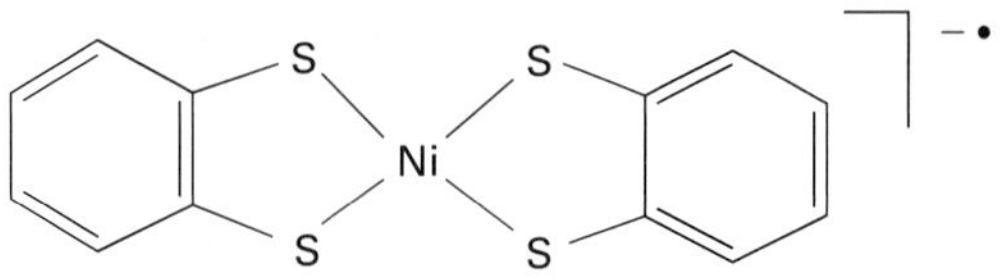

[Ni(bds)$_2$]

Fig. 6.5 Some typical inorganic electron acceptors used in charge-transfer complexes.

The Pt species ($\theta = 27$ K) possesses one-dimensional ... D$^{\cdot+}$A$^{\cdot-}$D$^{\cdot+}$A$^{\cdot-}$ ... chains but the nickel analogue ($\theta = 15$ K) features zig-zag one-dimensional ... D$^{\cdot+}$A$^{\cdot-}$D$^{\cdot+}$A$^{\cdot-}$ ... chains and longer M ... M separations. Therefore, the stronger intrachain coupling in the Pt analogue seems to lead to the enhanced magnetic coupling. There are no real, simple trends in these systems which illustrates the important role of interchain as well as intrachain interactions.

Insight into the effect of increasing the intrachain separation was obtained through the study of a tris(dithiolato)metallate salt [Cp$_2^*$Fe]$^{\cdot+}$ [Mo{S$_2$C$_2$(CF$_3$)$_2$}$_3$]$^{\cdot-}$. The salt possesses only parallel out-of-registry one-dimensional ... D$^{\cdot+}$A$^{\cdot-}$D$^{\cdot+}$A$^{\cdot-}$ ... chains (with no parallel chains in-registry) with intrachain Mo ... Mo separations of 1.424 nm. The enhanced shielding of the spin and reduction of the spin–spin interactions due to the bulky CF$_3$ groups results in a small θ value of 8.4 K. This supports the utility of one-dimensional ... D$^{\cdot+}$A$^{\cdot-}$D$^{\cdot+}$A$^{\cdot-}$ ... chain structures for achieving significant ferromagnetic coupling and bulk ferromagnetic behaviour as observed for [Cp$_2^*$Fe]$^{\cdot+}$ [TCNE]$^{\cdot-}$.

6.2.3 Donors featuring other alkylferrocenes

Charge-transfer complexes featuring TCNE and TCNQ and alkylated ferrocenes such as decaethylferrocene have been recently reported. The complexes show a similar one-dimensional DADA solid state structure to the $[Cp_2^*Fe]$ species but, due to the increased bulk of the ethylated donor, the intra- and interchain interactions feature larger intermolecular separations. These structural features lead to significantly decreased ferromagnetic coupling ($\theta = 6.8$ K, as opposed to 11.6 K for the decamethyl species) and to an absence of bulk magnetic ordering. Octamethylferrocene also demonstrates similar structural features with a classic one-dimensional DADA solid state structure. The magnetic behaviour is weakly ferromagnetic and attributed to the reduced symmetry of this donor molecule. Conductivities vary substantially with the nature of the donor and the degree of substitution of the metallocene exerts a fine-tuning effect on the structure of the anion stacks and an influence on the intramolecular interactions. Biferrocenes have been well studied because of the easy formation of the mixed-valence Fe^{II}–Fe^{III} species, and a discussion is featured in Chapter 5.

6.2.4 Other organometallic charge-transfer complexes

Based on components with higher values of S, and thus a predicted higher value of T_c, the bulk ferromagnet $[Cp_2^*Mn]^{\cdot +}$ $[TCNE]^{\cdot -}$ has been prepared. It is isostructural with the usual iron compound but has a T_c of 8.8 K against 4.8 K and the decamethylmanganocenium ion is a d^4 high-spin ($S = 1$) system compared to the d^5 low-spin ($S = \frac{1}{2}$) ferrocenium system. Another similar salt $[Cp_2^*Mn]^{\cdot +}$ $[DDQCl_2]^{\cdot -}$ has been prepared and exhibits unusual magnetic properties. The high temperature magnetic susceptibility gives a θ value of 28.6 K which suggests that the strongest exchange interactions (the interactions within individual chains) are ferromagnetic. However, above 3.8 K the magnetisation exceeds the expected calculated values, whereas dramatically different behaviour is observed below this temperature. Then, at c. 4 K the magnetisation abruptly drops by more than an order of magnitude, depending on the applied field, to lower than calculated values. There are clearly various transitions and phases within this compound and three distinct phase regions can be described: antiferromagnetic phase I, mixed phase II and the co-existing ferromagnetic–paramagnetic phase III.

Charge-transfer complexes of metal bis(arene)s and TCNQ have been reported but they do not show physical properties superior to the cyclopentadienyl sandwich systems. One exception is the reaction of bis(benzene)vanadium (which is isoelectronic with $[Cp_2^*Mn]$ (Fig. 6.6)) with TCNE which gives a material with interesting properties. The amorphous material dis-

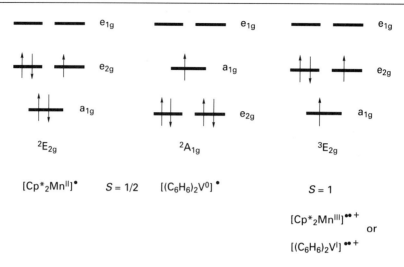

Fig. 6.6 The electronic configurations of $[Cp_2^*Mn]^{\cdot}$, $[Cp_2^*Mn^{III}]^{\cdot \cdot +}$, $[(C_6H_6)_2V^0]^{\cdot}$ and $[(C_6H_6)_2V^I]^{\cdot \cdot +}$.

plays bulk ferromagnetism at room temperature and its T_c even exceeds the decomposition temperature of the sample of about 350 K. Its strong magnetic behaviour is shown by its attraction to a permanent magnet and, to date, this is the only molecular/organic-based material with a critical temperature above room temperature. However, it is no longer an arene complex as it has an empirical composition of $[V(TCNE)_2 \cdot 0.5(CH_2Cl_2)]$. Extensive studies are now continuing on this complex and the other metallocenyl systems and are starting to show the potential for the formation of new materials with improved magnetic or electrical properties.

6.3 Metallocene-containing polymers

Metal-containing polymers have emerged as an important category of polymeric materials because of the significantly different properties they possess in comparison to conventional organic polymers. These include low temperature flexibility, high thermal and oxidative stability, flame retardancy, novel forms of chemical reactivity and interesting electrical and optical characteristics from the delocalisation of electrons. The particular attributes of the metallocene nucleus, in particular ferrocene, such as thermal stability and unusual redox behaviour, have encouraged the incorporation of these derivatives into polymer structures (Gonsalves and Chen, 1995; Manners, 1996).

There have been two basic synthetic routes towards these polymers: (i) formation of derivatives of organic polymers with pendant metallocenyl groups and (ii) the synthesis of metallocenyl compounds that contain polymerisable functional groups. These monomers can then be homo- or co-polymerised with conventional organic monomers. The inclusion of the metallocene in the side group structure of polymers has been very successful as polymer production can be accomplished with only minor modifications of previously established synthetic methodologies.

6.3.1 Addition and condensation polymers

One of the earliest examples in this field was the free radical addition polymerisation of vinylferrocene to form poly(vinylferrocene) (**VI.10**). It can

(**VI.10**)

be polymerised under radical, cationic and Ziegler–Natta conditions, but undergoes oxidation with peroxide initiators so azo initiators such as AIBN {azobis(iso-butyronitrile)} have been used extensively. The transition metal fragment exerts unusual steric and electronic effects during the polymerisation process due to the number of accessible oxidation states and large steric bulk. Organometallic acrylates and methacrylates containing the ferrocene nucleus also undergo radical-initiated homo- and co-polymerisation, but in these monomers the vinyl groups are somewhat removed from the influence of the ferrocene nucleus. In cationic-initiated polymerisation processes, the isopropenyl group has a distinct advantage over the vinyl group because of electronic effects. Therefore, a variety of η^5-cyclopentadienyl metal monomers that contain isopropenyl functional groups have been synthesised and then polymerised under different cationic-initiated conditions. In many cases, only low molecular weight products have been formed, presumably because of the unusually high stability of the α-ferrocenyl carbocation, which acts as a chain terminator.

Controlled polycondensation reactions using well-defined difunctional ferrocenes have yielded well-defined products of appreciable molecular weight. Ferrocene-containing polyamides and polyureas (Fig. 6.7) have recently been synthesised at ambient temperatures. Unfortunately, the low values of intrinsic viscosity are a drawback in the synthesis of high molecular weight polymers. Ferrocene is known to be an effective soot reducing agent in combustion and so derivatives have been incorporated into fire-retardant polymers. Indeed, poly(phosphate esters) have shown added advantages of accelerated char reduction thereby promoting extinction and reducing smoke formation.

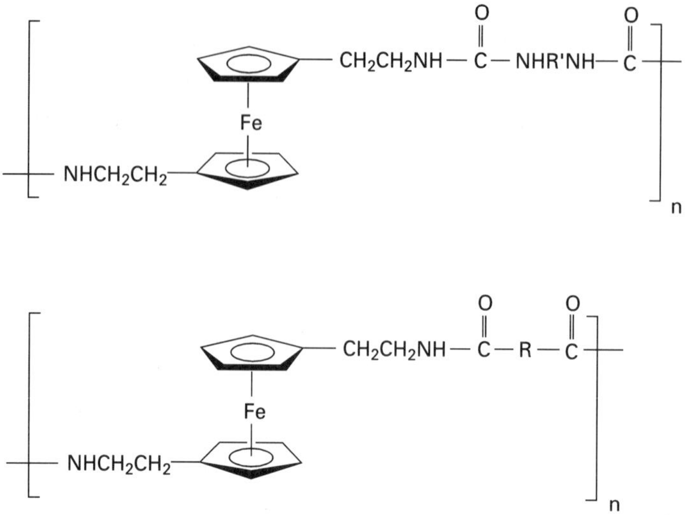

Fig. 6.7 Ferrocenyl polyamides and polyureas obtained from condensation polymerisation.

Some polymeric species feature the ferrocene units in close proximity and this enables the iron atoms to interact and sometimes yield delocalised, mixed-valent species after one-electron oxidation. These interesting materials have yet to be fully defined and shown to be of significantly high molecular weights but they show enormous potential. For example, poly(ferrocenylenes) **(VI.11)** with a molecular weight of < 4000 have been prepared by the conden-

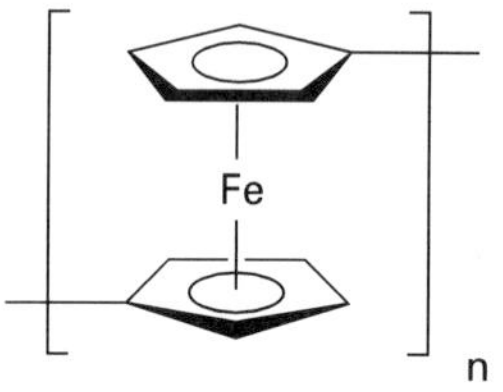

(VI.11)

sation reaction of 1,1'-dilithioferrocene·tmeda with 1,1'-diiodoferrocene and the reaction of 1,1'-dihaloferrocenes with magnesium also gives low molecular weight materials but with reasonable crystallinity. Oxidation of **VI.11** with TCNQ results in doped polymers that are delocalised on the Mössbauer timescale at room temperature and exhibit electrical conductivities of up to 10^{-2} S cm^{-1}.

6.3.2 Ring-opening polymerisation

The use of ring-opening polymerisation (ROP) has recently come to the fore, due mainly to the efforts of Manners and co-workers (1996). They have utilised ring-opening polymerisation of strained, ring-tilted silicon-, carbon-, germanium, phosphorus- and sulphur-bridged [1] and [2] ferrocenophanes to form many novel ferrocene-containing polymers (Scheme 6.7). (NB: The structures of these strained monomers were discussed in Chapter 5.) Thermal ROP involves heating the monomer to slightly beyond its melting point and the driving force for the polymerisation comes from the strain within the ferrocenophane. Soluble, high molecular weight polymers can be produced whose morphology and material properties (e.g. glass transition temperature, T_g) are dependent on the central bridging element(s) and the alkyl or aryl substituents. For example with the silicon-bridged species, when R = R′ = Me the polymer is an amber, film-forming thermoplastic (T_g = 33°C) but an amber, gummy amorphous material (T_g = − 26°C) is obtained when R = R′ = *n*-hexyl. Electronic spectra of the polymers indicate an electronic delocalisation along the polymer backbone and electrochemical studies provide evidence for interaction between the iron atoms. Recent reports on TCNE-oxidised low molecular weight poly(ferrocenylsilanes) show electron delocalisation on the Mössbauer timescale and ferromagnetic ordering at low temperatures.

An important extension of this work has been the introduction of 'living' anionic ROP at 25°C using initiators such as *n*-BuLi in THF. This works well with the silicon-bridged species and allows the synthesis of poly(ferrocenylsilanes) with controlled molecular weights and narrow polydispersities. As a

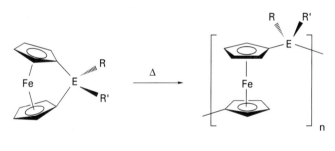

E = Si ; R, R' = alkyl, aryl, ferrocenyl
E = Ge ; R, R' = Me, Et, n-Bu, Ph

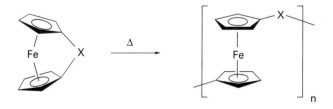

X = S, P(R), CH$_2$CH$_2$

Scheme 6.7 The formation of ferrocene-containing polymers by ring-opening polymerisation.

result of this and potential for the formation of block co-polymers and a 'fine-tuning' of the physical and material properties through the use of the almost limitless combinations of bridging atoms, substituents and metallocenes (ruthenium, cobalt and chromium 'sandwich' species have all been incorporated) this area seems likely to undergo exciting development in the future.

'Living' ring-opening polymerisation of norbornenes and norbornadienes is well known (Schrock, 1990) and incorporation of ferrocene into the backbone skeleton gives a new range of functionalised derivatives (e.g. **VI.12**). The

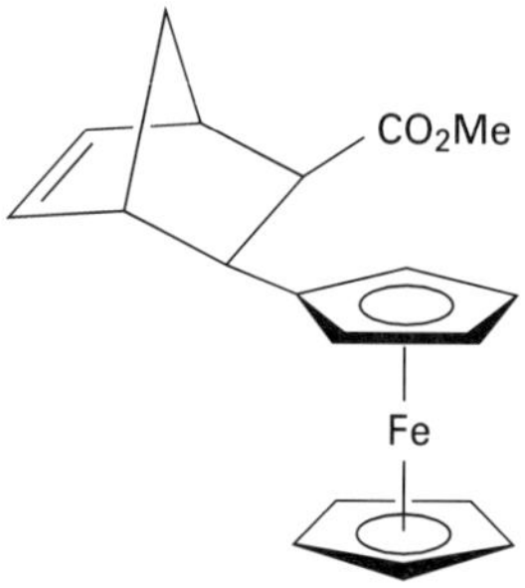

(**VI.12**)

norbornene monomer can be polymerised quantitatively using a transition metal complex initiator and possesses a reasonable polydispersity.

6.3.3 Polymers with ferrocene as a pendant group

A wide range of conventional organic polymers have been modified to include a ferrocenyl functional group. For example, addition of lithio- or dilithioferrocene to polystyrene leads to incorporation of a pendant ferrocene. Ferrocene can also be attached to carrier polymers (e.g. **VI.13**) by the

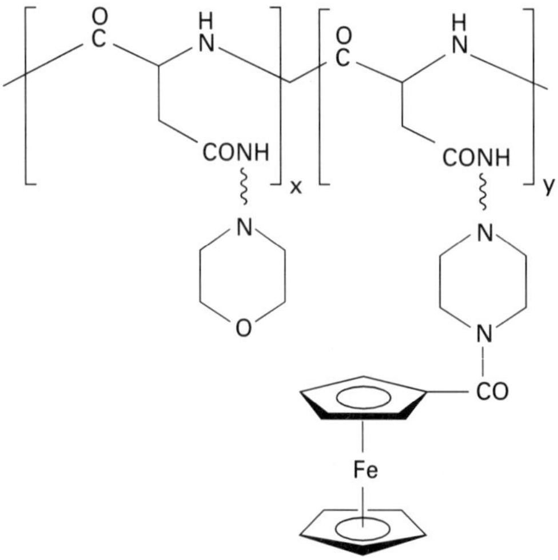

(**VI.13**)

reaction of the amine functions attached to the spacer with ferrocene carboxylic acids and carboxaldehydes. By incorporation of hydrosolubilising amine groups the polymers become water soluble and have been used as conjugates for biomedical applications requiring solubility in aqueous media.

6.3.4 Face-to-face and three-dimensional metallocene polymers

Using condensation reactions, rigid-rod metallocene polymers with a multi-stacked structure have been developed. Various methods have been employed, but the most successful has featured the treatment of a ferrocene monomer with $FeCl_2$ and $Na[N(SiMe_3)_2]$ to yield purple polymers (**VI.14**)

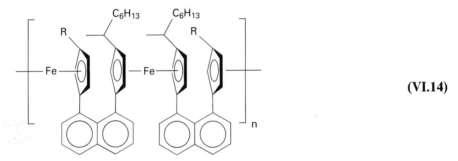

(**VI.14**)

R = H, 2 - octyl

with molecular weights of *c.* 18 000 (Rosenblum *et al.*, 1995). By using [Ni(acac)$_2$] instead of $FeCl_2$, novel mixed-metal co-polymers containing Ni and Fe can also be obtained though with lower molecular weights of < 3000. The electrical and magnetic properties of these species (and, in particular, I_2-doped materials) have shown conductivities of up to $6.7 \times 10^{-3} \, S \, cm^{-1}$. Structural work on well-defined oligomers indicates that the stacked metallocene units in the polymer form a helical structure.

In general, polymers containing ferrocene units connected by sulphide or disulphide bridges are insoluble in common organic solvents. Soluble polymers have however been obtained from *n*-butyl- or *tert*-butyl-substituted 1,2,3-trithia-[3] ferrocenophane monomers and the abstraction of the bridgehead sulphur by a simple phosphine. A similar reaction involving the doubly chalcogen-bridged ferrocenophane followed by atom abstraction polymerisation leads to a giant three-dimensional polymer with a molar mass $> 10^6$ (Galloway and Rauchfuss, 1993) (Fig. 6.8). The disulphide links in these species can be reductively cleaved, as in other disulphide-connected structures such as proteins, and the iron can be electrochemically oxidised to give ferrocenium units.

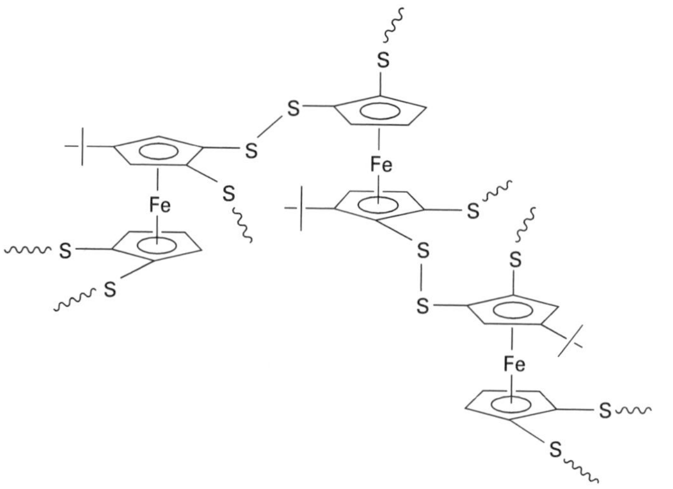

Fig. 6.8 A schematic representation of the three-dimensional polymer derived from {[Fe(t-BuC$_5$H$_2$)(C$_5$H$_3$)](S$_3$)$_2$}.

6.4 Metallocenes for non-linear optics

The interaction of the electromagnetic field of light (normally high intensity laser light) with a non-linear optical (NLO) material can result in the generation of new electromagnetic fields. As light passes through a species, its electric field interacts with inherent charges in the material causing the original beam to be altered in phase, frequency, amplitude or polarisation. The study of such interactions is the field of *non-linear optics* and describes deviations from linear behaviour as defined by the classical laws of optics. Such media are now creating intense interest not only because of their NLO properties but also because of attempts to control other characteristics such as solubility, processibility, optical clarity, absorption and thermal stability as these will obviously determine technological utility. Initial developments have already given rise to NLO devices, such as frequency doubling crystals in laser experimentation. However, there remain enormous possibilities in the fields of telecommunications, computer and optical signal processing devices through: (i) optical phase conjugation and image processing; (ii) optical switching (i.e. transmission of light depending on the refractive index of the material); (iii) optical data processing (for computers, where incredibly rapid movements are needed); and (iv) new frequency generation.

Research is gradually elucidating what actually governs second- and third-order NLO activity and, thereby, tailoring species to show greater effects is becoming possible. In recent years, organometallic compounds through their unique characteristics such as diversity of metals, oxidation states, ligands and geometries have found success and brought a new dimension to the area (Long, 1995, 1997).

6.4.1 Materials for second-order non-linear optics

The report in 1987 by Green *et al.* that a ferrocene derivative had an excellent second harmonic generation (SHG) efficiency demonstrated that organometallic compounds could have great potential in the field of second-order NLO (i.e. frequency-doubling). This has led to metallocenyl derivatives being the most thoroughly studied organometallic compounds to date, with regard to this field of study. The compound *cis*-[(C$_5$H$_5$)Fe(C$_5$H$_4$)CH=CH-*p*-C$_6$H$_4$NO$_2$] **(VI.15)** was analysed as it crystallises in a non-centrosymmetric

(VI.15)

space group (cf. the *trans*-isomer which shows no SHG signal due to possession of a centre of symmetry). Kurtz powder measurements at 1.064 μm for the compound gave an SHG efficiency 62 times that of the urea reference sample. It is thought that the facile redox ability exhibited by the metallocenyl species leads to large values of β via charge-transfer through the conjugated π-electron system. The donating strengths of the metallocenes are attributed to the low binding energy of metal electrons and their effectiveness is responsive to structural modifications which also influence their redox potentials. However, on the basis of binding energies and redox potentials, it was expected that the molecular hyperpolarisabilities would be larger than the observed values. The authors state that poor coupling between the metal centre and the substituents due to the π geometry lowers the donating ability of the metal centre.

Excellent SHG values have come from a series of salts **(VI.16)**, with the counterion being I⁻, Br⁻, Cl⁻, CF$_3$SO$_3$⁻, BF$_4$⁻, PF$_6$⁻, *p*-CH$_3$C$_6$H$_4$SO$_3$⁻, NO$_3$⁻ and B(C$_6$H$_5$)$_4$⁻. The value of the powder SHG signals depends on the

(VI.16)

counterion with I^- giving a figure of 220 times that of the urea standard. It is interesting to note that substitution of ruthenium for iron leads to smaller powder efficiencies. This is rationalised as being due to the less electron-rich and, therefore, less effective donor ruthenium and that the ruthenium compounds examined tended to crystallise in centrosymmetric orientations.

An interesting recent report has featured the construction of bimetallic donor-acceptor sesquifulvalene complexes with ferrocene and $(\eta^7$-cyclo-heptatrienyl)tricarbonyl chromium units **(VI.17)**. Hyper-Rayleigh scattering

(VI.17)

(HRS) techniques gave β values of 570×10^{-30} electrostatic units (e.s.u.) and 320×10^{-31} e.s.u. when Z = acetylene and alkene, respectively; these by far exceed those measured for mono- and bimetallic ferrocene derivatives to date and are in the region of the highest values ever reported for bimetallic complexes. The large β values are attributed to resonance enhancement and are, interestingly, not in agreement with theoretical calculations of the first hyperpolarisability β, which predict that sandwich compounds are not suitable as efficient NLO chromophores.

A potentially important development has been the report of some partially oxidised ferrocenyl complexes for non-linear optics. Using HRS techniques, the hyperpolarisabilities of a series of partially oxidised biferrocenes linked through conducting C=N linkages have been measured and found to vary linearly with the redox potential difference between the ferrocenyl moiety and the oxidant **(VI.18)**. As the —C=N— linkage acts as an acceptor, a D–A–D–A system can be designed. Oxidation of the neutral compounds results in an increase in the β values by an order of magnitude for the partially oxidised complexes. It appears that the non-linearity in these complexes results from differential electron transfer from the biferrocene moiety to the oxidant in the ground and excited states and effects a significant change in the dipole moment between the two states.

Elegant and thorough theoretical investigations have concluded that organometallic chromophores must possess a highly polarised, strongly coupled ligation sphere around a given metal for effective second-order NLO

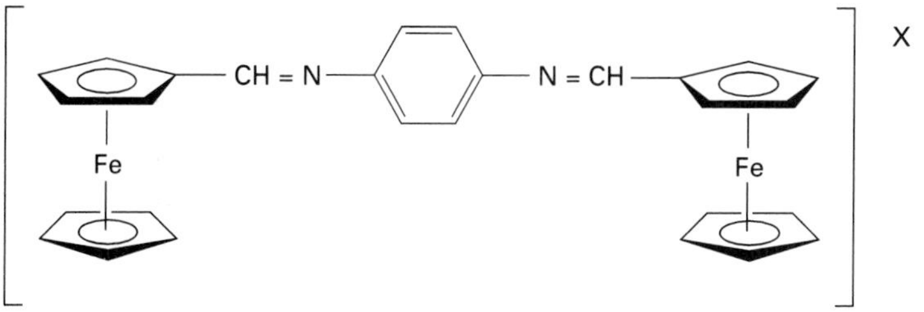

$$(VI.18)$$

$$X = (DDQ)_2$$

$$(TCNQ)_2$$

behaviour and that weakly-bound transition metal π-complexes do not display the dramatic increases in second-order NLO response with extra conjugation length observed for traditional organic chromophores. More importantly, it was shown that efficient quantum chemical methods can be used to understand hyperpolarisabilities and thus be utilised in designing new chromophoric units possessing optical NLO characteristics.

6.4.2 Materials for third-order non-linear optics

The structure/property relationships that govern third-order NLO polarisation are a little vague; however, polymeric materials with extended π-conjugation are known to be important and an increased effective conjugation, and hence large π-delocalisation length has been recognised as a way of achieving large third-order non-linearities. Organometallic polymers have extended π-electron delocalisation plus extra features which could enhance $\chi^{(3)}$. Transition metals are incorporated in such a way that some of their d orbitals interact with the conjugated π-electron orbitals of the organic repeating unit. Therefore, an extended, delocalised electron system within the polymer chain is formed. Electronic features can be manipulated by varying the ligands attached to the metal centre and many organometallic polymers have low-lying charge-transfer transitions not present in organic systems.

As was discussed in the previous section, metallocene derivatives are known to be important NLO materials. Less work has been completed featuring third-order materials than their second-order counterparts, but metallocenes allow direct comparisons to be made between organic and organometallic compounds. For example, some symmetric end-capped acetylenes have undergone third harmonic generation (THG) analysis. One of the compounds, 1,4-bi(ferrocenyl)butadiyne, showed a $\chi^{(3)}$ value of 225×10^{-14} e.s.u., which is more than double that of the organic species 1,4-diphenylbutadiyne. To help elucidate the NLO properties of organometallic structures, a series of aryl and vinyl derivatives of ferrocene have been synthesised (Fig. 6.9). The molecular second hyperpolarisabilities γ determined by degenerate four-wave mixing at 602 nm were overall fairly low. Nevertheless, some important points could be gleaned from the data: (i) γ increases strongly with the length of the conjugated π-electron system (a feature also found in organic polyenes) until a large limiting value is reached where additional conjugation is ineffective with regard to γ; (ii) the effective conjugation is determined by the length of the aryl–vinyl system, with the ferrocenyl group not being very significant in this respect; (iii) π-electron delocalisation through the ferrocene centre is less effective than through a double bond or phenyl group; and (iv) the d–d transitions of the metal in the ferrocene moiety do not make a significant contribution to the optical non-linearity as opposed to the d–π^* and π–π^* transitions.

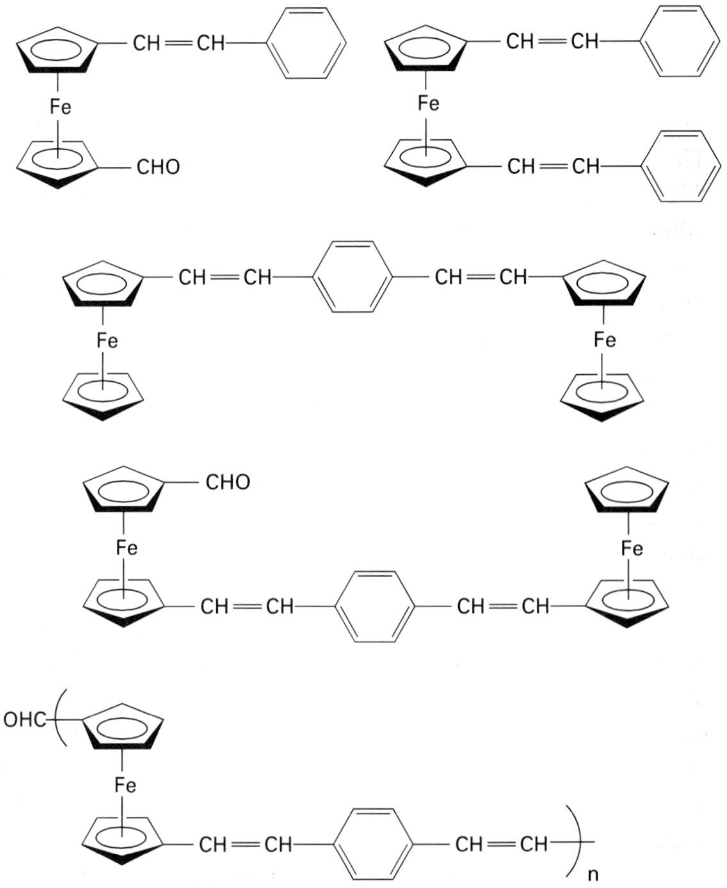

Fig. 6.9 The structures of some arylvinyl- and vinyl-ferrocene derivatives studied for third-order NLO.

The linear and third-order non-linear properties of a conjugated oxidised biferrocenylacetylide **(VI.19)** have been investigated. Remarkably, the com-

(VI.19)

pound exhibits strong electronic absorption in the near infrared region suggesting a high degree of delocalisation of the extended π-electron system. The magnitude and speed of the third-order non-linear optical response at 532 nm has been measured using time-resolved phase conjugation and at 1064 nm using 'Z-scan' techniques. Values of second molecular hyperpolarisability for both measured wavelengths were found to be comparable to those of other large conjugated molecular systems, i.e. superconducting tetrathiafulvalene (TTF) salts and 'doped' conjugated organic polymers.

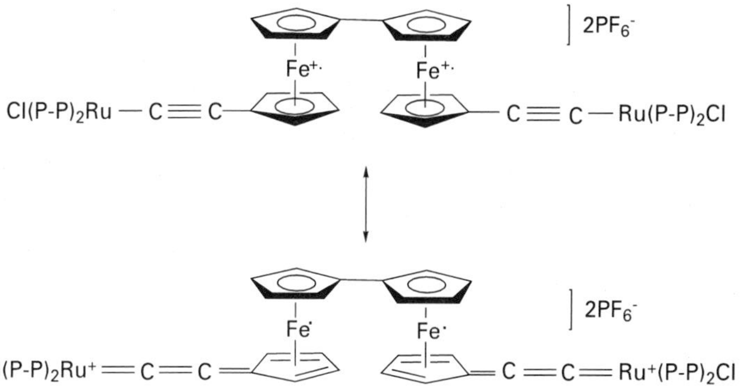

The third-order non-linear optical properties of a series of tetravalent group 4 (Ti, Zr, Hf) 'bent' metallocene compounds, including halides, acetylides and alkenylzirconocenes, have been studied. The rationale for studying group 4 metallocene complexes was that the orbital interaction between metal d orbitals and filled ligand orbitals could lead to a polarisable π-electron system and, therefore, good NLO properties. The LUMO of group 4 metallocene acetylides is very similar to that found for other group 4 metallocenes and the cylindrical symmetry of the π-system of the acetylide ligands makes them ideal for interaction with this metal-based LUMO. Moreover, acetylide ligands and d orbitals of the metal are often of similar symmetry. From THG techniques with a fundamental laser frequency of 1908 nm, the metallocene halide complexes do not have measurable optical non-linearities, but the acetylide and vinyl complexes have reasonably large non-linear optical coefficients, with the largest γ value being 154×10^{-36} e.s.u. for **VI.20**. This presumably arises from a conjugated π-system which involves the Cp–metal bonding network and similar symmetry orbitals of the vinyl and acetylide ligands.

257

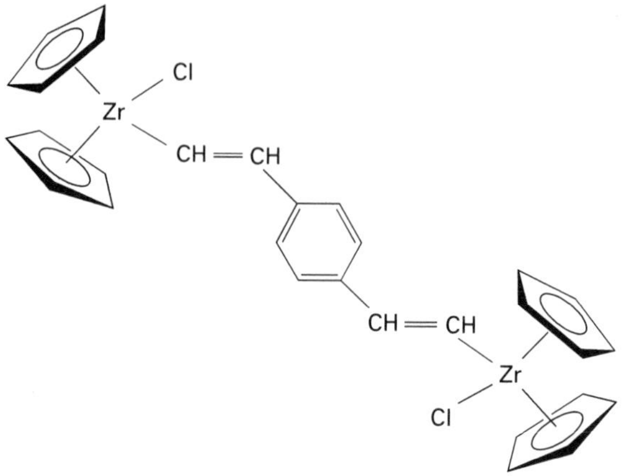

(VI.20)

6.5 Metallocenes in medicine

6.5.1 Ferrocene/ferrocenium couples in biosensors

The glucose sensor is one of the most important biosensors for glucose detection. This is necessary not only in the field of biotechnology for fermentation process control but also in medical services. It is used to measure blood glucose for diagnosis of the state of diabetes and hyperglycaemia. The oxidation of glucose by the enzyme glucose oxidase (GOD) to gluconolactone is sensitive and specific and, therefore, suitable for the determination of glucose with sensors. The current resulting from the biochemical reaction between the enzyme and glucose can be obtained by measuring the consumed oxygen or hydrogen peroxide produced in the reaction. However, the direct electrochemical measurement of the amount of oxidised product is impeded since the enzyme does not react with electrode surfaces directly and as there can be dissolved oxygen in the samples. A redox-active mediator can be used to facilitate the quantitative oxidation of glucose under catalytic conditions. By oxidising the reduced form of the mediator, the current can be obtained. Furthermore, the use of a mediator avoids the effect of any dissolved oxygen. Ferrocene acts as such a mediator enabling the electrochemical determination of glucose via a series of connected redox cycles (Scheme 6.8).

6.5.2 Metallocene anti-tumour agents

The anti-tumour properties of a series of bent metallocene dihalides and pseudohalides, $[Cp_2MX_2]$ (M = Ti, Mo, Nb, V; X = F, Cl, Br, I, NCS, N_3, Y), have been reported with regard to cell lines such as leukaemias P388 and L1210, colon 38 and Lewis lung carcinomas, B16 melanoma, Ehrlich ascites

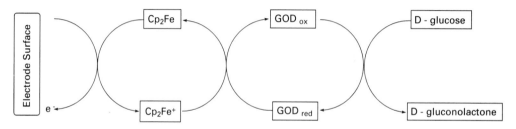

Scheme 6.8 The use of ferrocene as a mediator in a glucose sensor.

tumour, and several human colon and lung carcinomas transplanted into athymic mice. There seems to be a relationship between the activity and the position of the central metal in a diagonal relationship in the Periodic Table (Fig. 6.10) (Köpf-Maier and Köpf, 1987; Murray and Harding, 1994). Metallocenes featuring Ti, V, Nb or Mo all show excellent cancerostatic potency with distinct dose–activity relationships and cure rates of 100% at optimum doses. Within a group, the anti-tumour activity seems to decrease with increasing atomic weight, i.e. Ta and W metallocenes exhibit only sporadic activity for treatment of tumour-bearing mice whilst Zr and Hf metallocenes are inactive against Ehrlich ascites tumours. Within the series [Cp_2TiX_2], equally potent tumour inhibition was observed for a variety of halide and pseudohalide ligands and there are no distinguishing trends. However, modification of the cyclopentadienyl rings does seem to have an effect. Monosubstituted titanocene dichlorides effected reduced cure rates of 60–80%, 1,1'-disubstituted and 1,1'-bridged titanocene dichlorides (*ansa*-titanocenes) showed only sporadic cure rates of 10–30% and titanocene species containing pentamethylated cyclopentadienyl rings were cytostatically inactive against Ehrlich ascites tumour.

The mechanism of anti-tumour activity of metallocene dihalides is thought to result from the interaction of a hydrolysed metallocene species with DNA. Metals derived from the metallocenes accumulated in nucleic acid-rich regions of cells and nucleic acid synthesis, particularly DNA synthesis, is inhibited after *in vitro* and *in vivo* treatment with titanocene- or

Ti	V	
Zr	Nb	Mo
Hf	Ta	W

M = maximum activity
M = sporadic activity
M = no activity

Fig. 6.10 The relationship between the position of the central metal in the periodic table metallocene dihalides and their anti-tumour activity.

259

vanadocene-dichloride. Several adducts of metallocene dihalides with DNA have been isolated and characterised, but no such adducts have been detected with [Cp_2VCl_2], the most active metallocene *in vitro*. Studies into metallocene–DNA co-ordination chemistry have shown that [Cp_2MoCl_2] co-ordinates to both nucleobase (N) and phosphate (O) in a non-labile manner that effects major conformational changes but does not disrupt the Watson–Crick hydrogen bonding. What is clear is that the mechanism of interaction of metallocenes with DNA is very different to that of the well-known *cisplatin*.

The ionic metallocenium complexes (**VI.21**) also show anti-tumour activ-

$$\left[\begin{array}{c} \text{M} \end{array} \right]^{+} \quad X^{-}$$

$$\text{M = Fe, Co}$$
$$\text{X = FeCl}_4\text{, CCl}_3\text{COO,}$$
$$2,4,6\text{-(NO}_2)_3\text{C}_6\,\text{H}_2\text{O}$$

(VI.21)

ity. These salt-like, water-soluble species show marked anti-tumour properties against fluid Ehrlich ascites tumour and induced cure rates of 70–100% over a broad dose range. Other tumours inhibited by ferrocenium compounds are solid B16 melanoma, colon 38 carcinoma and Lewis lung carcinoma, and in general ferrocenium complexes exhibit a similar range of activity compared to the titanocene species. An important point is that the toxic properties of the metallocene cytostatic agents, especially those containing less heavy metals such as Ti or Fe, differ fundamentally from those of platinum compounds. This then opens up the possibility of combination therapy without dangerous toxic side effects.

6.6 Molecular recognition using metallocenes

The investigation of molecular systems that show specific and designed interactions with particular molecules or ions is an area of intense current interest. Metals have been used in the self-assembly of large molecular aggregates, novel topological arrays and in the design of novel receptor systems. This latter area has come to the fore by focusing on the metal's (and in particular, ferrocene's) ability to undergo metal-centred redox processes to generate oxidised or reduced compounds with differing properties. Thus, the incorporation of the redox-active metallocene enables a system to act as a molecular switch or receptor. For recent review articles consult Constable (1991), Beer (1992) and Hall (1995).

For example, a series of long-chain fatty acids or cholesterols featuring a substituted ferrocene unit have been synthesised by Gokel *et al.* (1991) (**VI.22**). The neutral molecule exhibited no aggregation phenomena in solu-

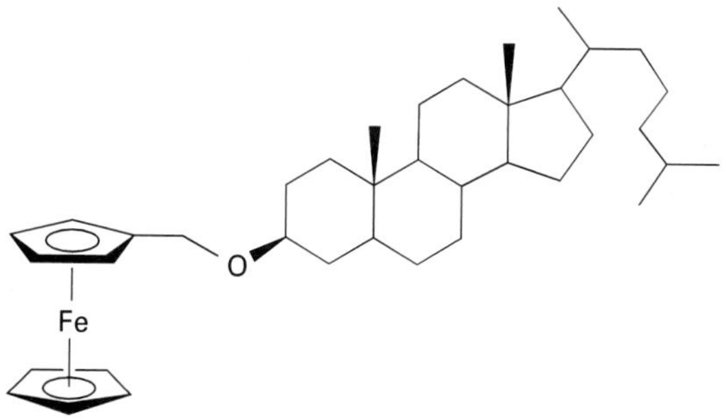

(**VI.22**)

tion but upon electrochemical or chemical oxidation to the cation approximately spherical vesicles with membranes of *c.* 4.5 nm thickness were formed. Clearly, the formation of the vesicles is redox-triggered (chemical reduction of the vesicles results in their destruction) and this illustrates the remarkable use of a redox-active metallocene as a molecular switch to control the formation of macromolecular assemblies. The same research group have also used ferrocene moieties to selectively bind a series of diamines. The biferrocenes formed (**VI.23**) contain a potential bonding cavity designed for diamines. The

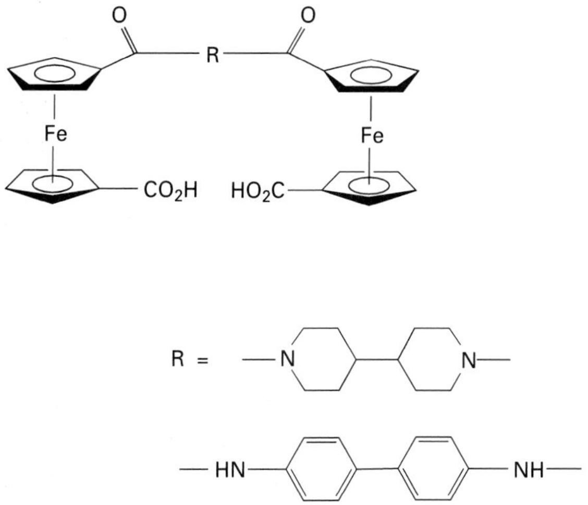

(**VI.23**)

$$R = $$

receptor– substrate interaction is between the carboxylic acid residues and the amino groups and the correct spatial arrangements of the acids may be achieved by the rotational motion of the two cyclopentadienyl rings of an individual ferrocene unit. The use of aromatic spacer groups between the ferrocenes allows π-stacking interactions between the substrate and the receptor which act as an additional form of recognition. In the correct conformation, the cavity size between the carboxylic acids is 0.7–0.8 nm and ^{1}H NMR chemical shift methods have detected significant substrate–receptor binding interactions with particular diamines.

The potential of metallocenes as molecular switches has been exploited in the design of molecular systems in which a metallocene bears a binding substituent such as a macrocycle, cryptand or calixarene. By suitable varia-tion of the size and structure of the host cavity, these species can be made to bind cationic, anionic or neutral guest species by means of electrostatic interactions either through space or through bonds between the binding sites and the metallocene unit. Therefore, selective binding can trigger a change in the redox potential of the metallocene unit and offer application in the fields of chemical sensors, cation transport across membranes, molecular electron-ics and mimics for metallo-enzymes. In order for ligand complexation to affect the electrochemistry of the redox centre, the guest species needs to be either in close proximity to the metal atom or co-ordinated by functional groups that are conjugated with the metallocene system. In general, there are two types of host compounds containing a metallocene unit: (i) the metal-locene is *appended* to either a macrocycle (**VI.24**) or a calixarene (**VI.25**); or (ii) the metallocene is *incorporated within* the macrocycle (**VI.26**). This has

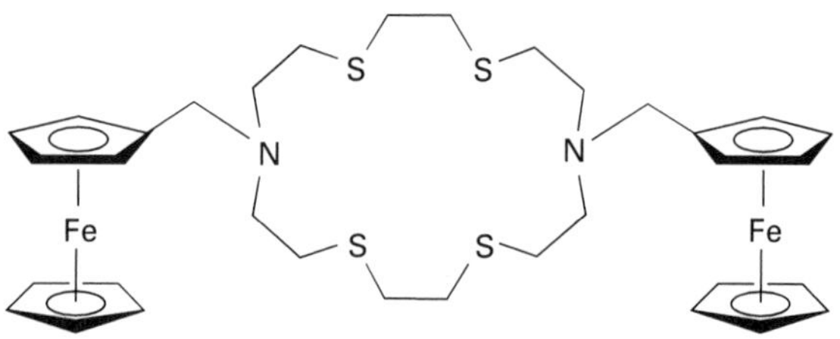

(VI.24)

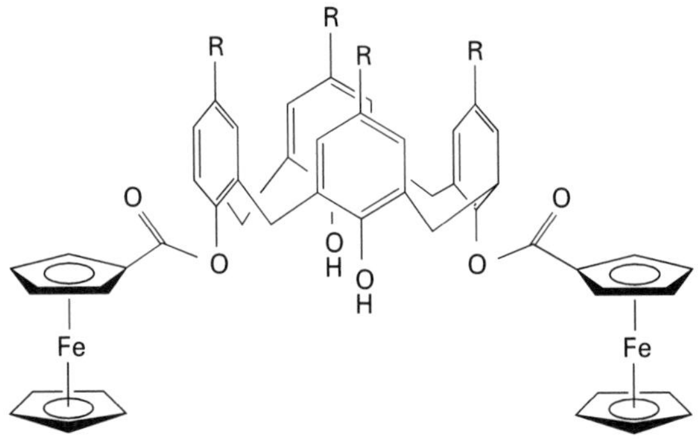

(VI.25)

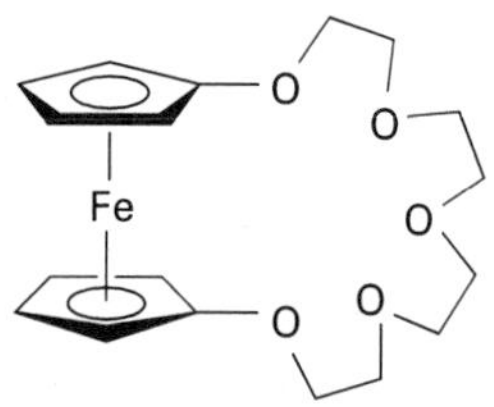

(VI.26)

given rise to a huge number of compounds featuring a range of ring sizes, type and number of heteroatoms and metallocenes which can be 'fine-tuned' for a particular purpose. For example, electrochemical studies showed that the redox couples of **VI.24** were perturbed to more positive potentials by 40 mV on co-ordination with Cu^{2+} (square planar co-ordination by the four sulphur atoms) whilst anodic shifts in the oxidation potential of **VI.26** result from the addition of alkali metal salts. Two distinct cyclic voltammetry waves corresponding to complexed and uncomplexed **VI.26** were observed for both Na^+ and Li^+ guest cations and the quantitative decrease in metal binding capacity on oxidation can be calculated as 40 and 742 for Na^+ and Li^+, respectively. The facility to switch cation binding on and off electro-chemically has been used to transport alkali metal cations across a membrane using **VI.26** as the carrier. In an attempt to mimic enzymes that selectively bind organic guest substrates as part of the process of catalysis, a variety of cyclophane host molecules have been designed. Receptors include cavitands that contain enforced, rigid, hydrophobic cavities with dimensions large enough to encapsulate simple neutral organic guests. These abiotic hosts have been modified to incorporate redox-active units (**VI.25**) to create sensory devices capable of detecting the inclusion of a neutral guest electron-ically and with the potential to catalyse reactions as the guest substrate.

Finally, an interesting series of neutral ferrocene anion receptors (e.g. **VI.27**) have been shown to selectively complex, electrochemically recognise

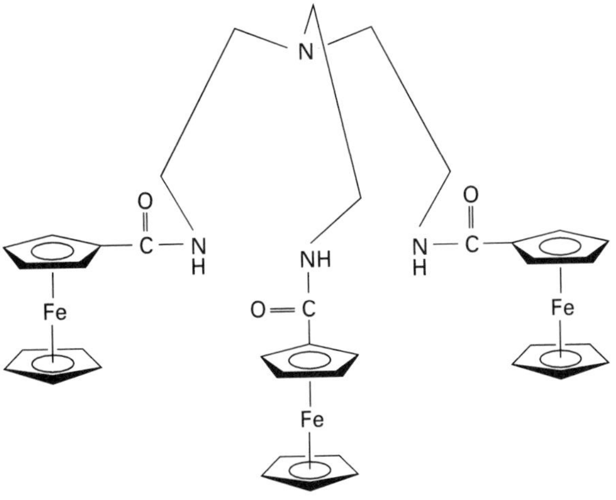

(VI.27)

and respond to a dihydrogen phosphate guest anion in the presence of excess amounts of hydrogen sulphate and chloride anions (Beer *et al.*, 1993). Shifts to higher frequency of *c.* 1–2 ppm for the amide protons were exhibited by the receptors on addition of $Bu_4^nN^+X^-$ ($X^- = H_2PO_4$, HSO_4, Cl) as were significant anion guest-induced cathodic perturbations of the ferrocenyl oxidation peak potentials for the receptors. Importantly, when an equimolar mixture of $H_2PO_4^-$, HSO_4^- and Cl^- was added to acetonitrile electrochemical solutions of the ligands, the redox couples shifted cathodically by an amount approximately the same as that induced by the $H_2PO_4^-$ anion alone.

6.7 Chiral metallocenes for catalysis

In asymmetric reactions, use of a chiral catalyst is desirable as long as the catalytic asymmetric reactions proceed with high stereoselectivity and produce the desired enantiomeric isomer in higher yield. Catalytic reactions by homogeneous transition metal complexes have developed rapidly and a gamut of reactions can be effected by transition metal catalysts. Chiral ferrocenyl derivatives have been recognised as useful chiral ligands in many asymmetric reactions, e.g. in stereoselective peptide synthesis, asymmetric transamination and asymmetric condensations. Many of the transition metal complexes used for catalytic reactions contain tertiary phosphines as ligands so optically active phosphine ligands can produce excellent chiral catalysts, and in this respect the formation and chemistry of chiral ferrocenylphosphines has come to the fore (Hayashi, 1995)

6.7.1 Chiral ferrocenylphosphines

Chiral ferrocenylphosphines (Fig. 6.11) have been used as chiral ligands for transition metal complexes that catalyse asymmetric reactions e.g. hydrogenation, Grignard cross-coupling, allylation, aldol reactions and hydrosilylation. The species have features that are unique when compared to other chiral phosphine ligands, i.e.: (i) they possess functional groups (X) at the ferrocenylmethyl position on the side chain; (ii) they have the ferrocene planar chirality that never undergoes racemisation; (iii) both mono- and bis-phosphines can be prepared from one chiral source, this being *N,N*-dimethyl-1-ferrocenylethylamine; and (iv) the characteristic orange colour assists during column chromatography. Most important is the presence of the functional groups, as they can be controlled by the ferrocenyl and methyl groups on the chiral carbon to point towards the reaction site on the catalyst

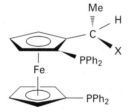

Fig. 6.11 The typical structure of chiral ferrocenylphosphines.

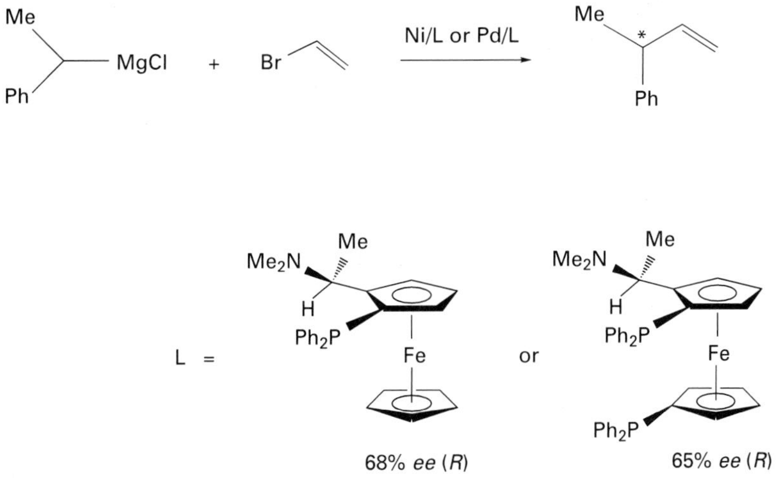

Scheme 6.9 An asymmetric cross-coupling reaction catalysed by a chiral ferrocenylphosphine ligand.

co-ordinated with phosphorus atoms on the ferrocenylphosphine ligand and will interact with a functional group on a substrate in a catalytic asymmetric reaction. Secondary interactions between the substrate and the functional groups on the phosphine ligand bring about high enantioselectivity in a variety of asymmetric catalytic reactions. The stereogenic, functionalised side chains can be modelled to fulfil a specific purpose and a well-designed ferrocenylphosphine ligand is reflected in, for example, the high efficiencies of nickel- or palladium-catalysed asymmetric cross-coupling or allylic substitution reactions.

The reaction of organometallic reagents (R—M) with alkenyl or aryl halides and related compounds (R'—X) to give cross-coupled products (R—R') is an extremely useful way of forming a carbon—carbon bond and is catalysed by nickel and palladium complexes. The use of chiral catalysts allows transformation of a racemic mixture of the secondary alkyl Grignard reagent into an optically active product by a kinetic resolution of the Grignard reagent. For example, the asymmetric cross-coupling of 1-phenylethylmagnesium chloride with vinyl bromide gives the optically active 3-phenyl-1-butene in the presence of a small amount of Ni or Pd catalyst co-ordinated with a chiral ferrocenylphosphine ligand (Scheme 6.9). Asymmetric cross-coupling with similar ligands has also been successfully applied to the formation of optically active allylsilanes. The reaction of α-(trimethylsilyl)benzylmagnesium bromide with vinyl bromide, (E)-bromostyrene and (E)-bromopropene using 0.5 mol % of a palladium complex (**VI.28**) produces high yields of the corresponding (R)-allylsilanes with 95, 85 and 95% ee, respectively. With other chiral phosphine ligands such as (4S,5S)-(+)-0-isopropylidene-2,3-dihydroxy-1,4-bis

(VI.28)

(diphenylphosphino)butane (DIOP) or prophos the optical purity of allylsilanes is less than 5% *ee*.

Chiral ferrocenylphosphines are extremely versatile chiral catalysts and have been used in allylic substitution reactions via π-allyl complexes, hydrogenation and hydrosilylation of olefins and ketones and aldol reactions of α-isocyanocarboxylates. An interesting example has been found to form very active and highly enantioselective catalysts for different reactions. **VI.29** is

(VI.29)

(R) - (S) - enantiomer

R = cyclohexyl

part of a class of chiral chelating diphosphines where the two ligating moieties can be varied independently from one another and allows the study of both the steric and electronic influence of the ligands on stereoselectivity. This is in contrast to most chiral diphosphines that, for synthetic reasons, feature identical phosphino groups. The ligand has been used in asymmetric rhodium-catalysed hydrogenation and hydroboration as well as in palladium-catalysed allylic alkylation reactions giving high enantioselectivities of up to 99%.

6.7.2 Chiral ferrocenyl alcohols and aminoalcohols

Following the success of chiral ferrocenylphosphines, the ferrocenyl moiety was incorporated into alcohol and aminoalcohol ligands for use in asymmetric synthesis. In particular, these species have been successfully used to catalyse enantioselective addition of dialkylzinc to aldehydes with high enantioselectivity (Butsugan *et al.*, 1995). Enantioselective addition of organometallic reagents to aldehydes is one of the most important and fundamental asymmetric reactions that afford optically active secondary alcohols as these species are components of many naturally occurring compounds, various biologically active compounds and industrial materials. In the catalytic asymmetric addition of dialkylzincs to various aldehydes (6.3), it is

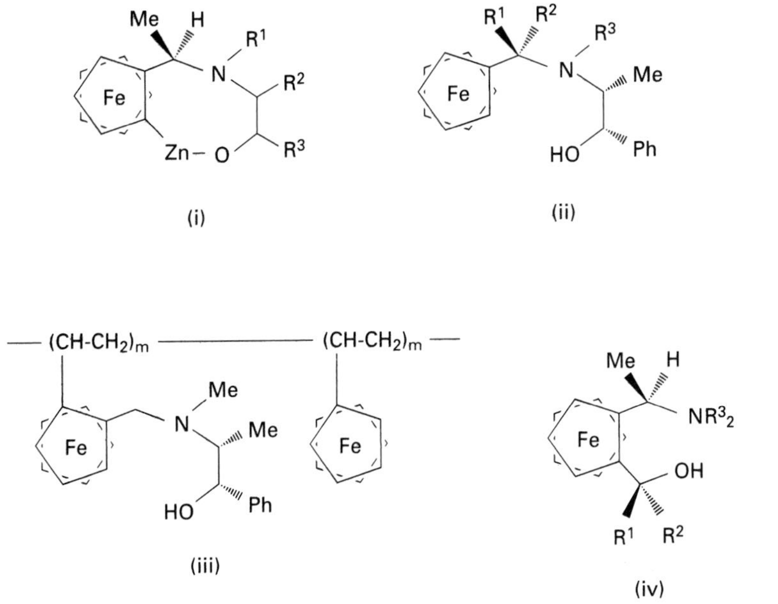

Fig. 6.12 Various chiral ferrocenyl aminoalcohol catalysts used in the alkylation of aldehydes with dialkylzincs.

known that a bulky substituent near the C—O bond and a tertiary amine moiety in the aminoalcohol catalyst are essential for high enantioselectivity and for acceleration of the alkylation.

$$\text{PhCHO} + \text{Et}_2\text{Zn} \xrightarrow[\text{catalyst}]{\text{chiral}} \overset{\displaystyle\text{OH}}{\underset{*}{\text{Ph}-\text{CH}}}-\text{Et} \tag{6.3}$$

Various ways of incorporation of the ferrocene moiety into aminoalcohol catalysts have resulted in four basic types of compound: (i) chiral ferrocenyl zincs bearing an aminoethanol auxiliary; (ii) N-(1-ferrocenylalkyl)-N-alkyl-norephedrines; (iii) chiral polymers bearing N-ferrocenylmethylephedrine; and (iv) chiral 1,2-disubstituted ferrocenyl aminoalcohols (Fig. 6.12). Particular success has been found using 1,2-disubstituted ferrocenyl aminoalcohols with planar chirality as these species display high enantioselectivity and high catalytic efficiency. They have been applied to the synthesis of optically active 3-alkylphthalides and vicinal thio- and seleno-alcohols.

6.7.3 Chiral ferrocenylchalcogenides

As an alternative to phosphines and amines, sulphides and selenides have been incorporated into chiral ferrocenes for asymmetric synthesis. Although this field is relatively new and well behind the development of chiral ferrocenylphosphine systems, there have been some promising results. For instance, the cross-coupling between allyl magnesium chloride and 1-chloro-1-phenylethane has been achieved with ligands containing thioethers and selenoethers instead of phosphines though the enantioselectivity (highest *ee* 45%) cannot compete with that of the phosphine systems. Some novel diferrocenyl dichalcogenide ligands (**VI.30**) have been shown to work effec-

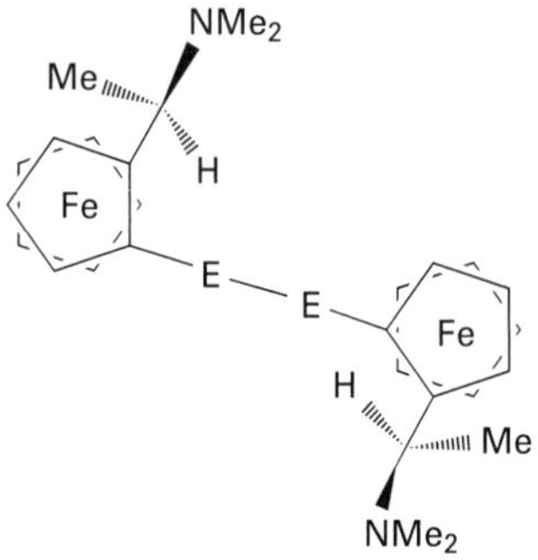

(VI.30)

tively as chiral species for the rhodium(I)-catalysed asymmetric hydrosilylation of several alkyl aryl ketones. Reasonable chemical yields and high enantiomeric excesses were found along with the fact that the product yield and the reaction rate were affected by the nature of the alkyl and aryl groups of the ketone.

6.7.4 Chiral titanocene catalysts

Apart from their role in olefin polymerisation (Section 6.1), chiral *ansa*-titanocenes (and zirconocenes) have been used in enantioselective C—C and C—H bond formation (Hoveyda and Morken, 1996) and in the asymmetric hydrogenation of imines (Scheme 6.10). The reduction affords amines with

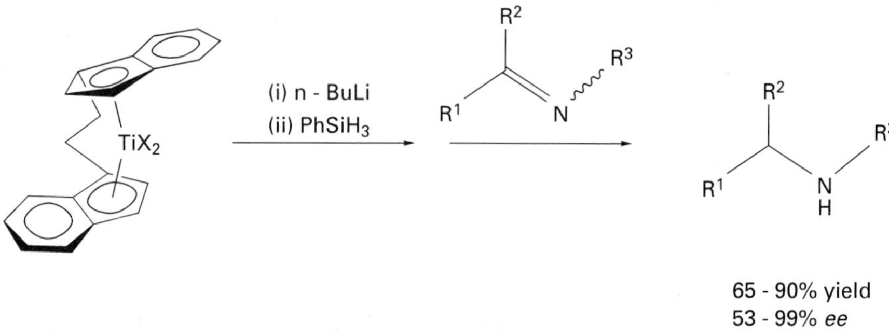

65 - 90% yield
53 - 99% *ee*

Scheme 6.10 The asymmetric reduction of imines using a chiral titanocene catalyst.

good to excellent enantioselectivity, with a feature of the catalytic system being that no co-ordinating group is necessary for high levels of enantioselectivity to be achieved. The catalyst discriminates purely on the basis of the 'shape' of the substrate. The intermediate active hydrogenation catalyst is presumed to be a titanium(III) hydride which forms two diastereomeric titanium amide complexes on reaction with an imine. Then hydrogenolysis of the transient amide complexes, via a σ-bond metathesis reaction, regenerates the titanium hydride and forms the two amine enantiomers. The catalyst is particularly effective for the reduction of cyclic imines, where enantiomeric excesses from 95 to 99% were achieved. For acyclic imines, lower *ee*'s were found and this was thought to be because of acyclic imines being mixtures of *anti* and *syn* isomers which interconvert during the reaction.

6.7.5 Chiral metallocene catalysts featuring calcium, samarium(II) and ytterbium(II)

To extend the number of optically active cyclopentadienyl ligands useful for co-ordination to electron-deficient metals and to reveal possible applications of their complexes in stereospecific catalytic reactions, a number of enantiomerically pure calcium, samarium(II) and ytterbium(II) complexes featuring different asymmetric donor-functionalised cyclopentadienyl ligands have been prepared (**VI.31**). There are very few enantiomerically pure organo-

M = Ca, Sm, Yb

E = OMe, NMe$_2$

(VI.31)

lanthanides or alkaline earth compounds, but because of the donating ability of the terminal functional group in the side chain, it is thought that π-complexes may be stabilised by additional intramolecular co-ordination and side arm participation may play an important role in catalytic processes.

6.8 Other uses of metallocenes

6.8.1 Metallocenes as flame retardents

Smoke development in accidental fires is of importance as more human lives are lost in fire disasters through the effects of smoke and gas evolution than flame action. Pronounced flame-retardent and smoke suppressing properties of ferrocene for a number of polymeric materials such as poly(vinylchloride) and polyurethane have been reported. Ferrocene is thought to act through conversion during combustion to the ferrocenium ion which may function as a Lewis acid, catalysing dehydrochlorination, cross-linking and char-forming reactions. Other suggestions are that the active smoke suppressant is the decomposition product α-Fe_2O_3 and intermediate iron oxychloride or chlorides and that ferrocene influences the gas phase chemistry, e.g. by inhibiting soot nucleation and growth. There is still debate about the mechanism of char formation but general agreement that the addition of ferrocene to PVC has three effects: (i) reduction of smoke formation; (ii) enhancement of char formation; and (iii) reduction of volatile aromatics with enhancement of volatile aliphatics. However, a major drawback of ferrocene is its volatility as considerable amounts are lost by sublimation at normal processing temperatures. Current research features the synthesis of ferrocene derivatives (e.g. ferrocenyl amides and esters) with good thermal stability, reduced volatility and significant flame-retardant/smoke-suppressant activity.

6.8.2 Metallocenes in organic synthesis

Since the early 1980s, there has been an increasing interest amongst organic chemists in the rare-earth elements because of their low toxicity and generally cheaper price compared to other synthetically useful metals such as palladium or rhodium. Whilst the majority of the synthetic applications involve the use of inorganic lanthanide compounds (e.g. SmI_2, $LnCl_3$) an increasing number of organic reactions using organolanthanides have now been investigated (Kagan, 1990). For instance, divalent lanthanides are very effective one-electron reducing agents and lanthanum(IV) derivatives are powerful oxidants. A large body of work on organic transformations has featured samarium diiodide, but di(cyclopentadienyl)samarium(II) has been increasingly studied. For example, aromatic acid chlorides can be efficiently reduced to α-diketones by samarocene in THF (6.4).

$$R-C(O)Cl + Cp_2Sm \xrightarrow{THF} R-C(O)-SmCp_2 \qquad (6.4)$$
$$R-C(O)-C(O)-R \xleftarrow{R-C(O)Cl}$$

The reaction proceeds via a transient acylsamarium species which then reacts as a nucleophile with RC(O)Cl. *In situ*-formed lanthanide(II) metallocene derivatives have been employed as reducing agents. For example, the system Cp_3Ln/NaH has been used to reduce 1-hexane, whilst dienes can be reduced regioselectively. Epoxides such as epoxypropane are smoothly deoxygenated when treated with the THF adduct of decamethylsamarocene

$[Cp_2^*Sm(THF)_2]$ and this species can also be used in a range of reductive carbonylation reactions.

Due to their strong reducing power, organolanthanide(II) complexes have been employed more frequently in organic synthesis than related lanthanide(III) species. Barbier reactions (6.5) have been carried out with samarium diiodide and di(cyclopentadienyl)samarium where radical intermediates have been postulated instead of samarium hydrocarbyls.

$$R^1 - C(O) - R^2 + R^3X + Cp_2Sm \xrightarrow{\text{THF}} R^1R^2R^3C - OH \qquad (6.5)$$

In these reactions, it was found that Cp_2Sm is superior to SmI_2 as the reactions proceeded much faster. Replacement of SmI_2 by Cp_2Sm also allowed isolation of organosamarium intermediates which were stable in THF solution. These solutions could then be reacted with various electrophiles such as D_2O, aldehydes or acid chlorides to afford the expected products (i.e. deuterated aldehydes, alcohols and ketones).

References

Beer, P.D. (1992) *Adv Inorg Chem*, **39**, 79.

Beer, P.D., Chen, Z., Goulden, A.J., Graydon, A., Stokes, S.E. and Wear, T. (1993) *J Chem Soc, Chem Commun*, 1834.

Brintzinger, H.H., Fischer, D., Mülhaupt, R., Rieger, B. and Waymouth, R.M. (1995) *Angew Chem*, **107**, 1255; *Angew Chem Int Ed Engl*, **34**, 1143.

Butsugan, Y., Araki, S. and Watanabe, M. (1995) in *Ferrocenes: Homogeneous Catalysis, Organic Synthesis, Materials Science* (eds A. Togni and T. Hayashi) pp. 143–169. VCH, Weinheim.

Coates, G.W. and Waymouth, R.M. (1995) in *Comprehensive Organometallic Chemistry* (eds E.W. Abel, F.G.A. Stone and G. Wilkinson) vol. 12, ch. 12. 1, pp. 1193–1207. Pergamon Press, Oxford.

Constable, E.C. (1991) *Angew Chem*, **103**, 418; *Angew Chem Int Ed Engl*, **30**, 407.

Corradini, P. and Guerra, G. (1991) *Prog Poly Sci*, **16**, 239.

Coughlin, E.B. and Bercaw, J.E. (1992) *J Am Chem Soc*, **114**, 7606.

Ewen, J.A. (1984) *J Am Chem Soc*, **106**, 6355.

Ewen, J.A., Jones, R.L., Razavi, A. and Ferrara, J.D. (1988) *J Am Chem Soc*, **110**, 6255.

Galloway, C.P. and Rauchfuss, T.B. (1993) *Angew Chem*, **105**, 1407; *Angew Chem Int Ed Engl*, **32**, 1319.

Gokel, G.W., Medina, J.G. Chen, Z. and Echegoyen, L. (1991) *J Am Chem Soc*, **113**, 365.

Gonsalves, K.E. and Chen, X. (1995) in *Ferrocenes: Homogeneous Catalysis, Organic Synthesis, Materials Science* (eds A. Togni and T. Hayashi) pp. 497–530. VCH, Weinheim.

Hall, C.D. (1995) in *Ferrocenes: Homogeneous Catalysis, Organic Synthesis, Materials Science* (eds A. Togni and T. Hayashi) pp. 279–316. VCH, Weinheim.

Hayashi, T. (1995) in *Ferrocenes: Homogeneous Catalysis, Organic Synthesis, Materials Science* (eds A. Togni and T. Hayashi) pp. 105–142. VCH, Weinheim.

Hoveyda, A.H. and Morken, J.P. (1996) *Angew Chem*, **108**, 1378; *Angew Chem Int Ed Engl*, **35**, 1262.

Kagan, H.B. (1990) *New J Chem*, **14**, 453.

Kaminsky, W. (1994) *Catal Today*, **20**, 257.

Köpf-Maier, P. and Köpf, H. (1987) *Chem Rev*, **87**, 1137.

Lango P., Grassi, A., Pellechia, C. and Zambelli, A. (1987) *Macromolecules*, **20**, 1019.

Long, N.J. (1995) *Angew Chem*, **107**, 37; *Angew Chem Int Ed Engl*, **34**, 21.

Long, N.J. (1997) in *Optoelectronic Properties of Inorganic Compounds* (eds J. Fackler Jr and D.M. Roundhill) Plenum, New York.

Manners, I. (1996) *Angew Chem*, **108**, 1712; *Angew Chem Int Ed Engl*, **35**, 1602.

Manners, I. (1996) *Polyhedron*, **15**, 4311.

Miller, J.S. and Epstein, A.J. (1994a) *Angew Chem*, **106**, 399.

Miller, J.S. and Epstein, A.J. (1994b) *Angew Chem Int Ed Engl*, **33**, 385.

Miller, J.S. and Epstein, A.J. (1996) *Chem Ind (London)*, 49.

Möhring, P.C. and Colville, N.J. (1994) *J Organomet Chem*, **479**, 1.

Murray, J.H. and Harding, M.M. (1994) *J Med Chem*, **37**, 1936.

Natta, G., Pasquon, I. and Zambelli, A. (1962) *J Am Chem Soc*, **84**, 1488.

Natta, G., Pino, P., Corradini, P., Danusso, F., Mantica, E., Mazzanti, G. and Moraglio, G. (1955) *J Am Chem Soc*, **77**, 1708.

Rosenblum, M., Nugent, H.M., Jang, K.-S. *et al.* (1995) *Macromolecules*, **28**, 6330.

Schrock, R.R. (1990) *Acc Chem Res*, **23**, 158.

Sinclair, K.B. and Wilson, R.B. (1994) *Chem Ind (London)*, 857.

Sinn, H. and Kaminsky, W. (1980) *Adv Organomet Chem*, **18**, 99.

Togni, A. (1995) in *Ferrocenes: Homogeneous Catalysis, Organic Synthesis, Materials Science* (eds A. Togni and T. Hayashi) pp. 433–469. VCH, Weinheim.

Togni, A. and Hayashi, T. (1995) (eds) *Ferrocenes: Homogeneous Catalysis, Organic Synthesis, Materials Science*. VCH, Weinheim.

Ziegler, K., Holzkamp, E., Breil, H. and Martin, H. (1955) *Angew Chem*, **67**, 426.

Bibliography

Abel, E.W., Stone, F.G.A. and Wilkinson, G. (eds) (1995) *Comprehensive Organometallic Chemistry II*, 14 volumes. Pergamon Press, Oxford.

Bochmann, M. (1994) *Organometallics 1*. Oxford University Press, Oxford.

Bochmann, M. (1994) *Organometallics 2*. Oxford University Press, Oxford.

Bruce, D.W. and O'Hare, D. (eds) (1996) *Inorganic Materials* 2nd edn. Wiley, Chichester.

Coates, G.E., Green, M.L.H. and Wade, K. (1968) *Organometallic Compounds* 3rd edn, 2 volumes. Methuen, London.

Cotton, F.A. and Wilkinson, G. (1988) *Advanced Inorganic Chemistry* 5th edn. Wiley, New York.

Crabtree, R.E. (1988) *The Organometallic Chemistry of the Transition Metals.* Wiley, New York.

Elschenbroich, Ch. and Salzer, A. (1992) *Organometallics* 2nd edn. VCH, Weinheim.

Greenwood, N.N. and Earnshaw, A. (1984) *Chemistry of the Elements.* Pergamon Press, Oxford.

Haiduc, I. and Zuckerman, J.J. (1985) *Basic Organometallic Chemistry.* Walter de Gruyter, Berlin.

Huheey, J.E. (1983) *Inorganic Chemistry* 3rd edn. Harper and Row, New York.

Macintyre, J.E. and Hodgson, A.J. (eds) (1995) *Dictionary of Organometallic Compounds*, 5 volumes. Chapman & Hall, London.

Marks, T.J. and Fischer, R.D. (eds) (1979) *Organometallic Chemistry of the f-elements.* Reidel, Dordrecht.

Pearson, A.J. (1985) *Metallo-organic Chemistry.* Wiley, New York.

Powell, P. (1988) *Organometallic Chemistry* 2nd edn. Chapman & Hall, London.

Purcell, K.F. and Kotz, J.C. (1977) *Inorganic Chemistry.* W.B. Saunders, Philadelphia.

Rosenblum, M. (1965) *The Chemistry of the Iron Group Metallocenes–Part 1.* Wiley, New York.

Shriver, D.F., Atkins, P.W. and Langford, C.H. (1990) *Inorganic Chemistry.* Oxford University Press, Oxford.

Togni, A. and Hayashi, T. (eds) (1995) *Ferrocenes: Homogeneous Catalysis, Organic Synthesis, Materials Science.* VCH, Weinheim.

Wilkinson, G., Stone, F.G.A. and Abel, E.W. (eds) (1982) *Comprehensive Organometallic Chemistry*, 9 volumes. Pergamon Press, Oxford.

Wilkinson, G., Gillard, R. D. and McCleverty, J.E. (eds) (1987) *Comprehensive Coordination Chemistry*, 7 volumes. Pergamon Press, Oxford.

Yamamoto, A. (1986) *Organotransition Metal Chemistry.* Wiley, New York.

Index

Index